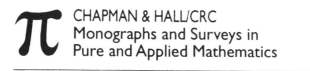

CHAPMAN & HALL/CRC
Monographs and Surveys in
Pure and Applied Mathematics 113

SELF-SIMILARITY

AND BEYOND

Exact Solutions
of Nonlinear Problems

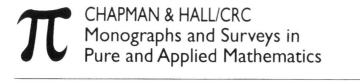

CHAPMAN & HALL/CRC
Monographs and Surveys in
Pure and Applied Mathematics **113**

SELF-SIMILARITY

AND BEYOND

Exact Solutions
of Nonlinear Problems

P. L. SACHDEV

CRC Press
Taylor & Francis Group
Boca Raton London New York

CRC Press is an imprint of the
Taylor & Francis Group, an **informa** business
A CHAPMAN & HALL BOOK

Chapman & Hall/CRC
Taylor & Francis Group
6000 Broken Sound Parkway NW, Suite 300
Boca Raton, FL 33487-2742

ISBN-13: 978-0-367-45548-4 (pbk)
ISBN-13: 978-1-58488-211-4 (hbk)

Library of Congress Card Number 00-034041

Library of Congress Cataloging-in-Publication Data

Sachdev, P.L.
 Self-similarity and beyond: exact solutions of nonlinear problems/P.L. Sachdev.
  p. cm.-- (Chapman & Hall/CRC monographs and surveys in pure and applied
 mathematics ; 113)
 Includes bibliographical references and index.
 ISBN 1-58488-211-5 (alk. paper)
 1. Differential equatins, Nonlinear--Numerical solutions. I. Title. II. Series.

QA377.S23 2000
515'.355--dc21   00-034041

Visit the Taylor & Francis Web site at
http://www.taylorandfrancis.com

and the CRC Press Web site at
http://www.crcpress.com

Preface

I have enjoyed finding exact solutions of nonlinear problems for several decades. I have also had the pleasure of association of a large number of students, postdoctoral fellows, and other colleagues, from both India and abroad, in this pursuit. The present monograph is an attempt to put down some of this experience. Nonlinear problems pose a challenge that is often difficult to resist; each new exact solution is a thing of joy.

In the writing of this book I have been much helped by my colleague Professor V. Philip, and former students Dr. B. Mayil Vaganan and Dr. Ch. Srinivasa Rao. I am particularly indebted to Dr. Rao for his unstinting help in the preparation of the manuscript. Mr. Renugopal, with considerable patience and care, put it in LaTex form.

My wife, Rita, provided invaluable support, care, and comfort as she has done in my earlier endeavours. Our sons, Deepak and Anurag, each contributed in their own ways.

I am grateful to the Council of Scientific and Industrial Research, India for financial support. I also wish to thank Dr. Sunil Nair, Commissioning Editor, Chapman & Hall, CRC Press, for his prompt action in seeing this project through.

P.L. Sachdev

Contents

CONTENTS

Chapter 1

Introduction

Nonlinear problems have always tantalized scientists and engineers: they fascinate, but oftentimes elude exact treatment. A great majority of nonlinear problems are described by systems of nonlinear partial differential equations (PDEs) together with appropriate initial/boundary conditions; these model some physical phenomena. In the early days of nonlinear science, since computers were not available, attempts were made to reduce the system of PDEs to ODEs by the so-called "similarity transformations." The ODEs could be solved by hand calculators. The scenario has since changed dramatically. The nonlinear PDE systems with appropriate initial/boundary conditions can now be solved effectively by means of sophisticated numerical methods and computers, with due attention to the accuracy of the solutions. The search for exact solutions is now motivated by the desire to understand the mathematical structure of the solutions and, hence, a deeper understanding of the physical phenomena described by them. Analysis, computation, and, not insignificantly, intuition all pave the way to their discovery.

The similarity solutions in earlier years were found by direct physical and dimensional arguments. The two most famous examples are the point explosion and implosion problems (Taylor (1950), Sedov (1959), Guderley (1942)). Simple scaling arguments to obtain similarity solutions, illustrating also the self-similar or invariant nature of the scaled solutions, were lucidly given by Zel'dovich and Raizer (1967). Their work was greatly amplified by Barenblatt (1996), who clearly explained the nature of self-similar solutions of the first and second kind. More importantly, Barenblatt brought out manifestly the role of these solutions as intermediate asymptotics; these solutions do not describe merely the behaviour of physical systems under certain conditions, they also describe the intermediate asymptotic behaviour of solutions of wider classes of problems in the ranges where they no longer depend on the details of the initial/boundary conditions, yet the system is still far from being in a limiting state.

1

The early investigators relied greatly upon the physics of the problem to arrive at the similarity form of the solution and, hence, the solution itself. This methodology underwent a severe change due to the work of Ovsyannikov (1962), who, using both finite and infinitesimal groups of transformations, gave an algorithmic approach to the finding of similarity solutions. This approach is now readily available in a practical form (Bluman and Kumei (1989)). A recent direct approach, not involving the use of the groups of finite and infinitesimal transformations, may be found even more convenient in the determination of similarity solutions; the final results via either approach are, however, essentially the same (Clarkson and Kruskal (1989); Hood (1995)). So the reduction to ODEs (if the PDEs originally involved two independent variables) is a routine matter, but then the ODEs have to seek their own initial/boundary conditions to be solved and used to explain some physical phenomenon. On the other hand, given a mathematical model, one must use both algorithmic and dimensional approaches suitably to discover if the problem is self-similar, solve the resulting ODEs subject to appropriate boundary conditions, and prove the asymptotic character of the solution. Since, in the process of reduction to self-similar form, the nonlinearity is fully preserved, the self-similar solution provides important clues to a wider class of solutions of the original PDE.

As a mathematical model is made more comprehensive to include other effects and extend its applicability, it may lose some of its symmetries, and the groups of infinitesimal or finite transformations to which the model is invariant may shrink. As a result, the self-similar form may either cease to exist or may become restricted. A simple example is the system of gasdynamic equations in plane geometry. As soon as the spherical or cylindrical geometry term is included in the equation of continuity, there is a diminution in the scale invariance (Zel'dovich and Raizer (1967)). Therefore, one must relinquish the self-similar hypothesis and assume a more general form of the solution; that is, one must go beyond self-similarity. In the gasdynamic context, several problems in nonplanar geometry, such as flow of a gas into vacuum or a piston motion leading to strong converging shock, are solved by assuming an infinite series in one of the independent variables, time, say, with coefficients depending on a similarity variable (Nageswara Yogi (1995); Van Dyke and Guttman (1982)). This results in an infinite (instead of finite) system of ODEs with appropriate boundary conditions; the zeroth order term in the series is the (known) solution in planar geometry. The series, of course, must be shown to converge in the physically relevant domain. The infinite system of ODEs, in a sense, reflects loss of some symmetry and, hence, greater complexity of the solution.

Another way to overcome the limitations imposed by invariance requirement is to exactly linearise the PDE system when possible, or choose a "natural" coordinate system such that the boundaries of the domain are level lines. The linearisation process immediately gives access to the principle of linear superposition and, hence, the ease of solution associated with it. Hodograph transformations for steady two-dimensional gasdynamic equations and Hopf-Cole transformation for the Burgers equation are well-known examples of exact linearisation. Linearisation, of course, imposes its own constraints, particularly with regard to initial and/or boundary conditions. An example of natural coordinated is again from gas dynamics where the shock trajectory and particle paths may be chosen as preferred coordinates. The transformed system is nonlinear, but has its own invariance properties leading to new classes of exact solutions of the original system of PDEs (Sachdev and Reddy (1982)).

There is yet another way of extending the class of similarity solutions. This is to embed the similarity solutions, suitably expanded, in a larger family; this family is obtained by varying the constants and introducing an infinite number of unknown functions into the expanded form of the similarity solution. These functions are then determined by substituting the assumed form of the solution into the PDEs and, hence, solving the resulting (infinite) system of ODEs appropriately. Thus, the similarity solution becomes a special (embedded) case of the larger family. What is the role and significance of the extended family of solutions must of course be carefully examined (Sachdev, Gupta, and Ahluwalia (1992); Sachdev and Mayil Vaganan (1993)). This embedding is analogous to that for nonlinear ODEs (see, for example, Hille (1970) and Bender and Orszag (1978) for the solution of Thomas-Fermi equation).

Exact asymptotic solutions can also be built up from the (known) linear solutions (Whitham (1974)). The scheme or form of the nonlinear solutions is chosen such that they extend far back (in time, say) the validity of the linear asymptotic solution. For example, for generalised Burgers equations, the exact solution of the planar Burgers equation for N wave neatly motivates the form of the solution for the former (Sachdev and Joseph (1994)). In exceptional circumstances, a "composite" solution may be written out which spans the infinitely long evolution of the N wave, barring a finite initial interval during which the initial (usually discontinuous) profile loosens its gradients (Sachdev, Joseph, and Nair (1994)).

The activist approach to nonlinear ODEs (Bender and Orszag (1978); Sachdev (1991)) suggests how one may build up large time approximate solutions of nonlinear PDEs by a balancing argument. For this purpose, one introduces some preferred variables, the similarity variable and time for instance, into the PDE and looks for possible solutions of truncated PDE made up of terms which balance in one of the independence variables. The simpler PDE thus obtained is usually more amenable to analysis than the original equation. The approximate solution so determined can be improved by taking into account the neglected lower order terms. Usually, a few terms in this analysis give a good description of the asymptotic solution (Grundy, Sachdev, and Dawson (1994); Dawson, Van Duijn, and Grundy (1996)).

We may revert and say that whenever similarity solutions exist, their existence theory greatly assists in the understanding of the original PDE system. These solutions also help in the quantitative estimation of how the solutions of certain classes of initial/boundary value problems evolve in time (Sachdev (1987)).

The role of numerical solution of nonlinear problems in discovering the analytic structure of the solution need hardly be emphasised; very often the numerical solution throws much light on what kind of analytic form one must explore. Besides, understanding the validity and place of exact/approximate analytic solution in the general context can be greatly enhanced by the numerical solution. In short, there must be a continuous interplay of analysis and computation if a nonlinear problem is to be successfully tackled.

The approaches outlined in the above go beyond self-similarity, but the exact solutions they yield are still generally asymptotic in nature; these solutions, per se, satisfy some special (singular) initial conditions but evolve to become intermediate asymptotics to which solutions of a certain larger but restricted class of initial/boundary value problems tend as time goes to infinity (Sachdev (1987)).

Chapter 2 deals with first-order PDEs, illustrating with the help of many examples the place of similarity solutions in the general solution. Exact similarity solutions via group theoretic methods and the direct similarity approach of Clarkson and Kruskal (1989) are discussed in Chapter 3, while travelling wave solutions are treated in Chapter 4. Exact linearisation of nonlinear PDEs, including via hodograph methods, is dealt with in Chapter 5. In Chapter 6, construction of more general solutions from special solutions of a given or a related problem is accomplished via nonlinearisation or embedding methods. Chapter 7 uses the balancing argument for nonlinear PDEs to find approximate solutions of nonlinear problems. The concluding chapter expounds series solutions for nonlinear PDEs with the help of several examples; the series are constructed in one of the independent variables, often the time, with the coefficients depending on the other independent variable.

The approach in the present monograph is entirely constructive in nature; there is very little by way of abstract analysis. The analytic and numerical solutions are often treated alongside. Most examples are drawn from real physical situations, mainly from fluid mechanics and nonlinear diffusion. The idea is to illustrate and bring out the main points. To highlight the goals of the present book we could do no better than quote from the last chapter on exact solutions in the book by Whitham (1974), "Doubtless much more of value will be discovered, and the different approaches have added enormously to the arsenal of 'mathematical methods.' Not least is the lesson that exact solutions are still around and one should not always turn too quickly to a search for the ϵ."

Chapter 2

First-Order Partial Differential Equations

2.1 Linear Partial Differential Equations of First Order

The most general first-order linear PDE in two independent variables x and t has the form

$$au_x + bu_t = cu + d \qquad (2.1.1)$$

where a, b, c, d are functions of x and t only. We single out the variable t (often "time" in physical problems) and write the first-order general PDE in the "normal" form

$$u_t + F(x, t, u, u_x) = 0.$$

The general solution of a first-order PDE involves an arbitrary function. In applications one is usually interested not in obtaining the general solution of a PDE, but a solution subject to some additional condition such as an initial condition (IC) or a boundary condition (BC) or both.

A basic problem for first-order PDEs is to solve

$$u_t + F(x, t, u, u_x) = 0, x \in R, t > 0 \qquad (2.1.2)$$

subject to the IC

$$u(x, 0) = u_0(x), x \in R \qquad (2.1.3)$$

where $u_0(x)$ is a given function. (The interval of interest for x may be finite.) This is called a Cauchy problem; it is a pure initial value problem. It may be viewed as a signal or wave at time $t = 0$. The initial signal or wave is a space distribution of u, and a "picture" of the wave may be obtained by drawing the graph of $u = u_0(x)$ in the xu-space. Then the PDE

5

(2.1.2) may be interpreted as the equation that describes the propagation of the wave as time increases.

We first consider the wave equation

$$u_t + cu_x = 0 \qquad (2.1.4)$$

with the IC

$$u(x,0) = u_0(x), \qquad (2.1.5)$$

where c is a constant.

If $x = x(t)$ defines a smooth curve C in the (x,t) plane, the total derivative of $u = u(x,t)$ along a curve is found by using the chain rule:

$$\frac{du}{dt} = \frac{\partial u}{\partial t} + \frac{\partial u}{\partial x}\frac{dx}{dt}.$$

The left-hand side of (2.1.4) is a total derivative of u along the curves defined by the equation $\dfrac{dx}{dt} = c$. Therefore, equation (2.1.4) is equivalent to the statement

$$\frac{du}{dt} = 0 \text{ along the curves } \frac{dx}{dt} = c. \qquad (2.1.6)$$

From (2.1.6) we find that

$$u = \text{ constant along the curves } x - ct = \xi \qquad (2.1.7)$$

where ξ is constant of integration. For different values of ξ we get a family of curves in the (x,t) plane. A curve of the family through an arbitrary point (x,t) intersects the x-axis at $(\xi,0)$. Since u is constant on this curve, its value $u(x,t)$ is equal to its value $u(\xi,0)$ at the initial time:

$$u = u(x,t) = u(\xi,0) = u_0(\xi) = u_0(x - ct) \qquad (2.1.8)$$

$u_0(x - ct)$ is the solution to the IVP (2.1.4) - (2.1.5).

The curves defined by (2.1.6) are called "characteristic curves" or simply characteristics of the PDE (2.1.4). A characteristic in the xt-space represents a moving wavelet in the x-space, $\dfrac{dx}{dt}$ being its speed. The greater the inclination of the line with the t-axis, the greater will be the speed of the corresponding wavelet. Signals or wavelets are propagated along the characteristics. Also, along the characteristics the PDE reduces to a system of ODEs (see (2.1.6)). At the initial time $t = 0$ the wave has the form $u_0(x)$. At a later time t the wave profile is $u_0(x - ct)$. This shows that in time t the initial profile is translated to the right a distance ct. Thus, c represents the speed of the wave.

Example 1

$$\frac{\partial u}{\partial t} + t^2 \frac{\partial u}{\partial x} = 0, \quad u(x,0) = f(x).$$

It is clear that $\dfrac{du}{dt} = 0$ along the characteristic curves $\dfrac{dx}{dt} = t^2$. On integration we get $x = \dfrac{t^3}{3} + \xi$ so that

$$u = \text{constant on } x = \xi + \frac{t^3}{3}.$$

Therefore,

$$u(x,t) = u(\xi,0) = f(\xi) = f\left(x - \frac{t^3}{3}\right).$$

The solution $u(x,t) = f\left(x - \dfrac{t^3}{3}\right)$ has a travelling wave form $u(x,t) = f(\eta), \eta = x - \dfrac{t^3}{3}$. The travelling wave moves with a nonconstant speed t^2 and a nonconstant acceleration $2t$.

The method of characteristics can also be applied to solve IVP for a nonhomogeneous PDE of the form $u_t + c(x,t)u_x = f(x,t), x \in R, \ t > 0$, $u(x,0) = u_0(x)$.

Example 2

$$\frac{\partial u}{\partial t} + c\frac{\partial u}{\partial x} = e^{-3x}, u(x,0) = f(x).$$

We note that

$$\frac{du}{dt} = e^{-3x} \quad \text{along} \quad \frac{dx}{dt} = c.$$

This pair of ODEs can be solved subject to the IC $x = \xi$, $u = f(\xi)$ at $t = 0$. We get

$$x = ct + \xi$$

and

$$\frac{du}{dt} = e^{-3(ct+\xi)}.$$

On integration we have

$$u(x,t) = \frac{e^{-3ct}}{-3c}e^{-3\xi} + g(\xi)$$

where g is the function of integration. Applying the IC we get

$$g(\xi) = \frac{e^{-3\xi}}{3c} + f(\xi).$$

Thus,

$$u(x,t) \;=\; \frac{e^{-3\xi}}{3c}(1 - e^{-3ct}) + f(\xi)$$

$$\;=\; \frac{e^{-3(x-ct)}}{3c}(1 - e^{-3ct}) + f(x - ct).$$

The solution here is of the similarity form $u(x,t) = \alpha(x,t) + \beta(\eta)$, where $\eta = x - ct$ is the similarity variable, a linear combination of the independent variables x and t.

Example 3

$$\frac{\partial u}{\partial t} + x\frac{\partial u}{\partial x} = t, \quad u(x,0) = f(x).$$

Here

$$\frac{du}{dt} = t \quad \text{along} \quad \frac{dx}{dt} = x,$$

which on integration yields

$$x = \xi e^t$$

and

$$u(x,t) = \frac{t^2}{2} + g(\xi).$$

At $t = 0$, $x = \xi$, $u = f(\xi)$; therefore, $g(\xi) = f(\xi)$. Thus

$$u = \frac{t^2}{2} + f(\xi) = \frac{t^2}{2} + f(xe^{-t}).$$

The solution here has the similarity form

$$u = \alpha(x,t) + \beta(\eta)$$

where $\eta = xe^{-t}$ is the similarity variable.

Example 4

$$xu_x + (x^2 + y)u_y + \left(\frac{y}{x} - x\right)u = 1.$$

The characteristics are given by

$$\frac{dx}{dt} = x, \quad \frac{dy}{dt} = x^2 + y, \quad \frac{du}{dt} + \left(\frac{y}{x} - x\right)u = 1,$$

the first two of which give the locus in the (x,y) plane, the so-called traces,

$$\frac{dy}{dx} - x + \frac{y}{x}$$

which on integration become

$$\frac{y}{x} - x = \text{constant.}$$

It is often easier to find the general solution of the PDE by introducing the variable describing the trace curves as a new independent variable: $\phi = \frac{y}{x} - x$. The given PDE then becomes

$$x \left(\frac{\partial u}{\partial x} \right)_\phi + \phi u = 1$$

which on integration with respect to ϕ gives

$$u = \phi^{-1} + x^{-\phi} f(\phi)$$

where f is an arbitrary function of ϕ.

2.2 Quasilinear Partial Differential Equations of First Order

The general first-order quasilinear equation has the form

$$a u_x + b u_t = c, \tag{2.2.1}$$

where a, b, and c are functions of x, t, and u. Quasilinear PDEs are simpler to treat than fully nonlinear ones for which u_x and u_t may not occur linearly. The solution $u = u(x, t)$ of (2.2.1) may be interpreted geometrically as a surface in (x, t, u) space, called an "integral surface."

The Cauchy problem for (2.2.1) requires that u assume prescribed values on some plane curve C. If s is a parametric on C, its representation is $x = x(s)$, $t = t(s)$. We may prescribe $u = u(s)$ on C. The ordered triple $(x(s), t(s), u(s))$ defines a curve Γ in the (x, t, u)-space; C is the projection of Γ onto the (x, t) plane. Thus, generally, the problem is to find the solution or an integral surface $u = u(x, t)$ containing the three-dimensional curve Γ. The direction cosines of the normal $\vec{n}$ to the surface $u(x, t) - u = 0$ are proportional to the components of grad $(u(x, t) - u) = (u_x, u_t, -1)$. If we define the vector $\vec{e} = (a, b, c)$, then the PDE (2.2.1) can be written as $\vec{e} \cdot \vec{n} = 0$. In other words, the vector direction (a, b, c) is tangential to the integral surface at each point. The direction (a, b, c) at any point on the surface is called the "characteristic direction." A space curve whose tangent at every point coincides with the characteristic direction is called a "characteristic curve" and is given by the equations

$$\frac{dx}{a} = \frac{dt}{b} = \frac{du}{c}. \tag{2.2.2}$$

The characteristics are curves in the (x, t, u)-space and lie on the integral surface. The projections of the characteristic curves onto the (x, t) plane are called "base characteristics" or "ground characteristics." Integration of (2.2.2) is not easy as a, b, c; now depend upon u as well. Prescribing u at one point of the characteristic enables one to determine u all along it. We assume that all the smoothness conditions on the functions a, b, and c are satisfied so that the system of ODEs (2.2.2) has a unique solution starting from a point on the initial curve. Lagrange proved that solution of Equation (2.2.1) is given by

$$F(\phi, \psi) = 0 \quad \text{or} \quad \phi = f(\psi),$$

where $\phi(x, t, u)$ and $\psi(x, t, u)$ are independent functions (that is, normals to the surfaces $\phi = $ constant and $\psi = $ constant are not parallel at any point of intersection) such that

$$a\phi_x + b\phi_t + c\phi_u = 0, \quad a\psi_x + b\psi_t + c\psi_u = 0 \qquad (2.2.3)$$

(The functions F and f are themselves arbitrary). $F(\phi, \psi) = 0$, called the "general integral," is an implicit relation between x, t, and u. Oftentimes it is possible to solve for u in terms of x and t. If $\phi = $ constant is a first integral of (2.2.2), it satisfies (2.2.3). A second integral of (2.2.2), $\psi = $ constant, also satisfies (2.2.3). Equation (2.2.2) represents the curves of intersection of the surfaces $\phi = c_1$ and $\phi = c_2$, where c_1 and c_2 are arbitrary constants. We thus have a two-parameter family of curves. If we impose the condition $F_1(c_1, c_2) = 0$ we get a one-parameter family of characteristics. An integral surface can be constructed by drawing characteristics from each point of the initial curve. Note that (2.2.2) may be written in the parametric form

$$\frac{dx}{d\tau} = a, \quad \frac{dt}{d\tau} = b, \quad \frac{du}{d\tau} = c \qquad (2.2.4)$$

where τ is a parameter measured along the characteristic.

One may also obtain a solution of (2.2.4) in the form $x = x(s, \tau), t = t(s, \tau)$, and $u = u(s, \tau)$, where s is a parameter measured along the initial curve. Solving for s and τ in terms of x and t from the first two equations and substituting in $u = u(s, \tau)$, one gets u as a function of x and t.

Example 1

Find the general solution of $(t + u)u_x + tu_t = x - t$. Also find the integral surface containing the curve $t = 1, u = 1 + x, -\infty < x < \infty$.

The characteristics of the given PDE are given by

$$\frac{dx}{t + u} = \frac{dt}{t} = \frac{du}{x - t}.$$

It is easy to see that
$$\frac{d(x+u)}{x+u} = \frac{dt}{t}.$$

On integration we have
$$\frac{x+u}{t} = c_1$$

where c_1 is a constant. Again
$$\frac{d(x-t)}{u} = \frac{du}{x-t},$$

implying
$$(x-t)^2 - u^2 = c_2,$$

where c_2 is another constant.

The general solution, therefore, is
$$(x-t)^2 - u^2 = f\left(\frac{x+u}{t}\right).$$

If the integral surface contains the given curve $t = 1, u = 1 + x$, we have
$$(x-1)^2 - (1+x)^2 = f(1+2x),$$

or
$$f(1+2x) = -4x$$

implying that
$$f(z) = -2(z-1)$$

and so
$$f\left(\frac{x+u}{t}\right) = -2\left(\frac{x+u}{t} - 1\right).$$

The solution therefore is
$$(x-t)^2 - u^2 = -\frac{2}{t}(x+u-t).$$

Solving for u, we have
$$u = \frac{1}{t} \pm \left(x - t + \frac{1}{t}\right).$$

The condition $u = 1 + x$ when $t = 1$ is satisfied only if we take the positive sign. Thus, the solution of the IVP is
$$u = \frac{2}{t} + x - t.$$

Clearly, the solution is defined only for $t > 0$.

While the general solution is quite implicit, the solution of IVP has the form $u = f(t) + g(\eta), \eta = x - t$, and may be found by similarity methods.

Example 2

Find the general solution of

$$(t^2 - u^2)u_x - xtu_t = xu.$$

Also find the integral surface containing the curve $x = t = u$, $x > 0$.

The characteristics of the given PDE are

$$\frac{dx}{t^2 - u^2} = \frac{dt}{-xt} = \frac{du}{xu}.$$

A first integral obtained from the second pair is $\phi(x, t, u) \equiv ut = c_1$, say. Each of the above ratios is equal to

$$\frac{xdx + tdt + udu}{x(t^2 - u^2) + t(-xt) + u(xu)} = \frac{xdx + tdt + udu}{0}.$$

Therefore, a second integral is $\psi(x, t, u) \equiv x^2 + t^2 + u^2 = c_2$, say. The general solution, therefore, is $\phi = f(\psi)$, that is,

$$ut = f(x^2 + t^2 + u^2).$$

Applying the initial condition $x = t = u$, we get

$$x^2 = f(3x^2),$$

giving

$$f(z) = \frac{z}{3}.$$

Therefore we get the special solution satisfying the IC as

$$ut = \frac{x^2 + t^2 + u^2}{3}.$$

Solving the quadratic in u we find that

$$u = \frac{3t - (5t^2 - 4x^2)^{1/2}}{2},$$

the root with the negative sign satisfying the given conditions.

Here, again, the general solution is rather implicit. The special solution satisfying given IC may be obtained by the similarity approach.

Conservation Laws

Considerable interest attaches to the quasilinear equations of the form

$$u_t + (f(u))_x = 0;$$

it is a divergence form or a conservation law. A simple model of traffic on a highway yields a conservation law of this type.

Consider a single-lane highway occupied by moving cars. We can define a density function $u(x, t)$ as the number of cars per unit length at the point x measured from some fixed point on the road at time t. The flux of vehicles $\phi(x, t)$ is the number of cars per unit time (say, hour) passing a fixed place x at time t. Here we regard u and ϕ as continuous functions of the distance x. If we consider an arbitrary section of the highway between $x = a$ and $x = b$, then the number of cars between $x = a$ and $x = b$ at time t is equal to $\int_a^b u(x, t)dx$. Assuming that there are neither entries nor exits on this section of the road, the time rate of change of the number of cars in the section $[a, b]$ equals the number of cars per unit time entering at $x = a$ minus the number of cars per unit time leaving at $x = b$. That is

$$\frac{d}{dt} \int_a^b u(x, t)dx = \phi(a, t) - \phi(b, t)$$

or

$$\int_a^b \frac{\partial u}{\partial t}dx = -\int_a^b \frac{\partial \phi}{\partial x}(x, t)dx.$$

This yields the conservation law

$$\frac{\partial u}{\partial t} + \frac{\partial \phi}{\partial x} = 0 \qquad (2.2.5)$$

since the interval $[a, b]$ is arbitrary. If we assume that the flux ϕ depends on the traffic density u, then the conservation equation becomes

$$\frac{\partial u}{\partial t} + \phi'(u)\frac{\partial u}{\partial x} = 0$$

or

$$\frac{\partial u}{\partial t} + c(u)\frac{\partial u}{\partial x} = 0$$

where $c(u) = \phi'(u)$.

Considering this, we see that $\dfrac{du}{dt} = 0$ along the characteristic $\dfrac{dx}{dt} = c(u)$. Unlike the linear case, the characteristic curves cannot in general be determined in advance since u is yet unknown. But, in the special case considered here, since u and $c(u)$ remain constant on a characteristic, the latter must be a straight line in the (x, t) plane. If, through an arbitrary point (x, t), we draw a characteristic back in time, it will cut the x-axis at the point $(\xi, 0)$. If $u = u_0(x)$ at $t = 0$, the equation of this characteristic is

$$x = \xi + c(u_0(\xi))t. \qquad (2.2.6)$$

Since u remains constant along this characteristic,

$$u(x, t) = u(\xi, 0) = u_0(\xi). \qquad (2.2.7)$$

As ξ varies, we get different characteristics. Equations (2.2.6) and (2.2.7) give the implicit solution $u(x,t) = u_0[x - c(u_0(\xi))t]$.

Shock waves

In the case of quasilinear equations, two characteristics may intersect. Consider the characteristics C_1 and C_2, starting from the points $x = \xi_1$ and $x = \xi_2$, respectively. Along C_1, $u(x,t) = u_0(\xi_1) = u_1$, say. Along C_2, $u(x,t) = u_0(\xi_2) = u_2$. The speeds of the characteristics are $c(u_1)$ and $c(u_2)$. If $c(u_1) > c(u_2)$, the angle characteristic ξ_1 makes with the t-axis is greater than that which the characteristic ξ_2 makes with it, and so they intersect. This means that, at the point of intersection P, u has simultaneously two values, u_1 and u_2. This is unphysical since u (usually a density in physical problems) cannot have two values at the same time. To overcome this difficulty we assume that the solution u has a jump discontinuity. It is found that the discontinuity in u propagates along special loci in space time. The trajectory $x = x_s(t)$ in the (x,t) plane along which the discontinuity, called a shock, propagates is referred to as the "shock path" or "shock trajectory;" $\dfrac{dx_s(t)}{dt}$ is the shock speed. The shock path is not a characteristic curve.

Let $u(x,0)$ be the initial distribution of u (some density). The dependence of c on u produces nonlinear distortion of the wave as it propagates. When $c'(u) > 0$ (c is an increasing function of u), higher values of u propagate faster than the lower ones. As a result, the initial wave profile distorts. The density distribution becomes steeper as time increases and the slope becomes infinite at some finite time, called the "breaking time."

We now determine how the discontinuity is formed and propagates. At the discontinuity the PDE itself does not apply (We assume that all the derivatives exist in the flow region). Equation $u_t + c(u)u_x = 0$ holds on either side. It may be written in the conservation form

$$u_t + \phi_x = 0$$

where $\phi'(u) = c(u)$. If $v(x,t)$ is the velocity at (x,t), then the flux $\phi(x,t) = u(x,t)v(x,t)$. Conservation of density at the discontinuity requires (relative inflow equals relative outflow)

$$u(x_s-,t)\left[v(x_s-,t) - \frac{dx_s}{dt}\right] = u(x_s+,t)\left[v(x_s+,t) - \frac{dx_s}{dt}\right].$$

Solving for $\dfrac{dx_s}{dt}$, we get the shock velocity as

$$\frac{dx_s}{dt} = \frac{\phi(x_s+, t) - \phi(x_s-, t)}{u(x_s+, t) - u(x_s-, t)}$$

$$= \frac{[\phi]}{[u]} \tag{2.2.8}$$

where $[\phi]$ and $[u]$ denote jumps in ϕ and u across the shock, respectively.
Consider the IVP

$$u_t + u u_x = 0 \tag{2.2.9}$$

$$u(x, 0) = \begin{array}{ll} 1 & x < 0 \\ 0 & x > 0 . \end{array}$$

Equation (2.2.9) can be written in the conservation form as

$$u_t + \phi_x = 0$$

where the flux $\phi = \dfrac{u^2}{2}$. The jump condition (2.2.8) becomes

$$\begin{aligned}
\frac{dx_s}{dt} &= \frac{[\phi]}{[u]} = \frac{\phi_+ - \phi_-}{u_+ - u_-} \\
&= \frac{\frac{u_+^2}{2} - \frac{u_-^2}{2}}{u_+ - u_-} \\
&= \frac{u_+ + u_-}{2}
\end{aligned}$$

where the subscripts $+$ and $-$ indicate that the quantity is evaluated at x_s+ and x_s-, respectively. Thus, the shock speed is the average of the values of u ahead of and behind the shock.

Again, (2.2.9) implies that $\dfrac{du}{dt} = 0$ along the characteristic $\dfrac{dx}{dt} = u$; in other words, $u = $ constant along the straight line characteristics having speed u. Characteristics starting from the x-axis have speed unity if $x < 0$ and zero if $x > 0$. So at $t = 0+$, the characteristics intersect and a shock is produced. The shock speed $\dfrac{dx_s}{dt} = \dfrac{0 + 1}{2} = \dfrac{1}{2}$, and hence the shock path is $x = \dfrac{t}{2}$. The initial discontinuity at $x = 0$ propagates along this path with speed $\dfrac{1}{2}$. A solution to the IVP is

$$u(x, t) = 1 \quad \text{if } x < \frac{1}{2}t; \quad u(x, t) = 0 \text{ if } x > \frac{1}{2}t.$$

In the present example there is a discontinuity in the initial data and a shock is formed immediately. Even when the initial condition $u(x, 0) = u_0(x)$ is continuous, a discontinuity may be formed in a finite time.

Consider the characteristics coming out of point $x = \xi$ on the initial line

$$x = \xi + F(\xi)t,$$

where $F(\xi) = c(u_0(\xi))$. Differentiating this equation with respect to t we get

$$0 = \xi_t + F(\xi) + F'(\xi)\xi_t \, t$$

or

$$\xi_t = \frac{-F(\xi)}{1 + F'(\xi)t}.$$

Since

$$u = u_0(\xi),$$

we have

$$\begin{aligned} u_t &= u_0'(\xi)\xi_t \\ &= \frac{-u_0'(\xi)F(\xi)}{1 + F'(\xi)t}. \end{aligned}$$

It is clear that for u_t (and hence u_x) to become infinite we must have $F'(\xi) < 0$. The breaking of the wave first occurs on the characteristic $\xi = \xi_B$ for which $F'(\xi) < 0$ and $|F'(\xi)|$ is a maximum. The time of first breaking of the wave is

$$t_B = -\frac{1}{F'(\xi_B)}.$$

Example 1

$$\begin{aligned} u_t + 2uu_x &= 0 \\ u(x,0) &= \begin{cases} 3 & x < 0 \\ 2 & x > 0 \end{cases} \end{aligned}$$

The given PDE in conservation form is

$$u_t + \phi_x = 0$$

where $\phi = u^2$. Here, $\dfrac{du}{dt} = 0$ along $\dfrac{dx}{dt} = 2u$, that is, u is constant along the straight line characteristics having speed $2u$. For $x < 0$ the speed of the characteristic is $\dfrac{dx}{dt} = 6$, an integration yields the equation of the characteristic as

$$x = 6t + \xi$$

where ξ is constant of integration. For $x > 0$ the characteristic speed is 4 and the corresponding characteristics are $x = 4t + \xi$. For $t > 0$ the

characteristics collide immediately and a shock wave is formed. The slope
of the shock is given by

$$\frac{dx_s}{dt} = \frac{[\phi]}{[u]} = \frac{\phi(3) - \phi(2)}{3 - 2} = 5.$$

The shock path is clearly $x = 5t$. The solution of the problem is $u(x, t) = 3$
for $x < 5t$ and $u(x, t) = 2$ for $x > 5t$.

We now consider examples of the form

$$u_t + c(x, t, u)u_x = f(x, t, u), x \in R, t > 0$$

$$u(x, 0) = u_0(x), x \in R.$$

Example 2

$$u_t - u^2 u_x = 3u, x \in R, t > 0$$

$$u(x, 0) = u_0(x), x \in R$$

Here, $\dfrac{du}{dt} = 3u$ along the characteristics $\dfrac{dx}{dt} = -u^2$. This system of ODEs
must be solved subject to the IC $u = u_0(\xi)$, $x = \xi$ at $t = 0$. We have, on
integration of the first, the result $u = ke^{3t}$ where k is constant of integration.
Since $u = u_0(\xi)$ at $t = 0$, we have

$$u = u_0(\xi)e^{3t}. \tag{2.2.10}$$

Now $\dfrac{dx}{dt} = -u_0^2(\xi)e^{6t}$. Therefore, using the initial condition $x = \xi$ at $t = 0$,
we get

$$x = \xi + \frac{u_0^2(\xi)}{6}(1 - e^{6t}). \tag{2.2.11}$$

Equations (2.2.10) and (2.2.11) constitute (an implicit) solution of the given
initial value problem.

Example 3

$$u_t + uu_x = -u, \quad x \in R, t > 0$$

$$u(x, 0) = -\frac{x}{2}, \quad x \in R$$

Here,

$$\frac{du}{dt} = -u \quad \text{along} \quad \frac{dx}{dt} = u.$$

Solving the first equation with 1C $x = \xi$, $u = -\dfrac{\xi}{2}$ at $t = 0$ we have

$$u = -\frac{\xi}{2}e^{-t}. \qquad (2.2.12)$$

Integrating $\dfrac{dx}{dt} = -\dfrac{\xi}{2}e^{-t}$ and using the initial conditions, we get

$$x = \frac{\xi}{2}(1 + e^{-t}). \qquad (2.2.13)$$

Substituting ξ from (2.2.13) into (2.2.12) we get the solution

$$u(x,t) = -\frac{xe^{-t}}{1 + e^{-t}}.$$

Example 4

Consider the IVP

$$u_t + uu_x = 0, \quad x \in R, t > 0$$

$$u(x,0) = 0, \text{ if } x < 0; \ u(x,0) = 1 \quad \text{if } x > 0.$$

Here, $\dfrac{du}{dt} = 0$ along characteristics $\dfrac{dx}{dt} = u$. Characteristics issuing from the x-axis have speed zero if $x < 0$ and 1 if $x > 0$. There is a void between $x = 0$ and $x = t$ for $t > 0$. We can imagine that all values of u between 0 and 1 are present initially at $x = 0$. In this void, continuous solution can be constructed which connects the solution $u = 1$ ahead to the solution $u = 0$ behind. We insert a fan of characteristics (which are straight lines here) passing through the origin. Each member of the fan has a different (constant) slope. The value of u these characteristics carry varies continuously from 0 to 1. That is, $u = c$ (constant), $0 < c < 1$, on the characteristic $x = ct$. Thus, the solution is

$$
\begin{aligned}
u(x,t) &= 0 \quad \text{for } x < 0 \\
&= \frac{x}{t} \quad \text{for } 0 < \frac{x}{t} < 1 \qquad (2.2.14) \\
&= 1 \quad \text{for } x > t.
\end{aligned}
$$

A solution of this form is called a "centred expansion wave"; it is clearly a similarity solution.

Example 5

$$x(y^2 + u)u_x - y(x^2 + u)u_y = (x^2 - y^2)u$$

$$u = 1 \quad \text{on } x + y = 0$$

The characteristic equations are

$$\frac{dx}{x(y^2 + u)} = \frac{dy}{-y(x^2 + u)} = \frac{du}{(x^2 - y^2)u}$$

which, on some manipulation, give

$$\frac{dx}{x} + \frac{dy}{y} + \frac{du}{u} = 0 \qquad (2.2.15)$$

and

$$x\,dx + y\,dy - du = 0. \qquad (2.2.16)$$

Equations (2.2.15) and (2.2.16) integrate to give

$$xyu = C_1$$

and

$$x^2 + y^2 - 2u = C_2$$

where C_1 and C_2 are arbitrary constants. The general solution therefore is

$$x^2 + y^2 - 2u = f(xyu). \qquad (2.2.17)$$

The initial data $u = 1$ on $x + y = 0$ gives $f(-x^2) = 2x^2 - 2$ or $f(x^2) = -2x^2 - 2$. Thus, the general solution (2.2.17) in this case reduces to

$$2xyu + x^2 + y^2 - 2u + 2 = 0.$$

Example 6

$$xu_x + yu_y = x\exp(-u)$$
$$u = 0 \text{ on } y = x^2$$

The characteristic equations are

$$\frac{dx}{x} = \frac{dy}{y} = \frac{du}{x\exp(-u)} \qquad (2.2.18)$$

and have the first integrals

$$\frac{y}{x} = C_1$$

and

$$e^u = x + C_2$$

from the first and second and first and third of (2.2.18), respectively. C_1 and C_2 are arbitrary constants. The general solution of the given PDE therefore is

$$e^u = x + g(y/x), \qquad (2.2.19)$$

which is a similarity form for the dependent variable $U = e^u$. If we use the given 1C, we get $g(x) = 1 - x$, and so (2.2.19) in this case becomes

$$e^u = x + 1 - \frac{y}{x}$$

or

$$u = \ln \left(x + 1 - \frac{y}{x} \right).$$

Direct Similarity Approach for First-Order PDEs

Although we discuss self-similar solutions in detail in Chapter 3, here we give two examples to illustrate the simple approach of Clarkson and Kruskal (1989) which is direct and does not require group theoretic ideas.

Example 1

$$u_t + uu_x = 0. \tag{2.2.20}$$

We assume that (2.2.20) has solution of the form

$$u(x, t) = \alpha(x, t) + \beta(x, t)H(\eta), \eta = \eta(x, t), \beta(x, t) \neq 0. \tag{2.2.21}$$

Differentiating (2.2.21) to get u_t and u_x and, hence, substituting in (2.2.20), we have

$$\beta^2 \eta_x HH' \quad + \quad \beta\beta_x H^2 + \beta(\eta_t + \alpha\eta_x)H'$$

$$+ \quad (\beta_t + \alpha\beta_x + \beta\alpha_x)H + (\alpha_t + \alpha\alpha_x) = 0. \tag{2.2.22}$$

Equation (2.2.22) becomes an ODE for the determination of the similarity function $H(\eta)$ if

$$\beta\beta_x \quad = \quad \beta^2\eta_x\Gamma_1(\eta) \tag{2.2.23}$$
$$\beta(\eta_t + \alpha\eta_x) \quad - \quad \beta^2\eta_x\Gamma_2(\eta) \tag{2.2.24}$$
$$\beta_t + \alpha\beta_x + \beta\alpha_x \quad = \quad \beta^2\eta_x\Gamma_3(\eta) \tag{2.2.25}$$
$$\alpha_t + \alpha\alpha_x \quad = \quad \beta^2\eta_x\Gamma_4(\eta). \tag{2.2.26}$$

Equation (2.2.22) then becomes

$$HH' + \Gamma_1(\eta)H^2 + \Gamma_2(\eta)H' + \Gamma_3(\eta)H + \Gamma_4(\eta) = 0. \tag{2.2.27}$$

We solve (2.2.23) - (2.2.26) to obtain the unknown functions $\alpha, \beta, \eta, \Gamma_1, \Gamma_2, \Gamma_3$, and Γ_4. In the process of solution the following remarks are found useful.

Remark 1

If $\alpha(x, t)$ has the form $\alpha(x, t) = \hat{\alpha}(x, t) + \beta(x, t)\Omega(\eta)$, then we may set $\Omega \equiv 0$.

Remark 2

If $\beta(x,t)$ is to be determined from an equation of the form $\beta(x,t) = \hat{\beta}(x,t)\Omega(\eta)$, then we may put $\Omega \equiv 1$.

Setting $\Gamma_1(\eta) = \dfrac{\Omega_1'(\eta)}{\Omega_1(\eta)}$ in (2.2.23) and integrating with respect to x, we obtain

$$\beta = B(t)\Omega_1(\eta). \tag{2.2.28}$$

Using Remark 2 in (2.2.28), we set $\Omega_1(\eta) \equiv 1$ so that

$$\Gamma_1(\eta) = 0 \tag{2.2.29}$$

and from (2.2.23)

$$\beta = B(t). \tag{2.2.30}$$

Substituting (2.2.30) in (2.2.25) we get

$$\alpha_x = B(t)\eta_x\Gamma_3(\eta) - \frac{B'(t)}{B(t)}. \tag{2.2.31}$$

Setting $\Gamma_3(\eta) = \Omega_3'(\eta)$ in (2.2.31) and integrating with respect to x, we obtain

$$\alpha = B(t)\Omega_3(\eta) - \frac{B'(t)}{B(t)}x + A(t). \tag{2.2.32}$$

Making use of Remark 1 in (2.2.32) we set $\Omega_3 \equiv 0$, and so

$$\Gamma_3 \equiv 0 \tag{2.2.33}$$

from (2.2.31) and (2.2.32). Equation (2.2.32) now reduces to

$$\alpha = A(t) - \frac{B'(t)}{B(t)}x. \tag{2.2.34}$$

Using (2.2.30), (2.2.24) is written as a first-order PDE for η,

$$\eta_t + [\alpha - B(t)\Gamma_2(\eta)]\eta_x = 0,$$

with the characteristic equations

$$\frac{dt}{1} = \frac{dx}{\alpha - B\Gamma_2(\eta)} = \frac{d\eta}{0}. \tag{2.2.35}$$

The second equation in (2.2.35) gives

$$\eta - \text{constant} = S, \text{ say}, \tag{2.2.36}$$

as the similarity variable. Setting $\Gamma_2(\eta) = l$, where l is a constant, the first equation in (2.2.35) becomes

$$\frac{dx}{dt} + \frac{B'}{B}x = A(t) - lB(t). \tag{2.2.37}$$

Here we have used (2.2.34) for α. The general solution of (2.2.37) is

$$\eta = xB(t) - \int A(t)B(t)dt + l\int B^2(t)dt \qquad (2.2.38)$$

where η, the constant of integration, serves as the similarity variable. Using (2.2.30) and (2.2.34), the solution form (2.2.21) becomes

$$u(x,t) = A(t) - x\frac{B'(t)}{B(t)} + B(t)H(\eta). \qquad (2.2.39)$$

Using (2.2.30), (2.2.34), and (2.2.38) in (2.2.26), we get

$$\left(A' - \frac{B'}{B}A\right) + x\left(2\frac{B'^2}{B^2} - \frac{B''}{B}\right) = B^3\Gamma_4(\eta). \qquad (2.2.40)$$

Equations (2.2.38) and (2.2.40) imply that

$$\Gamma_4(\eta) = m\eta + k \qquad (2.2.41)$$

where m and k are arbitrary constants. Substituting (2.2.38) and (2.2.41) in (2.2.40), we get

$$\left(A' - \frac{B'}{B}A\right) + x\left(\frac{2B'^2}{B^2} - \frac{B''}{B}\right) = B^3\left[mxB - m\int ABdt + ml\int B^2dt + k\right].$$

Equating coefficients of x and terms free of x on both sides of this equation, we get

$$BB'' - 2B'^2 + mB^6 = 0 \qquad (2.2.42)$$

$$A' - \frac{B'}{B}A = B^3\left[k - m\int ABdt + ml\int B^2dt\right]. \qquad (2.2.43)$$

Equations (2.2.42) and (2.2.43) are solved for two special cases.

(i) $m = 0$

In this case, equation (2.2.42) becomes

$$BB'' - 2B'^2 = 0, \qquad (2.2.44)$$

giving

$$B(t) = bt^{-1} \qquad (2.2.45)$$

where b is an arbitrary constant.

Using (2.2.45) and $m = 0$ in (2.2.43), we have

$$\frac{dA}{dt} + \frac{1}{t}A = \frac{kb^3}{t^3}$$

yielding a special solution

$$A(t) = -kb^3 t^{-2}. \tag{2.2.46}$$

On using (2.2.45) and (2.2.46) in (2.2.38) and (2.2.39), we have the similarity variable and the solution as

$$\eta(x,t) = bxt^{-1} - \frac{kb^4}{2}t^{-2} - lb^2 t^{-1} \tag{2.2.47}$$

and

$$u(x,t) = -kb^3 t^{-2} + xt^{-1} + bt^{-1} H(\eta), \tag{2.2.48}$$

respectively. Using (2.2.29), (2.2.33), (2.2.41), and $\Gamma_2(\eta) = l$ in (2.2.27), we find that $H(\eta)$ satisfies the first order ODE

$$(H + l)H' + k = 0 \tag{2.2.49}$$

with the solution

$$\frac{H^2}{2} + lH + k\eta = p$$

where p is the constant of integration; solving for H we have

$$H(\eta) = -l \pm \sqrt{l^2 + 2p - 2k\eta}. \tag{2.2.50}$$

Substituting (2.2.50) and (2.2.47) into (2.2.48), we get an explicit solution of (2.2.20) as

$$u(x,t) = xt^{-1} - kb^3 t^{-2} + bt^{-1} \left[-l \pm \sqrt{l^2 + 2p - 2kxbt^{-1} + 2klb^2 t^{-1} + k^2 b^4 t^{-2}} \right].$$

(ii) $l = k = 0$

Another special solution of (2.2.42) is

$$B(t) = qt^{-1/2} \tag{2.2.51}$$

provided

$$4q^4 m + 1 = 0 \tag{2.2.52}$$

where q is a constant. Correspondingly, a solution of (2.2.43) may be found to be

$$A(t) - c - al^{-1} \tag{2.2.53}$$

where a and c are arbitrary constants. Making use of (2.2.51)–(2.2.53) in (2.2.38) and (2.2.39), we get the following similarity reduction of (2.2.20):

$$\eta(x,t) = q(x - 2a)t^{-1/2} - 2qct^{1/2} \tag{2.2.54}$$

$$u(x,t) = c - at^{-1} + \frac{x}{2}t^{-1} + qt^{-1/2}H(\eta). \tag{2.2.55}$$

The ODE for this special case $l = k = 0$ governing $H(\eta)$ is obtained by using the results $\Gamma_i(\eta) = 0$, $i = 1, 2, 3$, $\Gamma_4(\eta) = m\eta$, $4q^4 m + 1 = 0$, and (2.2.54) in (2.2.27):

$$HH' - \frac{1}{4q^4}\eta = 0$$

which immediately integrates to give

$$H(\eta) = \pm\sqrt{\frac{\eta^2}{4q^4} + r} \tag{2.2.56}$$

where r is constant of integration. Using (2.2.56) and (2.2.54) in (2.2.55), we get another explicit solution of (2.2.20):

$$u(x, t) = c - at^{-1} + \frac{x}{2}t^{-1} \pm qt^{-1/2}\sqrt{r + \frac{1}{4q^2}[(x - 2a)t^{-1/2} - 2ct^{1/2}]^2}. \tag{2.2.57}$$

Example 2

$$(t + u)u_x + tu_t = x - t \tag{2.2.58}$$

Assume a solution of the form

$$u(x, t) = \alpha(x, t) + \beta(x, t)H(\eta), \eta = \eta(x, t), \beta(x, t) \neq 0. \tag{2.2.59}$$

Differentiating (2.2.59) to get u_x and u_t and substituting in (2.2.58) and rearranging terms, we get

$$\beta^2\eta_x HH' \quad + \quad \beta(t\eta_x + \alpha\eta_x + t\eta_t)H' + \beta\beta_x H^2$$

$$+(t\beta_x + \alpha\beta_x + t\beta_t + \alpha_x\beta)H$$

$$+(t\alpha_x + \alpha\alpha_x + t\alpha_t - x + t) = 0. \tag{2.2.60}$$

Equation (2.2.60) will be an ODE for $H(\eta)$ only if

$$\beta[(t + \alpha)\eta_x + t\eta_t] = \beta^2\eta_x\Gamma_1(\eta) \tag{2.2.61}$$

$$\beta\beta_x = \beta^2\eta_x\Gamma_2(\eta) \tag{2.2.62}$$

$$(t + \alpha)\beta_x + t\beta_t + \alpha_x\beta = \beta^2\eta_x\Gamma_3(\eta) \tag{2.2.63}$$

$$(t + \alpha)\alpha_x + t\alpha_t - x + t = \beta^2\eta_x\Gamma_4(\eta). \tag{2.2.64}$$

It then takes the form

$$HH' + \Gamma_1(\eta)H' + \Gamma_2(\eta)H^2 + \Gamma_3(\eta)H + \Gamma_4(\eta) = 0. \tag{2.2.65}$$

In (2.2.62), let $\Gamma_2(\eta) = \dfrac{\Omega_2'(\eta)}{\Omega_2(\eta)}$ so that it integrates and gives

$$\beta(x,t) = B(t)\Omega_2(\eta). \tag{2.2.66}$$

Using Remark 2 in Example 1, we may put $\Omega_2 \equiv 1$ in (2.2.66) and obtain

$$\beta(x,t) = B(t) \tag{2.2.67}$$

and

$$\Gamma_2(\eta) \equiv 0. \tag{2.2.68}$$

In (2.2.63) we put $\Gamma_3(\eta) = \Omega_3'(\eta)$, use (2.2.67), and integrate with respect to x to get

$$\alpha(x,t) = \left[A(t) - xt\frac{B'(t)}{B(t)}\right] + B(t)\Omega_3(\eta). \tag{2.2.69}$$

Using Remark 1 of Example 1, we may put $\Omega_3 \equiv 0$ in (2.2.69) and have

$$\alpha(x,t) = \left[A(t) - xt\frac{B'(t)}{B(t)}\right]. \tag{2.2.70}$$

Since $\Gamma_3(\eta) = \Omega_3'(\eta)$, we also have

$$\Gamma_3(\eta) \equiv 0. \tag{2.2.71}$$

Substituting (2.2.70) in (2.2.61), we get

$$\left[t + A(t) - xt\frac{B'(t)}{B(t)} - B(t)\Gamma_1(\eta)\right]\eta_x + t\eta_t = 0. \tag{2.2.72}$$

The characteristics of (2.2.72) are

$$\frac{dx}{t + A(t) - xt\frac{B'}{B} - B\Gamma_1(\eta)} = \frac{dt}{t} = \frac{d\eta}{0}. \tag{2.2.73}$$

A first integral from (2.2.73) is clearly $\eta = $ constant; this is the similarity variable.

Setting

$$\Gamma_1(\eta) = l, \text{ a constant} \tag{2.2.74}$$

in the first equation of (2.2.73), we have

$$\frac{dx}{dt} + \frac{B'(t)}{B(t)}x = 1 + \frac{A(t) - lB(t)}{t} . \tag{2.2.75}$$

The solution of (2.2.75) gives the similarity variable

$$\eta = xB(t) - \int\left[1 + \frac{A(t) - lB(t)}{t}\right]B(t)dt . \tag{2.2.76}$$

Substituting (2.2.76) into (2.2.64) and using (2.2.70), we get

$$
- \left[t + A(t) - xt\frac{B'(t)}{B(t)} \right] \frac{tB'(t)}{B(t)}
$$
$$
+ t \left[A'(t) - x\frac{B'(t)}{B(t)} - xt\frac{B''(t)}{B(t)} + xt\frac{B'^2(t)}{B^2(t)} \right]
$$
$$
- x + t = B^3(t)\Gamma_4(\eta). \tag{2.2.77}
$$

Equations (2.2.76) and (2.2.77) imply that

$$
\Gamma_4(\eta) = m\eta + k \tag{2.2.78}
$$

where m and k are constants. Equations (2.2.76) and (2.2.78), when used in (2.2.77), give

$$
-t^2\frac{B'}{B} - t\frac{AB'}{B} + xt^2\frac{B'^2}{B^2} + tA' - xt\frac{B'}{B} - xt^2\frac{B''}{B} + xt^2\frac{B'^2}{B^2} - x + t
$$
$$
= B^3 \left[mxB - m\int \left\{ 1 + \frac{A - lB}{t} \right\} B\,dt + k \right]. \tag{2.2.79}
$$

Equating coefficients of x and terms free of x on both sides of (2.2.79), we get

$$
2t^2 B'^2 - tBB' - t^2 BB'' - B^2 = mB^6, \tag{2.2.80}
$$
$$
-t^2 B' - tAB' + tA'B + tB = kB^4 - mB^4 \int \left[1 + \frac{A - lB}{t} \right] B\,dt. \tag{2.2.81}
$$

For the special case $m = 0$, (2.2.80) gives

$$
B(t) = bt. \tag{2.2.82}
$$

Substituting (2.2.82) in (2.2.81), we get

$$
\frac{dA}{dt} - \frac{1}{t}A = kb^3 t^2.
$$

A special solution of this linear equation is

$$
A(t) = \frac{k}{2}b^3 t^3. \tag{2.2.83}
$$

On using (2.2.82), (2.2.83), (2.2.70), and (2.2.67) in (2.2.76) and (2.2.59), we get

$$
\eta = bxt - \frac{bt^2}{2} - \frac{k}{8}b^4 t^4 + \frac{lb^2}{2}t^2 \tag{2.2.84}
$$
$$
u = \frac{k}{2}b^3 t^3 - x + btH(\eta). \tag{2.2.85}
$$

On using (2.2.67), (2.2.70), and (2.2.82), Equation (2.2.60) becomes

$$HH' + lH' + k = 0 \qquad (2.2.86)$$

and integrates to give

$$H = -l \pm \sqrt{l^2 + 2p - 2k\eta} \qquad (2.2.87)$$

where p is the constant of integration. Using (2.2.87) and (2.2.84) in (2.2.85), we get a similarity solution of the given PDE. It may be explicitly written as

$$u(x, t) = \frac{k}{2} b^3 t^3 - x - lbt$$

$$\pm bt \sqrt{l^2 + 2p - 2kbxt + kbt^2 - klb^2 t^2 + \frac{k^2}{4} b^4 t^4}.$$

We have obtained some special exact solutions of (2.2.58) via the direct similarity approach. A richer class of solutions may be obtained if the intermediate equations can be solved more generally.

2.3   Reduction of $u_t + u^n u_x + H(x, t, u) = 0$ to the form $U_t + U^n U_x = 0$

A large number of physical models are described by special cases of the generalised Burgers equation (GBE) (see Chapter 6)

$$u_t + u^n u_x + H(x, t, u) = \frac{\delta}{2} u_{xx}, \qquad (2.3.1)$$

where δ is the coefficient of viscosity. The inviscid limit of (2.3.1) as $\delta \to 0$ is

$$u_t + u^n u_x + H(x, t, u) = 0. \qquad (2.3.2)$$

The term $H(x, t, u)$ in (2.3.2) may represent the effects of damping, geometrical spreading, or sources of some sort. Equation (2.3.2) plays an important role in the analytical theory of GBEs.

We seek the most general transformation of the type

$$\tau = \tau(x, t) \qquad (2.3.3)$$
$$y = y(x, t) \qquad (2.3.4)$$
$$U(y, \tau) = f(x, t)u(x, t) \qquad (2.3.5)$$

which reduces (2.3.2) to the form

$$U_\tau + U^n U_y = 0. \qquad (2.3.6)$$

A more general form $U = F(x, t, u)$ is not considered in order that Rankine-Hugoniot conditions for (2.3.2) and (2.3.6) remain the same.

We assume that $f(x, t) > 0$ and

$$J = \begin{vmatrix} y_t & y_x \\ \tau_t & \tau_x \end{vmatrix} \neq 0. \tag{2.3.7}$$

Differentiating (2.3.5) with respect to x and t, we get

$$U_y y_x + U_\tau \tau_x = f u_x + f_x u$$

$$U_y y_t + U_\tau \tau_t = f u_t + f_t u. \tag{2.3.8}$$

Solving for U_τ and U_y from (2.3.8), we have

$$U_\tau = -\frac{1}{J}\left[(y_x f_t - y_t f_x)u + y_x f u_t - y_t f u_x\right]$$

$$\tag{2.3.9}$$

$$U_y = -\frac{1}{J}\left[(\tau_t f_x - \tau_x f_t)u + \tau_t f u_x - \tau_x f u_t\right].$$

Substituting (2.3.9) in (2.3.6), we get

$$-\frac{1}{J}[(y_x f_t - y_t f_x)u + y_x f u_t - y_t f u_x] - \frac{f^n u^n}{J}[(\tau_t f_x - \tau_x f_t)u + \tau_t f u_x - \tau_x f u_t] = 0$$

or

$$u_t + \left(\frac{f_t}{f} - \frac{y_t f_x}{y_x f}\right)u - \frac{y_t}{y_x}u_x$$

$$+ f^n u^n \left[\left(\frac{\tau_t f_x}{y_x f} - \frac{\tau_x f_t}{y_x f}\right)u + \frac{\tau_t}{y_x}u_x - \frac{\tau_x}{y_x}u_t\right] = 0. \tag{2.3.10}$$

For (2.3.10) to be of the form (2.3.2), we must have

$$y_t = 0, \quad \tau_x = 0, \quad \text{and} \quad \frac{f^n \tau_t}{y_x} = 1. \tag{2.3.11}$$

Equation (2.3.10) then takes the form

$$u_t + u^n u_x + \frac{f_t}{f}u + \frac{f_x}{f}u^{n+1} = 0. \tag{2.3.12}$$

From (2.3.11) we see that y is a function of x alone and τ is a function of t alone. Equation (2.3.11$_3$) then becomes

$$f = \left[\frac{y'(x)}{\tau'(t)}\right]^{1/n}. \tag{2.3.13}$$

Let

$$G(t) = \frac{f_t}{f} = -\frac{1}{n}\frac{\frac{d^2\tau}{dt^2}}{\frac{d\tau}{dt}} = -\frac{d}{dt}\ln\left[\left(\frac{d\tau}{dt}\right)^{\frac{1}{n}}\right] \qquad (2.3.14)$$

and

$$F(x) = \frac{f_x}{f} = \frac{1}{n}\frac{\frac{d^2 y}{dx^2}}{\frac{dy}{dx}}$$

$$= \frac{d}{dx}\ln\left[\left(\frac{dy}{dx}\right)^{\frac{1}{n}}\right]. \qquad (2.3.15)$$

Equation (2.3.12) can now be written as

$$u_t + u^n u_x + G(t)u + F(x)u^{n+1} = 0 \qquad (2.3.16)$$

where $G(t)$ and $F(x)$ are given by (2.3.14) and (2.3.15). Thus, $H(x,t,u)$ in (2.3.2) must be of the form $G(t)u + F(x)u^{n+1}$. Conversely, for given $G(t)$ and $F(x)$, the relations (2.3.14) and (2.3.15) determine the transformation functions τ and y in (2.3.3) and (2.3.4).

Equation (2.3.14) may be written as

$$\tau(t) = \int^t \left(\exp\left(\int^s G(s_1)ds_1\right)\right)^{-n} ds. \qquad (2.3.17)$$

Similarly, from (2.3.15)

$$y(x) = \int^x \left(\exp\left(\int^s F(s_1)ds_1\right)\right)^n ds. \qquad (2.3.18)$$

Therefore,

$$f(x,t) = \exp\left(\int^t G(s)ds\right)\exp\left(\int^x F(s_1)ds_1\right). \qquad (2.3.19)$$

Thus, we have the following result: the most general equation of the form (2.3.2) that can be reduced to (2.3.6) by the transformation (2.3.3)-(2.3.5) is (2.3.16); the transformation itself is given by (2.3.17) - (2.3.19).

Equations of the form (2.3.16) appear in many physical applications. Nimmo and Crighton (1986) considered the case $n = 1$ with $F(x) \equiv 0$ and $G(t) = \left(\frac{j}{2t} + \alpha\right), j = 0, 1, 2$. In this case, (2.3.16) takes the form

$$u_t + uu_x + \left(\frac{j}{2t} + \alpha\right)u = 0. \qquad (2.3.20)$$

From (2.3.17), (2.3.18), and (2.3.19) we get the transformation

$$\tau = \int^t \left\{\exp\left(\int^s \left(\frac{j}{2s_1} + \alpha\right)ds_1\right)\right\}^{-1} ds$$

$$= \int^t s^{-j/2} e^{-\alpha s} ds$$

$$y = \int^x (e^0)^1 ds = x; U(y, \tau) = f(x, t)u \qquad (2.3.21)$$

where

$$f(x, t) = \exp \int^t \left(\frac{j}{2s} + \alpha \right) ds. \exp \int^x 0 ds_1 = t^{j/2} e^{\alpha t}.$$

This changes (2.3.20) to the form

$$U_t + U U_y = 0.$$

Lefloch (1988) considered the special case of (2.3.16) for $n = 1$, $G(t) \equiv 0$, and $F(x) = \dfrac{\beta}{x}$:

$$u_t + u u_x + \frac{\beta}{x} u^2 = 0. \qquad (2.3.22)$$

The transformation which reduces (2.3.22) to $U_\tau + U U_y = 0$ is

$$y = \int^x \left(\exp \left(\int^s \frac{\beta}{s_1} ds_1 \right) \right) ds$$

$$= \frac{x^{\beta+1}}{\beta + 1}$$

$$\tau = \int^t \left(\exp \int^s 0. ds_1 \right)^{-1} ds$$

$$= t$$

$$U = f(x, t)u = x^\beta u$$

since

$$f(x, t) = \exp \left(\int^t 0. ds \right) \cdot \exp \left(\int^x \frac{\beta}{s_1} ds_1 \right) = x^\beta.$$

The inviscid limit of Burgers-Fisher equation

$$u_t + u u_x + u(u - 1) = \frac{\delta}{2} u_{xx} \qquad (2.3.23)$$

is

$$u_t + u u_x + u^2 - u = 0. \qquad (2.3.24)$$

This is a special case of (2.3.16) with $n = 1$, $F(x) = 1$, and $G(t) = -1$. The transformation which changes (2.3.24) to $U_\tau + U U_y = 0$ is

$$y = \int^x \left(\exp \left(\int^s 1 ds_1 \right) \right) ds = \int^x e^s ds = e^x$$

$$\tau = \int^t \left(\exp\left(\int^s (-1) ds_1 \right) \right)^{-1} ds = \int^t (e^{-s})^{-1} ds = e^t$$

$$U(y, \tau) = f(x, t)u = e^{x-t} u(x, t)$$

since

$$f(x, t) = \exp\left(\int^t (-1) ds \right) \exp\left(\int^x 1.ds_1 \right) = e^{-t} \cdot e^x.$$

Murray (1970) considered the equation $u_t + g(u)u_x + \lambda u^\alpha = 0$ where $g'(u) > 0$ for $u > 0$ and $\lambda > 0$ is a constant (see Section 2.4). We consider a special case $g(u) = u$ and $\alpha = 2$, namely $u_t + uu_x + \lambda u^2 = 0$. This is (2.3.16) with $F(x) = \lambda, G(t) = 0$ and $n = 1$:

$$u_t + uu_x + \lambda u^2 = 0. \tag{2.3.25}$$

The transformation which reduces (2.3.25) to $U_\tau + UU_y = 0$ is

$$y = \int^x \left(\exp\left(\int^s \lambda ds_1 \right) \right) ds = \int^x e^{\lambda s} ds = \frac{e^{\lambda x}}{\lambda}$$

$$\tau = \int^t \left(\exp\left(\int^s 0.ds_1 \right) \right)^{-1} ds = \int^t ds = t$$

$$U = f(x, t)u = e^{\lambda x} u$$

since

$$f(x, t) = \exp\left(\int^t 0.ds \right) \cdot \exp\left(\int^x \lambda ds_1 \right) = e^{\lambda x}.$$

In the problem of propagation of waves in tubes we get the following equation for right-running waves (Shih (1974)):

$$u_t + \left(a_0 + \frac{\gamma + 1}{2} u \right) u_x + \frac{F}{4D} u^2 = 0 \tag{2.3.26}$$

where F, D are a constants (see also Crighton (1979)).
    With

$$t' = \frac{F}{4D} t \quad \text{and} \quad x' = \frac{2}{\gamma + 1} \frac{F}{4D} x,$$

Equation (2.3.26) reduces (after dropping primes) to

$$u_t + (a + u)u_x + u^2 = 0, a = \frac{2a_0}{\gamma + 1}. \tag{2.3.27}$$

With $\bar{x} = x - at$, (2.3.27) changes to

$$u_t + uu_{\bar{x}} + u^2 = 0. \tag{2.3.28}$$

Equation (2.3.28) is a special case of (2.3.25) with $\lambda = 1$. Therefore, the transformation $\bar{t} = t$, $y = e^{\bar{x}}$, $U = e^{\bar{x}} u(x,t) = yu(\ln y, t)$ reduces (2.3.28) to the form $U_{\bar{t}} + UU_y = 0$; here we assume that $y > 0, t > 0$.

We carried out a detailed analysis for the reduction of $u_t + uu_x = 0$ to an ODE by the direct approach of Clarkson and Kruskal (1989) in Section 2.2. A similar analysis may be done for (2.3.6) for $n \geq 2$ to find its symmetries and, hence, the solution.

2.4　Initial Value Problem for $u_t + g(u)u_x + \lambda h(u) = 0$

An obvious generalization of the equation $u_t + u^n u_x = 0$ discussed in detail in Section 2.3 is

$$u_t + g(u)u_x + \lambda h(u) = 0 \tag{2.4.1}$$

where $\lambda \geq 0$ is a parameter and $g(u)$ and $h(u)$ are nonnegative functions of u such that $g_u(u) > 0, h_u(u) > 0$ for $u > 0$.

Many model equations in applications are special cases of (2.4.1). In particular, when $h(u)$ can be negative for some u, interesting phenomena appear; they occur in a model for the Gunn effect (Murray (1970)) (see also Section 2.3). While it is not possible to give an explicit general discussion of (2.4.1), much progress can be made when $h(u) = O(u^\alpha)$, $\alpha > 0$, $0 < u \ll 1$. Indeed, Murray (1970) has shown that in this case, a finite initial disturbance zero outside a finite range in x decays (i) within a finite time and finite distance for $0 < \alpha < 1$ and is unique under certain conditions, (ii) within an infinite time like $O(\exp -\lambda t)$ and in a finite distance for $\alpha = 1$, and (iii) within an infinite time and distance like $O(t^{-1/(\alpha-1)})$ for $1 < \alpha \leq 3$ and $O(t^{-1/2})$ for $\alpha \geq 3$. The asymptotic speed of propagation of the discontinuity was given in each case together with its role in the decay process. We follow Murray (1970) closely in this section. After giving some results regarding the general Equation (2.4.1), we give a detailed analysis for the simpler case $u_t + (u + a)u_x + \lambda u = 0$, which displays many interesting features and is itself a descriptor of some physical phenomenon. It is a limiting case of the Burgers equation with damping, $u_t + (u+a)u_x + \lambda u = \dfrac{\delta}{2}u_{xx}$, as $\delta \to 0$, and plays an important role in its analysis. In the following section we shall discuss more recent work of Bukiet, Pelesko, Li, and Sachdev (1996), where special cases of (2.4.1) admitting similarity form of solutions would be studied. In this work, a numerical scheme for (2.4.1) was developed and the asymptotic nature of the exact solutions confirmed.

An initial-boundary value problem for (2.4.1) is posed as follows:

$$u(0,t) = 0, \quad t > 0$$

$$u(x,0) = u_0(x) = \begin{cases} 0 & x < 0 \\ f(x) & 0 < x < X \\ 0 & X < x \end{cases} \qquad (2.4.2)$$

where $f(x)$ is such that

$$0 \le f(x) \le 1. \qquad (2.4.3)$$

With $g(u)$ a monotonic increasing function, weak or discontinuous solutions of (2.4.1) occur when $\lambda = 0$ for some value of $t > 0$, even for smooth functions $u_0(x)$ (see Section 2.2). If a discontinuity exists at $t = 0$, its propagation and decay are considered from the beginning.

Let the path of the shock discontinuity in the (x, t)-plane be given by

$$x = x_s(t). \qquad (2.4.4)$$

The Rankine-Hugoniot condition which holds across the shock is

$$\frac{dx_s}{dt} = \frac{1}{u_1 - u_2} \int_{u_2}^{u_1} g(u)du \qquad (2.4.5)$$

where $u_1(t)$ and $u_2(t)$ are the values of $u(x, t)$ at x_s- and x_s+, respectively. This can be obtained by applying the Gauss theorem to (2.4.1) across the shock. For simplicity we require $u = 0$ to be a solution of (2.4.1), implying that $h(0) = 0$. Equation (2.4.1) shows that, along the characteristics, we have

$$\frac{dx}{dt} = g(u)$$

$$\frac{du}{dt} + \lambda h(u) = 0. \qquad (2.4.6)$$

In parametric form we have

$$\frac{dx}{d\sigma} = g(u), \frac{dt}{d\sigma} = 1$$

$$\frac{du}{d\sigma} = -\lambda h(u) \qquad (2.4.7)$$

where σ is a parameter measured along the characteristics.

The solution of (2.4.7) may be obtained as

$$x(\sigma) = \xi + \int_0^\sigma g[u(x(\tau), \tau)]d\tau$$

$$t(\sigma) = \sigma \qquad (2.4.8)$$

$$\int_{f(\xi)}^u \frac{ds}{h(s)} = -\lambda\sigma.$$

Here, $t = 0$ when $\sigma = 0$, and ξ is the value of x at $t = 0$. Let t_c be the critical time beyond which the solution (2.4.8) ceases to be single-valued and a shock is formed.

Let

$$\int_{f(\xi)}^{u} \frac{ds}{h(s)} = H(u) - H(f(\xi)) \tag{2.4.9}$$

so that

$$H'(u) = \frac{1}{h(u)}. \tag{2.4.10}$$

The integration of (2.4.8) yields

$$H(u) = H(f(\xi)) - \lambda\sigma \tag{2.4.11}$$

$$
\begin{aligned}
u(\sigma) &= H^{-1}[H(f(\xi)) - \lambda\sigma] \\
&= G[H(f(\xi)) - \lambda\sigma]
\end{aligned}
\tag{2.4.12}
$$

where the inverse function $G = H^{-1}$ exists since H is monotonic. On using (2.4.12), we get from (2.4.8$_1$)

$$x = \xi + \int_0^t g[G\{H(f(\xi)) - \lambda\tau\}]d\tau. \tag{2.4.13}$$

To find when the solution ceases to be single-valued, we differentiate (2.4.13) with respect to ξ and equate the result to zero. We find that the earliest time t_c at which the shock is formed satisfies

$$0 = 1 + \int_0^{t_c} g'[G\{H(f(\xi)) - \lambda\tau\}]G'\{H(f(\xi)) - \lambda\tau\}H'(f(\xi))f'(\xi)d\tau,$$

that is,

$$
\begin{aligned}
1 &= \int_0^{t_c} \frac{d}{d\tau} g[G\{H(f(\xi)) - \lambda\tau\}]\frac{1}{h(f(\xi))}f'(\xi)d\tau \\
&= \frac{1}{\lambda}\frac{f'(\xi)}{h(f(\xi))}[g(G\{H(f(\xi)) - \lambda t_c\}) - g(f(\xi))]. \tag{2.4.14}
\end{aligned}
$$

Here we have made use of the fact that $GH(f(\xi)) = f(\xi)$. When $\lambda = 0$, (2.4.12) gives $u(\sigma) = GH(f(\xi)) = f(\xi)$. Therefore, from (2.4.12) and (2.4.14) we get

$$t_c = \left\{[-g'(f(\xi))\, f'(\xi)]^{-1}\right\}_{min}. \tag{2.4.15}$$

Now we consider in some detail the special case

$$u_t + (u + a)u_x + \lambda u = 0, a \geq 0, \lambda > 0; \tag{2.4.16}$$

here $g(u) = u + a$ and $h(u) = u$, and so

$$H(u) \;=\; \int \frac{du}{u} = \ln u$$

$$G(u) \;=\; H^{-1}(u) = e^u$$

$$H(f(\xi)) - \lambda\sigma \;=\; \ln(f(\xi)) - \lambda\sigma$$

$$u(\sigma) \;=\; G\{H(f(\xi)) - \lambda\sigma\} = e^{\ln(f(\xi)) - \lambda\sigma}$$

$$\;=\; f(\xi)e^{-\lambda\sigma}. \tag{2.4.17a}$$

Using (2.4.8) we have

$$x(\sigma) \;=\; \xi + \int_0^\sigma [u(x(\tau), \tau) + a]d\tau \tag{2.4.17b}$$

$$t \;=\; \sigma.$$

With smooth initial data, a shock will form at the time $t = t_c$ obtained from (2.4.14):

$$1 = \frac{1}{\lambda}\frac{f'(\xi)}{f(\xi)}[f(\xi)e^{-\lambda t_c} + a - (f(\xi) + a)]$$

or

$$\lambda = f'(\xi)(e^{-\lambda t_c} - 1).$$

The earliest time for shock formation, therefore, is

$$t_c = \frac{1}{\lambda}\left[\ln\frac{f'(\xi)}{f'(\xi) + \lambda}\right]_{\min}. \tag{2.4.18}$$

For t_c to be positive we must have $-f'(\xi) > \lambda$ for some $0 \le \xi < X$.

If a t_c does not exist, then the solution of IVP is given by (2.4.17a) for all $t \ge 0$. It decays exponentially as $t \to \infty$. When $a = 0$, (2.4.17b) gives

$$x \;=\; \xi + \int_0^\sigma u(x(\tau), \tau)d\tau$$

$$\;=\; \xi + \int_0^\sigma f(\xi)e^{-\lambda\tau}d\tau$$

$$\;=\; \xi + \frac{f(\xi)}{\lambda}(1 - e^{-\lambda\sigma}).$$

Recalling that $\sigma = t$, we have

$$x \to \xi + \frac{f(\xi)}{\lambda} \text{ as } t \to \infty.$$

From this it follows that in the limit $t \to \infty$,

$$x \leq x_m = \left[\xi + \frac{1}{\lambda}f(\xi)\right]_{\max}, \quad 0 \leq \xi \leq X. \tag{2.4.19}$$

Thus, the solution $u(x,t)$ decays to zero in a finite distance but exponentially in time. The solution does not decay in a finite distance if $a > 0$.

We consider a form of initial condition $u_0(x)$ with a shock present at $x = X$ having $u_2 = 0$ for all $t \geq 0$. The characteristic solution (2.4.17a)-(2.4.17b) holds for all x and t, including $x = x_s\pm$, that is, $u(x_s-,t) = u_1(t)$, $u(x_s+,t) = u_2(t) = 0$. The shock speed is given by (2.4.5):

$$\frac{dx_s}{dt} = \frac{1}{u_1 - 0}\int_0^{u_1}(u+a)du = \frac{1}{2}u_1(t) + a. \tag{2.4.20}$$

Put $x = x_s(\equiv x_s-)$ in (2.4.17b) to get

$$x_s = \xi + \int_0^t [u(x(\tau),\tau) + a]d\tau, \sigma = t \tag{2.4.21}$$

just behind the shock.

On differentiation x_s with respect to t, we have

$$\begin{aligned}
\frac{dx_s}{dt} &= \frac{d}{dt}\xi(x_s,t) + u(x_s,t) + a + \int_0^t \frac{\partial}{\partial t}\left[f(\xi(x_s,t))e^{-\lambda\tau} + a\right]d\tau \\
&= \frac{d\xi}{dt} + u_1(t) + a - \frac{1}{\lambda}(e^{-\lambda t} - 1)\frac{df(\xi)}{dt}. \tag{2.4.22}
\end{aligned}$$

From (2.4.17a) we get

$$f(\xi(x_s,t)) = e^{\lambda t}u_1(t) \tag{2.4.23}$$

and

$$\frac{d}{dt}f(\xi(x_s,t)) = e^{\lambda t}\left[\frac{du_1}{dt} + \lambda u_1\right].$$

Since $f(x)$ is a monotonic increasing function for $0 \leq x < X$, its inverse exists. Let $f^{-1} = F$. Then (2.4.23) becomes

$$\xi(x_s,t) = F(e^{\lambda t}u_1(t)). \tag{2.4.24}$$

Differentiating (2.4.24), we get

$$\frac{d}{dt}\xi(x_s,t) = F'(e^{\lambda t}u_1)\left[\frac{du_1}{dt} + \lambda u_1\right]e^{\lambda t}. \tag{2.4.25}$$

Equating (2.4.20) and (2.4.22) and using (2.4.25) and (2.4.23) therein, we get

$$\frac{1}{2}u_1(t) + a = F'(e^{\lambda t}u_1)\left[\frac{du_1}{dt} + \lambda u_1\right]e^{\lambda t} + u_1(t) + a$$
$$- \frac{1}{\lambda}(e^{-\lambda t} - 1)e^{\lambda t}\left[\frac{du_1}{dt} + \lambda u_1\right]$$

or

$$\frac{1}{\lambda}\frac{du_1}{dt} = (1 - e^{\lambda t})^{-1}\left\{u_1\left(e^{\lambda t} - \frac{1}{2}\right) + e^{\lambda t}F'(e^{\lambda t}u_1)\left[\frac{du_1}{dt} + \lambda u_1\right]\right\}.$$
$$(2.4.26)$$

The solutions $u_1(t)$ of equation (2.4.26) will now be studied. Considering $f(x)$ as in Figure 2.1 and letting $\delta \to 0$, we get the top-hat situation as shown in Figure 2.2. For $\delta = 0$ we have initially $u = 0$ for $x < 0$ and $u = 1$ for $0 < x < X$. Thus, we have a centered simple wave at $x = 0$. Therefore, for $t \geq 0$, (2.4.17a) holds with $u = u_1$ and $f(\xi) = 1$ and, since $\sigma = t$, we have

$$u_1(t) = e^{-\lambda t}. \qquad (2.4.27)$$

Equation (2.4.20) now becomes

$$\frac{dx_s}{dt} = \frac{1}{2}e^{-\lambda t} + a$$

which, on integration from 0 to t, gives

$$x_s(t) = X + at + \frac{1}{2\lambda}(1 - e^{-\lambda t}) \qquad (2.4.28)$$

where $x_s(0) = X$.

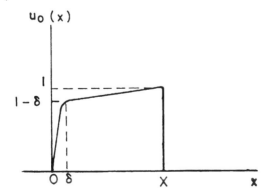

Figure 2.1. Typical initial profile for $u(x, 0)$.

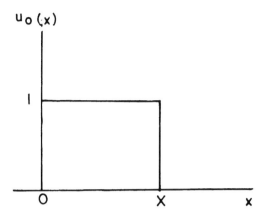

Figure 2.2. Top hat initial condition $u(x, 0)$.

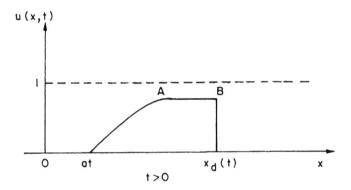

Figure 2.3. $u(x, t)$ when the point A never overtakes B, and for all x such that $u(x, t) > 0$, $x - X-$ at $< 1/(?\lambda)$ for all $t \geq 0$.

Equation (2.4.27) can be obtained also from (2.4.26) by letting $F' \to \infty$. The solution at this stage is shown in Figure 2.3.

This solution is valid for $t \leq t_0$ where t_0 is the time at which the first characteristic of the centred wave at $x = 0$ catches up with the shock, that is, when the point A in the Figure 2.3 catches up with the shock at B. From (2.4.12) we get the distance travelled by A by putting $\xi = 0$ and $f(\xi) = 1$. At $t = t_0$ (2.4.17b) gives

$$
\begin{aligned}
x(t_0) &= \int_0^{t_0} (e^{-\lambda \tau} + a) d\tau \\
&= at_0 + \frac{1}{\lambda}(1 - e^{-\lambda t_0}).
\end{aligned}
\tag{2.4.29}
$$

From (2.4.28) we get

$$
\begin{aligned}
x_s(t_0) &= X + at_0 + \frac{1}{2\lambda}(1 - e^{-\lambda t_0}) \\
&= at_0 + \frac{1}{\lambda}(1 - e^{-\lambda t_0})
\end{aligned}
$$

on using (2.4.29). Therefore,

$$
e^{-\lambda t_0} = 1 - 2\lambda X. \tag{2.4.30}
$$

The equation satisfied by $u_1(t)$ for $t \geq t_0$ is given by (2.4.26) (assuming t_0 exists) with $F' = 0$:

$$
\frac{1}{\lambda}\frac{du_1}{dt} = (1 - e^{-\lambda t})^{-1} u_1 \left(e^{\lambda t} - \frac{1}{2}\right). \tag{2.4.31}
$$

This solution must match the solution obtained from (2.4.27) at $t = t_0$. Therefore, on using (2.4.30), we have

$$
u_1(t_0) = e^{-\lambda t_0} = 1 - 2\lambda X. \tag{2.4.32}
$$

We consider two cases arising from (2.4.30).

i) $2\lambda X > 1$.

In this case, t_0 in (2.4.30) does not exist, so $u_1(t)$ given by (2.4.27) is valid for all $t \geq 0$, $x_s(t)$ is given by (2.4.28), and u is found parametrically from (2.4.17) as

$$
\begin{aligned}
u(\xi, t) &= f(\xi)e^{-\lambda t} \\
x &= \int_0^t [f(\xi)e^{-\lambda \tau} + a]d\tau \\
&= at + \frac{f(\xi)}{\lambda}(1 - e^{-\lambda t}). \tag{2.4.33}
\end{aligned}
$$

ii) $2\lambda X < 1$.

A finite t_0 exists, and $u_1(t)$ is given by (2.4.27) for $t < t_0$. For $t > t_0$, we solve (2.4.31) subject to (2.4.32) and obtain

$$
x_s(t) = X + at + \frac{1}{2\lambda}(1 - e^{-\lambda t_0})^{1/2}[2(1 - e^{-\lambda t})^{1/2} - (1 - e^{-\lambda t_0})^{1/2}] \tag{2.4.34}
$$

where we have used the condition $x_s(t_0)$ from (2.4.28). It follows from (2.4.34) that

$$
\begin{aligned}
x_s(t) - X - at \;&\leq\; \frac{1}{2\lambda}(1 - e^{-\lambda t_0})^{1/2}\left[2 - (1 - e^{-\lambda t_0})^{1/2}\right] \\
&=\; \frac{1}{2\lambda}(2\lambda X)^{1/2}[2 - (2\lambda X)^{1/2}] \\
&=\; \left(\frac{2X}{\lambda}\right)^{1/2} - X \qquad\qquad (2.4.35)
\end{aligned}
$$

where we have used (2.4.32). For $t \geq t_0$, the solution for $u(x,t)$ for $at \leq x < x_s(t)$ is given parametrically by (2.4.33).

As mentioned in the introduction to this section, Murray (1970) also considered the general Equation (2.4.1) with $h(u) = u^\alpha, \alpha > 0, u \ll 1$, including the limiting case $\lambda \to 0$.

Bukiet, Pelesko, Li, and Sachdev (1996) devised a characteristic-based numerical scheme for first-order PDEs and verified the asymptotic results of Murray with reference to the following initial conditions for the special case of (2.4.1), namely

$$
u_t + (\gamma u^\beta)_x + \lambda u^\alpha = 0. \qquad\qquad (2.4.36)
$$

(i) Smooth IC

$$
u(x,0) = \begin{cases} 0 & x < 0 \\ \sin^2(\pi x) & 0 \leq x \leq 1 \\ 0 & 1 < x \end{cases} \qquad\qquad (2.4.37)
$$

The parameters in (2.4.36) were chosen to be $\gamma = 1/2, \beta = 2, \lambda = \pi/2$, and $\alpha = 1$.

(ii) Top hat IC

$$
u(x,0) = \begin{cases} 0 & x < 0 \\ h & 0 < x < X \\ 0 & X < x \end{cases} \qquad\qquad (2.4.38)
$$

The parameters in (2.4.36) for this IC were

$$
\gamma = \frac{1}{2}, \beta = 2, \lambda = 1, \text{ and } \alpha = 1.5, 2.5, 4.
$$

For the continuous IC (2.4.37), the formation of the shock and its subsequent propagation were studied numerically. Asymptotic decay law agreed with the analytic formulae of Murray (1970).

For the top hat IC (2.4.38), different cases were considered: when the rarefaction wave catches up to the shock and when it does not. Again, the analytic results of Murray (1970) were confirmed numerically.

2.5 Initial Value Problem for $u_t + u^\alpha u_x + \lambda u^\beta = 0$

In this section we continue the analysis of Section 2.4, but restrict ourselves to Equation (2.5.1) below, which is a special case of (2.4.1) with $g(u) = u^\alpha, h(u) = u^\beta$. With this choice it becomes possible to find explicit solutions for many cases either by the method of characteristics or by reduction to an ODE via similarity analysis. Apart from finding explicit solutions, the concern here is to demonstrate the limiting nature of the similarity solution. We follow the work of Bukiet, Pelesko, Li, and Sachdev (1996). An important contribution of this paper is the development of a characteristic-based numerical scheme for nonlinear scalar hyperbolic equations, which involves the solving of ODEs. The solution thus computed displays sharp, well-defined shocks when they exist. The analytic solutions found here demonstrate the efficacy of the numerical scheme developed by Bukiet et al. (1996).

Consider the equation

$$u_t + u^\alpha u_x + \lambda u^\beta = 0 \tag{2.5.1}$$

with the top hat initial data

$$u(x,0) = \begin{cases} 0 & x < 0 \\ h & 0 < x < X \\ 0 & X < x \end{cases} \tag{2.5.2}$$

where h, α and β are positive constants; $\lambda > 0$ is the dissipative constant.

If $\alpha = 0$, the solution is a decaying travelling wave moving to the right with speed 1. For $\lambda = 0$, (2.5.1) reduces to

$$u_t + u^\alpha u_x = 0, \tag{2.5.3}$$

and so

$$\frac{du}{dt} = 0 \text{ along the characteristic curves } \frac{dx}{dt} = u^\alpha. \tag{2.5.4}$$

That is, u is constant along the characteristics $\dfrac{dx}{dt} = u^\alpha$ which are straight lines in the (x, t) plane. The initial condition $u = 0$ for $x < 0$ and $u = h$ for $x > 0$ give rise to a rarefaction wave centred at $x = 0$. All values from 0 to h propagate along the characteristics of the rarefaction wave. Since $\alpha > 0$, the characteristic with value h for u has the highest speed: $\dfrac{dx}{dt} = h^\alpha$. The equation of this characteristic is obtained by integrating (2.5.4):

$$x_F = h^\alpha t \tag{2.5.5}$$

where we have used the condition $x_F = 0$ at $t = 0$, x_F denoting the front of the rarefaction wave. A shock originates at $x = X$ (since $u = h$ for $x < X$ and $u = 0$ for $x > X$). The motion of this shock is given by

$$
\begin{aligned}
\frac{dx_s}{dt} &= \frac{1}{0 - u_b} \int_{u_b}^{0} u^\alpha du \\
&= \frac{u_b^\alpha}{\alpha + 1}.
\end{aligned}
\tag{2.5.6}
$$

Here, u_a and u_b are values of u immediately ahead of and behind the shock, respectively. Before the rarefaction catches up with the shock, the value of u_b is h. Therefore,

$$
\frac{dx_s}{dt} = \frac{h^\alpha}{\alpha + 1}
$$

or

$$
x_s = \frac{h^\alpha}{\alpha + 1} t + X, \quad 0 \le t \le t_0,
\tag{2.5.7}
$$

since $x_s = X$ at $t = 0$; t_0 is the time at which the rarefaction catches up with the shock. Thus the solution for $x_F < x < x_s$ is $u = h$. In order to solve for u in the rarefaction $(0 \le x \le x_F)$, we use (2.5.4). Thus,

$$
\frac{dx}{dt} = u^\alpha, \quad u = C \text{ (constant)},
$$

therefore,

$$
x = C^\alpha t,
$$

since $x = 0, t = 0$ in the rarefaction wave. We readily have

$$
u = \left(\frac{x}{t}\right)^{\frac{1}{\alpha}}.
\tag{2.5.8}
$$

From (2.5.5) and (2.5.7) we find that the rarefaction wave catches up with the shock at the time t_0 when $x_s = x_F$, that is,

$$
\frac{h^\alpha}{\alpha + 1} t_0 + X = h^\alpha t_0
$$

or

$$
t_0 = \frac{(\alpha + 1)X}{\alpha h^\alpha}.
\tag{2.5.9}
$$

At this time the position of the shock is

$$
\begin{aligned}
x_s &= \frac{h^\alpha}{\alpha + 1} \frac{(\alpha + 1)X}{\alpha h^\alpha} + X \\
&= \frac{\alpha + 1}{\alpha} X
\end{aligned}
\tag{2.5.10}
$$

where we have used (2.5.7) and (2.5.9).

For $t > t_0$, the motion of the shock can be found from Equations (2.5.6) and (2.5.8). The value of u_b is now less than h. On using (2.5.8), (2.5.6) becomes

$$\frac{dx_s}{dt} = \frac{1}{\alpha + 1}\left(\frac{x_s}{t}\right)$$

which, on integration, yields

$$x_s = Kt^{\frac{1}{\alpha+1}} \tag{2.5.11}$$

where K is a constant. From (2.5.10) we have $x_s = \dfrac{\alpha+1}{\alpha}X$ at $t = t_0$; therefore,

$$\frac{\alpha+1}{\alpha}X = Kt_0^{\frac{1}{\alpha+1}}. \tag{2.5.12}$$

Equation (2.5.12) gives the value of K as

$$\begin{aligned}
K &= \frac{\alpha+1}{\alpha}\frac{X}{t_0^{\frac{1}{\alpha+1}}} \\
&= (\alpha+1)\left[\frac{1}{\alpha+1}\left(\frac{hX}{\alpha}\right)^\alpha\right]^{\frac{1}{\alpha+1}}
\end{aligned}$$

where we have made use of (2.5.9). Using this value of K in (2.5.11), we have

$$x_s = (\alpha+1)\left[\frac{t}{\alpha+1}\left(\frac{hX}{\alpha}\right)^\alpha\right]^{\frac{1}{\alpha+1}}. \tag{2.5.13}$$

The results can now be summarised as follows:

$$u(x,t) = \begin{cases} \left(\dfrac{x}{t}\right)^{1/\alpha} & 0 \le x \le x_F \\ h & x_F < x < x_s, \ t \le t_0 \\ 0 & x > x_s \end{cases} \tag{2.5.14}$$

and

$$u(x,t) = \begin{cases} \left(\dfrac{x}{t}\right)^{1/\alpha} & 0 \le x \le x_s \\ 0 & x > x_s, \ t \ge t_0. \end{cases} \tag{2.5.15}$$

ii) $\lambda \neq 0$, $\beta = 1$. Here we have the equation

$$u_t + u^\alpha u_x + \lambda u = 0 \tag{2.5.16}$$

so that

$$\frac{du}{dt} + \lambda u = 0 \text{ along the characteristic curves } \frac{dx}{dt} = u^\alpha. \tag{2.5.17}$$

Integration of (2.5.17) gives

$$u = u_0 e^{-\lambda t}$$

where $u = u_0$ at $t = 0$. Since u varies from 0 to h in the rarefaction wave, decay of the height of the top hat is given by $\max u = he^{-\lambda t}$. Using this result in the characteristic direction in (2.5.17), the position of the wavefront is found to be

$$x_F = \frac{h^\alpha}{\lambda\alpha}(1 - e^{-\lambda\alpha t}), \qquad (2.5.18)$$

where we have inserted the 1C $x_F = 0$ at $t = 0$. The equation for the motion of the shock wave is

$$\frac{dx_s}{dt} = \frac{\int_{u_b}^0 u^\alpha du}{0 - u_b} = \frac{(he^{-\lambda t})^\alpha}{\alpha + 1} \qquad (2.5.19)$$

which, on integration and use of 1C $x_s = X$ at $t = 0$, gives

$$x_s = X + \frac{h^\alpha}{\lambda\alpha(\alpha + 1)}(1 - e^{-\lambda\alpha t}). \qquad (2.5.20)$$

Equation (2.5.20) gives the shock trajectory. The time t_0 at which the front of the rarefaction catches up with the shock is obtained by equating (2.5.18) and (2.5.20),

$$\frac{h^\alpha}{\lambda\alpha}(1 - e^{-\lambda\alpha t_0}) = X + \frac{h^\alpha}{\lambda\alpha(\alpha + 1)}(1 - e^{-\lambda\alpha t_0}), \qquad (2.5.21)$$

yielding

$$t_0 = -\frac{1}{\lambda\alpha}\ln\left[1 - \frac{(\alpha + 1)\lambda X}{h^\alpha}\right]. \qquad (2.5.22)$$

Thus, the rarefaction wave catches up with the shock only if

$$1 - \frac{(\alpha + 1)\lambda X}{h^\alpha} > 0,$$

that is,

$$h > [(\alpha + 1)\lambda X]^{\frac{1}{\alpha}}. \qquad (2.5.23)$$

If t_0 exists, the location of the shock at this time is given by

$$\begin{aligned} x_s &= X + \frac{h^\alpha}{\lambda\alpha(\alpha + 1)} \cdot \frac{\lambda(\alpha + 1)X}{h^\alpha}, \\ &= \frac{(\alpha + 1)}{\alpha}X, \end{aligned}$$

where we have used (2.5.20) and (2.5.22). Suppose a characteristic in the rarefaction has a value $u = c$ at time t. Then, as in (2.5.17), we have

$$\frac{dx}{dt} = (ce^{-\lambda t})^\alpha = c^\alpha e^{-\lambda \alpha t}$$

which, on integration and use of the IC $x = 0$, $t = 0$, gives

$$x = \frac{c^\alpha}{\lambda \alpha}(1 - e^{-\lambda \alpha t}).$$

The solution in the rarefaction wave is

$$
\begin{aligned}
u(x,t) &= ce^{-\lambda t} \\
&= \left(\frac{\lambda \alpha x}{e^{\lambda \alpha t} - 1}\right)^{\frac{1}{\alpha}}.
\end{aligned}
\tag{2.5.24}
$$

After $t = t_0$, the motion of the shock is given by

$$
\begin{aligned}
\frac{dx_s}{dt} &= \frac{u_b^\alpha}{\alpha + 1} = \frac{\lambda \alpha x_s}{(\alpha + 1)(e^{\lambda \alpha t} - 1)} \\
&= \frac{\lambda \alpha e^{-\lambda \alpha t} x_s}{(\alpha + 1)(1 - e^{-\lambda \alpha t})}.
\end{aligned}
\tag{2.5.25}
$$

On integrating (2.5.25), using the condition $x_s = \dfrac{\alpha + 1}{\alpha}X$ at $t = t_0$ and recalling (2.5.22), we get

$$x_s = \frac{\alpha + 1}{\alpha} \frac{(hX)^{\alpha/(\alpha+1)}}{[\lambda(\alpha + 1)]^{1/(\alpha+1)}}(1 - e^{-\lambda \alpha t})^{1/(\alpha+1)}, t > t_0. \tag{2.5.26}$$

It is clear that, whether the rarefaction catches up with the shock or not, the shock decays in infinite time but in finite distance.

If the rarefaction does not catch up with the shock, it follows from (2.5.20) that the location of the shock, as $t \to \infty$, is

$$x_{s,\infty} = X + \frac{h^\alpha}{\lambda \alpha(\alpha + 1)}. \tag{2.5.27}$$

If the rarefaction does catch up with the shock, that is, if t_0 exists, then the location of the shock from (2.5.26) in the limit t tending to infinity is

$$x_{s,\infty} = \frac{\alpha + 1}{\alpha} \frac{1}{[\lambda(\alpha + 1)]^{1/(\alpha+1)}}(hX)^{\alpha/(\alpha+1)}. \tag{2.5.28}$$

To summarise, if $\beta = 1$,

$$
u(x,t) = \begin{cases} \left(\dfrac{\lambda \alpha x}{e^{\lambda \alpha t} - 1}\right)^{1/\alpha} & 0 \le x \le x_F \\ he^{-\lambda t} & x_F < x < x_s, t \le t_0 \\ 0 & x > x_s \end{cases}
\tag{2.5.29}
$$

where t is less than t_0 and

$$u(x,t) = \begin{cases} \left(\dfrac{\lambda \alpha x e^{-\lambda \alpha t}}{1 - e^{-\lambda \alpha t}} \right)^{1/\alpha} & 0 \le x \le x_s \\ 0 & x > x_s, \ t \ge t_0 \end{cases} \qquad (2.5.30)$$

where $t \ge t_0$.

(iii) $\beta \neq 1, \beta \neq \alpha \neq 1$

Here the solution with top hat initial data will be found. Along the characteristic curves we have

$$\frac{dx}{dt} = u^\alpha, \quad \frac{du}{dt} + \lambda u^\beta = 0. \qquad (2.5.31)$$

Integrating (2.5.31) subject to conditions $u = h$ at $t = 0$, we get

$$u = [h^{1-\beta} - \lambda(1 - \beta)t]^{1/(1-\beta)}. \qquad (2.5.32)$$

The wavefront, whose location is $x_F = 0$ at $t = 0$, is obtained by integrating $\dfrac{dX_F}{dt} = u_F$, etc. We have

$$x_F = \frac{1}{\lambda(\alpha + 1 - \beta)} \left[h^{\alpha+1-\beta} - \{ h^{1-\beta} - \lambda(1 - \beta)t \}^{\frac{\alpha+1-\beta}{1-\beta}} \right]. \qquad (2.5.33)$$

Now the equation of motion of the shock wave is found from (2.5.32) as

$$\frac{dx_s}{dt} = \frac{u_b^\alpha}{\alpha + 1} = \frac{[h^{1-\beta} - \lambda(1 - \beta)t]^{\alpha/(1-\beta)}}{\alpha + 1}. \qquad (2.5.34)$$

Integrating (2.5.34) and using the condition $x_s = X$ at $t = 0$, we have

$$x_s = X + \frac{1}{\lambda(\alpha + 1)(\alpha + 1 - \beta)} \left[h^{\alpha+1-\beta} - \{ h^{1-\beta} - \lambda(1 - \beta)t \}^{\frac{\alpha+1-\beta}{1-\beta}} \right]. \qquad (2.5.35)$$

The time t_0 at which the front of the rarefaction catches up with the shock is found by equating (2.5.33) and (2.5.35):

$$t_0 = \frac{\left[h^{\alpha+1-\beta} - \dfrac{\lambda(\alpha+1)(\alpha+1-\beta)}{\alpha} X \right]^{\frac{1-\beta}{\alpha+1-\beta}} - h^{1-\beta}}{\lambda(\beta - 1)}. \qquad (2.5.36)$$

For t_0 to exist we must have the right side of (2.5.36) greater than zero; besides, the expression in square brackets must be positive. This requires that

$$h > \left[\frac{\lambda(\alpha + 1)(\alpha + 1 - \beta)X}{\alpha} \right]^{1/(\alpha+1-\beta)}. \qquad (2.5.37)$$

From (2.5.35) and (2.5.36) the shock position at $t = t_0$ is found to be

$$
\begin{aligned}
x_s &= X + \frac{1}{\lambda(\alpha+1)(\alpha+1-\beta)} \cdot \frac{\lambda(\alpha+1)(\alpha+1-\beta)}{\alpha} X \\
&= \frac{\alpha+1}{\alpha} X.
\end{aligned}
$$

This is in agreement with the result (2.5.10).

The characteristic solution in an implicit form may easily be found to be

$$
x = \frac{1}{\lambda(\alpha+1-\beta)} \left\{ \left[u^{1-\beta} + \lambda(1-\beta)t \right]^{\frac{\alpha+1-\beta}{1-\beta}} - u^{\alpha+1-\beta} \right\}. \tag{2.5.38}
$$

(iv) $\beta = \alpha + 1$

As for the derivation of (2.5.33), we have in this case

$$
x_F = \frac{1}{\lambda(1-\beta)} \ln \left[\frac{h^{1-\beta}}{h^{1-\beta} - \lambda(1-\beta)t} \right]. \tag{2.5.39}
$$

The equation of motion of the shock wave is

$$
\frac{dx_s}{dt} = \frac{1}{h^{1-\beta} - \lambda(1-\beta)t}
$$

which, on integration and use of 1C $x_s = X$ at $t = 0$, gives

$$
x_s = X + \frac{1}{\lambda\beta(1-\beta)} \ln \left[\frac{h^{1-\beta}}{h^{1-\beta} - \lambda(1-\beta)t} \right]. \tag{2.5.40}
$$

The time t_0 at which the front of the rarefaction wave catches up with the shock is obtained by equating (2.5.39) and (2.5.40):

$$
\lambda\beta(1-\beta)X = (\beta-1) \ln \left[\frac{h^{1-\beta}}{h^{1-\beta} - \lambda(1-\beta)t_0} \right] \tag{2.5.41}
$$

or

$$
t_0 = \frac{h^{1-\beta}(e^{\lambda\beta X} - 1)}{\lambda(\beta-1)}. \tag{2.5.42}
$$

The value t_0 in (2.5.42) always exists since $\beta = \alpha+1 > 1$, α being positive. From (2.5.40) and (2.5.42) the location of the shock at $t = t_0$ is

$$
x_s = \frac{(\alpha+1)X}{\alpha}.
$$

Some smooth solutions of $u_t + u^\alpha u_x + \lambda u^\beta = 0$

We investigate the conditions under which nonnegative, bounded C^1 solutions of (2.5.1) exist on a semi-infinite domain: $\epsilon < x \leq \infty$, $\delta < t \leq \infty$, where ϵ and δ are positive. These are some classes of initial and boundary conditions for which shock waves do not arise. First we study solutions of the form $u(x,t) = F(x)$ and $u(x,t) = G(t)$, that is, solutions which are functions of one variable only. Consider solutions of the form $u(x,t) = F(x)$. On substitution of this into (2.5.1) we get

$$F^\alpha F' + \lambda F^\beta = 0. \tag{2.5.43}$$

Therefore,

$$u(x,t) = F(x) = [(\alpha + 1 - \beta)(C - \lambda x)]^{\frac{1}{\alpha+1-\beta}} \tag{2.5.44}$$

where C is constant of integration.

If $\beta > \alpha + 1$ and $C < 0$ (so that the expression within the square brackets in (2.5.44) is positive and the exponent is negative), the solution $u(x,t)$ is bounded. If $\beta = \alpha + 1$, (2.5.43) integrates to give the bounded solution

$$u(x,t) = F(x) = Ce^{-\lambda x} \tag{2.5.45}$$

where C is the constant of integration.

If $\alpha + 1 > \beta$, there are no bounded solutions of the form $u(x,t) = F(x)$.

Consider now solutions of the form $u(x,t) = G(t)$. Substitution into (2.5.1) gives

$$G' + \lambda G^\beta = 0. \tag{2.5.46}$$

If $\beta < 1$, there exist bounded solutions

$$u(x,t) = G(t) \geq \begin{cases} [(1-\beta)(C-\lambda t)]^{\frac{1}{1-\beta}} & t < \frac{C}{\lambda} \\ \\ 0 & t \geq \frac{C}{\lambda} \end{cases} \tag{2.5.47}$$

where the constant of integration C is greater than zero.

These solutions decay to zero in a finite time C/λ. If $\beta = 1$, Equation (2.5.46) integrates to give

$$u(x,t) = G(t) = Ce^{-\lambda t}, \tag{2.5.48}$$

a solution which decays to zero in infinite time.

If $\beta > 1$, (2.5.46) may be integrated to yield

$$u(x,t) = G(t) = \frac{1}{[(\beta - 1)(\lambda t + C)]^{1/(\beta-1)}} \tag{2.5.49}$$

where $C > 0$ is constant of integration. The solution (2.5.49) is bounded and decays to zero in infinite time.

Special exact solutions of the equation
$u_t + u^\alpha u_x + \lambda u^\beta = 0$

Consider the one-parameter family of stretching transformations (see Chapter 3)

$$\overline{x} = \epsilon^a x, \overline{t} = \epsilon^b t, \overline{u} = \epsilon^c u$$

$$\frac{\partial u}{\partial t} = \epsilon^{b-c}\frac{\partial \overline{u}}{\partial \overline{t}}, \frac{\partial u}{\partial x} = \epsilon^{a-c}\frac{\partial \overline{u}}{\partial \overline{x}}. \tag{2.5.50}$$

Substituting (2.5.50) into (2.5.1), we get

$$\epsilon^{b-c}\overline{u}_{\overline{t}} + \epsilon^{a-c}\overline{u}_{\overline{x}} \cdot \epsilon^{-ca}\overline{u}^\alpha + \lambda\epsilon^{-c\beta}\overline{u}^\beta = 0.$$

For invariance of (2.5.1) we must have

$$b = c(1 - \beta), \quad a = c(1 + \alpha - \beta).$$

For $\beta \neq 1, 1 + \alpha$ we have solutions of the form

$$u(x, t) = t^{c/b}H(\eta) = t^{1/(1-\beta)}H(\eta)$$

where

$$\eta = xt^{-a/b} = xt^{(\beta-\alpha-1)/(1-\beta)}. \tag{2.5.51}$$

Substituting (2.5.51) into (2.5.1), we get

$$H' = \frac{H - (\beta - 1)\lambda H^\beta}{(\alpha + 1 - \beta)\eta + (\beta - 1)H^\alpha}. \tag{2.5.52}$$

For $\alpha + 1 > \beta$ and $\beta > 1$, $H(\eta)$ approaches the constant solution

$$H^* = [\lambda(\beta - 1)]^{1/(1-\beta)} \tag{2.5.53}$$

of (2.5.52) as $\eta \to \infty$.

There are no bounded solutions of (2.5.52) as $\eta \to \infty$ for $\alpha + 1 < \beta$, since, in this case, either the denominator in (2.5.53) becomes zero at some finite point or H' is proportional to $1/\eta$ for large η so that $H \approx O(\ln \eta)$ as $\eta \to \infty$. There are also no bounded solutions for $\beta < 1$ since u grows with time.

If $\beta \neq 1, \alpha + 1$ we also have solutions of the form

$$u(x, t) = x^{c/a}H(\eta) = x^{1/(1+\alpha-\beta)}H(\eta)$$

$$\eta = xt^{(\beta-\alpha-1)/(1-\beta)}. \tag{2.5.54}$$

Substituting (2.5.54) in (2.5.1) we get, after some simplification,

$$H' = \frac{(\beta - 1)[(\alpha + 1 - \beta)\lambda H^\beta + H^{\alpha+1}]}{\eta(\alpha + 1 - \beta)[(\beta - \alpha - 1)\eta^{\frac{\beta-1}{\beta-\alpha-1}} + (1 - \beta)H^\alpha]}. \tag{2.5.55}$$

For $\alpha + 1 > \beta$ and $\beta > 1$, $H(\eta) \to 0$ exponentially as $\eta \to \infty$. So bounded solutions exist. There are no bounded solutions to (2.5.55) for $\alpha + 1 < \beta$ or $\beta < 1$ since the denominator in the RHS in these cases vanishes at some finite point η. Some additional similarity solutions of (2.5.16) exist if $\alpha = 1$.

Bukiet et al. (1996) have studied numerically most of the special cases of (2.5.1) which have exact solutions. Specific values of parameters in the PDEs, IC, and BC (wherever applicable) were chosen. In each case the numerical results agreed closely with the analytic ones. However, the main point of their study was to show the superiority of the proposed numerical scheme over other schemes such as the Lax-Wendroff scheme. The shocks when they formed were accurately located and their progression in time predicted. The shocks were sharp, showing no spurious oscillations. The only drawback of this scheme is that, in its present form, it is applicable only to scalar hyperbolic PDEs.

We conclude this chapter by referring to a result due to Logan (1987) regarding the general first-order, nonlinear PDE

$$F(x, t, u, p, q) = 0 \qquad (2.5.56)$$

where $p = u_x$ and $q = u_t$.

For this purpose we need the following definition. Equation (2.5.56) is constant conformally invariant under the one-parameter family of stretching T_c,

$$\bar{x} = \epsilon^a x, \quad \bar{t} = \epsilon^b t, \quad \bar{u} = \epsilon^c u \qquad (2.5.57)$$

if, and only if,

$$F(\bar{x}, \bar{t}, \bar{u}, \bar{u}_{\bar{x}}, \bar{u}_{\bar{t}}) = A(\epsilon) F(x, t, u, u_x, u_t) \qquad (2.5.58)$$

for all ϵ in I, for some function A with $A(1) = 1$. If $A(\epsilon) \equiv 1$, then we say that (2.5.56) is absolutely invariant.

If Equation (2.5.56) is constant conformally invariant under the one-parameter family of stretching transformation (2.5.57), then it can be changed to an ODE of the form

$$H(s, f, f') = 0 \qquad (2.5.59)$$

where

$$u = t^{c/b} f(s), \quad s = (x^b / t^a). \qquad (2.5.60)$$

Chapter 3

Exact Similarity Solutions of Nonlinear PDEs

PART A

3.1 Reduction of PDEs by Infinitesimal Transformations

The similarity method is one of the standard methods for obtaining exact solutions of PDEs. The number of independent variables in a PDE is reduced by one by making use of appropriate combinations of the original independent variables as new independent variables, called "similarity variables." The similarity variables can themselves be identified by using the invariance properties of PDEs when subjected to finite or infinitesimal transformations.

Consider the one-parameter (ϵ) group of finite transformations

$$\bar{x} = \epsilon^2 x, \quad \bar{t} = \epsilon t, \quad \bar{u} = \epsilon u. \tag{3.1.1}$$

If we apply the transformations (3.1.1) to the scalar nonlinear transport equation

$$u_t + u u_x = 0 \tag{3.1.2}$$

we find that it simply becomes

$$\bar{u}_{\bar{t}} + \bar{u} \bar{u}_{\bar{x}} = 0, \tag{3.1.3}$$

that is, the form of Equation (3.1.2) is unaltered. We say that Equation (3.1.2) is invariant under the transformations (3.1.1).

We explain below the theory of infinitesimal transformations given originally by Ovsiannikov (1982) and explained more clearly by Bluman and Cole (1974).

Lie Group Transformations

Let (x, y) be the coordinates of a point in the plane. We consider a set of one-parameter transformations, T, of the plane onto itself:

$$T: \quad \begin{aligned} x_1 &= \phi(x, y; \alpha) \\ y_1 &= \psi(x, y; \alpha) \end{aligned} \qquad (3.1.4)$$

where α is a parameter which can vary continuously over a given range. An example of such a transformation is the one-parameter family of rotations, R, about the origin in the xy-plane which takes a point $P(x, y)$ to another point $P_1(x_1, y_1)$ through an angle α:

$$R: \quad \begin{aligned} x_1 &= x \cos \alpha - y \sin \alpha \\ y_1 &= x \sin \alpha + y \cos \alpha, \quad 0 \le \alpha < 2\pi. \end{aligned} \qquad (3.1.5)$$

This is a one-parameter group of rotations. We consider the set of all one-parameter transformations, T, which form a continuous group (Lie group). Here the identity transformation $(x, y) \rightarrow (x, y)$ may be characterised by the parameter value $\alpha = 0$:

$$x = \phi(x, y; 0), \quad y = \psi(x, y; 0). \qquad (3.1.6)$$

The Infinitesimal Transformation

For many purposes it suffices to study only transformations close to the identity. We assume that the functions ϕ and ψ in (3.1.4) are differentiable a sufficient number of times with respect to α. If ϵ is an infinitesimally small value of α, then the transformation

$$\begin{aligned} x_1 &= \phi(x, y; \epsilon) \\ y_1 &= \psi(x, y; \epsilon) \end{aligned} \qquad (3.1.7)$$

differs only infinitesimally from the identity transformation.

We can expand (3.1.7) and write it as

$$\begin{aligned} x_1 &= \phi(x, y; 0) + \epsilon \left. \frac{\partial \phi}{\partial \alpha}(x, y; \alpha) \right|_{\alpha=0} \\ &\quad + \frac{\epsilon^2}{2!} \left. \frac{\partial^2 \phi}{\partial \alpha^2}(x, y; \alpha) \right|_{\alpha=0} + \cdots \\ y_1 &= \psi(x, y; 0) + \epsilon \left. \frac{\partial \psi}{\partial \alpha}(x, y; \alpha) \right|_{\alpha=0} \\ &\quad + \frac{\epsilon^2}{2!} \left. \frac{\partial^2 \psi}{\partial \alpha^2}(x, y; \alpha) \right|_{\alpha=0} + \cdots \end{aligned} \qquad (3.1.8)$$

and have the infinitesimal transformation

$$x_1 = x + \epsilon X + O(\epsilon^2)$$

$$y_1 \ = \ y + \epsilon Y + O(\epsilon^2) \tag{3.1.9}$$

where

$$
\begin{aligned}
X \ &= \ X(x,y) = \left. \frac{\partial \phi}{\partial \alpha}(x,y;\alpha) \right|_{\alpha=0} \\
Y \ &= \ Y(x,y) = \left. \frac{\partial \psi}{\partial \alpha}(x,y;\alpha) \right|_{\alpha=0}
\end{aligned}
\tag{3.1.10}
$$

In the rotation group (3.1.5) with $\alpha = \epsilon \ll 1$, we have

$$
\begin{aligned}
x_1 \ &= \ x\cos\epsilon - y\sin\epsilon \approx x - \epsilon y + O(\epsilon^2), \\
y_1 \ &= \ x\sin\epsilon + y\cos\epsilon \approx y + \epsilon x + O(\epsilon^2).
\end{aligned}
\tag{3.1.11}
$$

Here,

$$X = -y, \ Y = x. \tag{3.1.12}$$

Application to PDEs

We first study a system, S, consisting of the general second-order PDE in the dependent variable u and independent variables x and t, namely

$$H(x,t,u,u_x,u_t,u_{xx},u_{xt},u_{tt}) = 0 \tag{3.1.13}$$

and appropriate boundary conditions, left unspecified here. Now consider a one-parameter group of infinitesimal transformations which takes the (x,t,u)-space into itself so that a given solution, $u = \phi(x,t)$ of (3.1.13), is transformed into another solution, $\bar{u} = \bar{\phi}(\bar{x},\bar{t})$:

$$
\begin{aligned}
\bar{x} \ &= \ x + \epsilon X, \\
\bar{t} \ &= \ t + \epsilon T, \\
\bar{u} \ &= \ u + \epsilon U.
\end{aligned}
\tag{3.1.14}
$$

We require that Equation (3.1.13) is invariant under the transformations (3.1.14). That is,

$$H(\bar{x},\bar{t},\bar{u},\bar{u}_{\bar{x}},\bar{u}_{\bar{t}},\bar{u}_{\bar{x}\bar{x}},\bar{u}_{\bar{x}\bar{t}},\bar{u}_{\bar{t}\bar{t}}) = 0. \tag{3.1.15}$$

We have

$$
\begin{aligned}
\bar{u} \ &= \ u + \epsilon U = \phi(\bar{x},\bar{t}) = \phi(x + \epsilon X, t + \epsilon T) \\
&= \ \phi(x,t) + \epsilon X \frac{\partial \phi}{\partial x} + \epsilon T \frac{\partial \phi}{\partial t} + O(\epsilon^2).
\end{aligned}
\tag{3.1.16}
$$

Equating terms of order ϵ on both sides, we get

$$X \frac{\partial u}{\partial x} + T \frac{\partial u}{\partial t} = U. \tag{3.1.17}$$

Equation (3.1.17) is called the "invariant surface condition." Its characteristic equations are

$$\frac{dx}{X} = \frac{dt}{T} = \frac{du}{U}. \tag{3.1.18}$$

The general solution of (3.1.17) is obtained by finding two independent integrals

$$\begin{aligned} \eta(x,t,u) &= \quad \text{constant} \\ h(x,t,u) &= \quad \text{constant}. \end{aligned} \tag{3.1.19}$$

The general solution of (3.1.17), therefore, is

$$h(x,t,u) = f(\eta) \tag{3.1.20}$$

where f is an arbitrary function of its argument. If X/T is a function of x and t alone, then the integral of the first equality in (3.1.18) will be of the form

$$\eta = \eta(x,t) = \text{ constant}. \tag{3.1.21}$$

Here, η is the so-called "similarity variable," and the general solution of (3.1.17) is

$$u = F(x,t,f(\eta)). \tag{3.1.22}$$

The function $f(\eta)$ is determined by substituting the form (3.1.22) into the given PDE (3.1.13). In the case of two independent variables, the resulting equation will be an ODE in $f(\eta)$. Each solution of the ODE yields a similarity solution for the PDE.

To find the generators X, T, U of the group of transformations, we have to determine how the derivatives transform under the infinitesimal transformation (3.1.14).

In the following calculations, x, t, and u are regarded as independent variables. Thus,

$$\frac{\partial X}{\partial x} = X_x + X_u u_x, \quad \frac{\partial X}{\partial t} = X_t + X_u u_t. \tag{3.1.23}$$

From (3.1.14) we have

$$\begin{aligned} \frac{\partial x}{\partial \bar{x}} &= \frac{\partial}{\partial \bar{x}}[\bar{x} - \epsilon X(x,t,u) + O(\epsilon^2)] \\ &= 1 - \epsilon(X_x + X_u u_x)\frac{\partial x}{\partial \bar{x}} + O(\epsilon^2) \end{aligned} \tag{3.1.24}$$

or

$$\frac{\partial x}{\partial \bar{x}} = 1 - \epsilon(X_x + X_u u_x) + O(\epsilon^2).$$

Similarly, we find that

$$\frac{\partial x}{\partial \bar{t}} = -\epsilon(X_t + X_u u_t) + O(\epsilon^2) \tag{3.1.25}$$

$$\frac{\partial t}{\partial \bar{x}} = -\epsilon(T_x + T_u u_x) + O(\epsilon^2) \tag{3.1.26}$$

$$\frac{\partial t}{\partial \bar{t}} = 1 - \epsilon(T_t + T_u u_t) + O(\epsilon^2). \tag{3.1.27}$$

Also,

$$\frac{\partial u}{\partial \bar{x}} = [1 - \epsilon(X_x + X_u u_x)]u_x - \epsilon(T_x + T_u u_x)u_t.$$

We now have the operator

$$\frac{\partial}{\partial \bar{x}} = [1 - \epsilon(X_x + X_u u_x)]\frac{\partial}{\partial x} - \epsilon(T_x + T_u u_x)\frac{\partial}{\partial t}$$
$$+ [\{1 - \epsilon(X_x + X_u u_x)\}u_x - \epsilon(T_x + T_u u_x)u_t]\frac{\partial}{\partial u} \tag{3.1.28}$$

keeping terms up to order ϵ. Thus,

$$\frac{\partial \bar{u}}{\partial \bar{x}} = \frac{\partial}{\partial \bar{x}}[u + \epsilon U(x, t, u)]$$
$$= [1 - \epsilon(X_x + X_u u_x)]u_x - \epsilon(T_x + T_u u_x)u_t$$
$$+ \epsilon[U_x + U_u u_x] + O(\epsilon^2)$$

or

$$\frac{\partial \bar{u}}{\partial \bar{x}} = u_x + \epsilon \tilde{U}_x + O(\epsilon^2) \tag{3.1.29}$$

$$\tilde{U}_x = U_x + U_u u_x - X_x u_x - X_u u_x^2 - T_x u_t - T_u u_x u_t. \tag{3.1.30}$$

Similarly, we get

$$\frac{\partial \bar{u}}{\partial \bar{t}} = u_t + \epsilon \tilde{U}_t + O(\epsilon^2) \tag{3.1.31}$$

$$\tilde{U}_t = U_t + U_u u_t - X_t u_x - X_u u_t u_x - T_t u_t - T_u u_t^2. \tag{3.1.32}$$

From (3.1.29) we have $\dfrac{\partial^2 \bar{u}}{\partial \bar{x}^2} = \dfrac{\partial}{\partial \bar{x}}[u_x + \epsilon \tilde{U}_x]$. Using (3.1.28), we find, on simplification, the second derivative

$$\frac{\partial^2 \bar{u}}{\partial \bar{x}^2} = u_{xx} + \epsilon \tilde{U}_{xx} + O(\epsilon^2) \tag{3.1.33}$$

where

$$
\begin{aligned}
\tilde{U}_{xx} = & \ U_{xx} + (2U_{xu} - X_{xx})u_x - T_{xx}u_t + (U_{uu} - 2X_{xu})u_x^2 \\
& - 2T_{xu}u_xu_t - X_{uu}u_x^3 - T_{uu}u_x^2u_t + (U_u - 2X_x)u_{xx} \\
& - 2T_xu_{xt} - 3X_uu_{xx}u_x - T_uu_{xx}u_t - 2T_uu_{xt}u_x. \quad (3.1.34)
\end{aligned}
$$

In a similar manner we get

$$
\frac{\partial^2 \bar{u}}{\partial \bar{x} \partial \bar{t}} = u_{xt} + \epsilon \tilde{U}_{xt} + O(\epsilon^2) \qquad (3.1.35)
$$

where

$$
\begin{aligned}
\tilde{U}_{xt} = & \ U_{xt} + (U_{tu} - X_{tx})u_x + (U_{xu} - T_{tx})u_t \\
& + (U_{uu} - X_{xu} - T_{tu})u_xu_t - X_{tu}u_x^2 - X_{uu}u_x^2u_t - T_{xu}u_t^2 \\
& - T_{uu}u_xu_t^2 + (U_u - X_x - T_t)u_{xt} - X_tu_{xx} - 2X_uu_xu_{xt} \\
& - X_uu_tu_{xx} - T_xu_{tt} - T_uu_xu_{tt} - 2T_uu_tu_{xt} \qquad (3.1.36)
\end{aligned}
$$

and

$$
\frac{\partial^2 \bar{u}}{\partial \bar{t}^2} = u_{tt} + \epsilon \tilde{U}_{tt} + O(\epsilon^2) \qquad (3.1.37)
$$

where $\tilde{U}_{tt}$ is obtained from $\tilde{U}_{xx}$ by simply interchanging x and t as well as X and T.

The twice-extended group is a group of transformations of the eight variables

$$
\{x, t, u, u_x, u_t, u_{xx}, u_{xt}, u_{tt}\}.
$$

The infinitesimal operator is

$$
\begin{aligned}
U^* = & \ X\frac{\partial}{\partial x} + T\frac{\partial}{\partial t} + U\frac{\partial}{\partial u} + \tilde{U}_x\frac{\partial}{\partial u_x} + \tilde{U}_t\frac{\partial}{\partial u_t} \\
& + \tilde{U}_{xx}\frac{\partial}{\partial u_{xx}} + \tilde{U}_{xt}\frac{\partial}{\partial u_{xt}} + \tilde{U}_{tt}\frac{\partial}{\partial u_{tt}}. \qquad (3.1.38)
\end{aligned}
$$

The PDE (3.1.13) will be invariant under the group of transformations (3.1.14) if

$$
H(\bar{x}, \bar{t}, \bar{u}, \bar{u}_{\bar{x}}, \bar{u}_{\bar{t}}, \bar{u}_{\bar{x}\bar{x}}, u_{\bar{x}\bar{t}}, \bar{u}_{\bar{t}\bar{t}}) = 0
$$

for any solution $u = H(x, t)$ of (3.1.13). The condition for this is

$$
U^*H = 0 \quad \text{when} \quad H = 0, \qquad (3.1.39)
$$

where U^* is given by (3.1.38). Substituting (3.1.29) - (3.1.37) into (3.1.39), using (3.1.13) and equating to zero the coefficients of like derivative terms in u, we get an overdetermined system of linear PDEs to obtain the generators X, T, and U. After finding X, T, and U, the invariant surface condition (3.1.17) is used to find the similarity form of the solution. These results can be generalised to systems of PDEs involving arbitrary number of dependent and independent variables.

In the absence of any boundary conditions, the invariant surface condition helps us to find a class of solutions of the PDE.

The above remarks may be illustrated by considering the one-dimensional heat equation

$$u_t = u_{xx}. \tag{3.1.40}$$

If we apply the transformation (3.1.14) to (3.1.40), the latter becomes (to order ϵ)

$$u_t + \epsilon \tilde{U}_t = u_{xx} + \epsilon \tilde{U}_{xx}, \tag{3.1.41}$$

from which it follows that

$$\tilde{U}_t = \tilde{U}_{xx}. \tag{3.1.42}$$

In the classical similarity method as explained by Bluman and Cole (1974), u_{xx} in $\tilde{U}_{xx}$ is replaced by u_t, and the coefficients of like derivative terms, u_t, u_x, u_{tx}, etc., are equated to zero. The equation $\tilde{U}_t = \tilde{U}_{xx}$, with (3.1.32) and (3.1.34) in view, may be written out as

$$
\begin{aligned}
&U_t + U_u u_t - X_t u_x - X_u u_t u_x - T_t u_t - T_u u_t^2 = U_{xx} + (2U_{xu} - X_{xx})u_x \\
&-T_{xx}u_t + (U_{uu} - 2X_{xu})u_x^2 - 2T_{xu}u_x u_t - X_{uu}u_x^3 - T_{uu}u_x^2 u_t \\
&+(U_u - 2X_x)u_t - 2T_x u_{xt} - 3X_u u_t u_x - T_u u_t^2 - 2T_u u_{xt}u_x. \tag{3.1.43}
\end{aligned}
$$

Equating to zero the coefficients of different products of derivatives of u, etc. gives an overdetermined system for the generators, which, however, must be solved in a consistent manner. For example, from the coefficients of $u_x u_{xt}, u_x u_t$, and u_x^2, we get

$$T_u = 0, \ X_u = 0 \ \text{ and } \ U_{uu} = 0. \tag{3.1.44}$$

It may be noted that, here, $H = u_t - u_{xx}$, and the infinitesimal operator U^* in (3.1.38) in the present case is

$$U^* = \tilde{U}_t \frac{\partial}{\partial u_t} + \tilde{U}_{xx}\frac{\partial}{\partial u_{xx}}.$$

Therefore,

$$
\begin{aligned}
U^* H &= \left(\tilde{U}_t \frac{\partial}{\partial u_t} + \tilde{U}_{xx}\frac{\partial}{\partial u_{xx}} \right)(u_t - u_{xx}) \\
&= \tilde{U}_t - \tilde{U}_{xx} \tag{3.1.45}
\end{aligned}
$$

so that $U^* H = 0$ implies that $\tilde{U}_t - \tilde{U}_{xx} = 0$. Thus, the condition (3.1.39) is satisfied. The method is illustrated with full details through the following example.

Example

We consider the nonlinear diffusion equation

$$\frac{\partial u}{\partial t} = \frac{\partial}{\partial x}\left(D(u)\frac{\partial u}{\partial x}\right) \tag{3.1.46}$$

which may be written as

$$\frac{\partial u}{\partial t} = D(u)\frac{\partial^2 u}{\partial x^2} + D'(u)\left(\frac{\partial u}{\partial x}\right)^2. \tag{3.1.47}$$

We seek a one parameter group of infinitesimal transformations which takes the (x, t, u)-space into itself and under which Equation (3.1.47) is invariant:

$$\bar{x} = x + \epsilon X, \quad \bar{t} = t + \epsilon T, \quad \bar{u} = u + \epsilon U \tag{3.1.48}$$

where the generators X, T, and U are functions of x, t, and u. Invariance of Equation (3.1.47) under (3.1.48) gives (after substituting for u_{xx} from (3.1.47))

$$U_t + U_u u_t - X_t u_x - X_u u_t u_x - T_t u_t - T_u u_t^2 = UD'(u)\left[\frac{u_t}{D(u)} - \frac{D'(u)}{D(u)}u_x^2\right]$$

$$+ D(u)\left[U_{xx} + (2U_{xu} - X_{xx})u_x - T_{xx}u_t + (U_{uu} - 2X_{xu})u_x^2 - 2T_{xu}u_x u_t\right.$$

$$- X_{uu}u_x^3 - T_{uu}u_x^2 u_t + (U_u - 2X_x)\frac{u_t}{D(u)} - (U_u - 2X_x)\frac{D'(u)}{D(u)}u_x^2 - 2T_x u_{xt}$$

$$\left. - 3X_u u_x \frac{u_t}{D(u)} + 3X_u \frac{D'(u)}{D(u)}u_x^3 - T_u \frac{u_t^2}{D(u)} + T_u \frac{D'(u)}{D(u)}u_x^2 u_t - 2T_u u_{xt}u_x\right]$$

$$+ D'(u)[2U_x u_x + 2U_u u_x^2 - 2X_x u_x^2 - 2X_u u_x^3 - 2T_x u_x u_t - 2T_u u_x^2 u_t]$$

$$+ UD''(u)u_x^2. \tag{3.1.49}$$

Equating to zero the coefficients of $u_{xt}u_x, u_{xt}, u_x^3, u_x^2 u_t, u_x^2, u_t^2, u_x u_t, u_x, u_t$, and u^0, we get the determining equations:

$$T_u = 0 \tag{3.1.50}$$

$$T_x = 0 \tag{3.1.51}$$

$$D(u)X_{uu} - D'(u)X_u = 0 \tag{3.1.52}$$

$$D(u)T_{uu} + D'(u)T_u = 0 \tag{3.1.53}$$

$$\left[U_u - 2X_x + U\frac{D'(u)}{D(u)}\right]_u = 0 \tag{3.1.54}$$

$$T_u - T_u = 0 \tag{3.1.55}$$

$$X_u + D(u)T_{xu} + D'(u)T_x = 0 \qquad (3.1.56)$$

$$X_t - D(u)X_{xx} + 2D(u)U_{xu} + 2D'(u)U_x = 0 \qquad (3.1.57)$$

$$T_t - D(u)T_{xx} - 2X_x + U\frac{D'(u)}{D(u)} = 0 \qquad (3.1.58)$$

$$U_t - D(u)U_{xx} = 0. \qquad (3.1.59)$$

From (3.1.50), (3.1.51) and (3.1.56) we have

$$X = X(x,t), \quad T = T(t). \qquad (3.1.60)$$

From (3.1.54) and (3.1.60) we get

$$U_u + U\frac{D'(u)}{D(u)} = \phi(x,t) \qquad (3.1.61)$$

where ϕ is an arbitrary function of x and t. Equations (3.1.58) and (3.1.60) yield

$$T'(t) - 2X_x + U\frac{D'(u)}{D(u)} = 0. \qquad (3.1.62)$$

Using (3.1.61) in (3.1.62) we get

$$U_u = T'(t) - 2X_x + \phi(x,t). \qquad (3.1.63)$$

It follows that

$$U_{uu} = 0. \qquad (3.1.64)$$

From (3.1.58) and (3.1.60) we get

$$U = \frac{D(u)}{D'(u)}(2X_x - T'(t)). \qquad (3.1.65)$$

Since $U_{uu} = 0$, we have either

$$2X_x = T'(t) \qquad (3.1.66)$$

or

$$\left[\frac{D(u)}{D'(u)}\right]'' = 0. \qquad (3.1.67)$$

It follows that

$$D(u) = \alpha(u + \beta)^m, \qquad (3.1.68)$$

where α, β, and m are arbitrary constants. Equations (3.1.66) and (3.1.65) imply that

$$U = 0. \qquad (3.1.69)$$

Now (3.1.66) gives

$$X = \frac{1}{2}T'(t)x + f(t) \qquad (3.1.70)$$

so that (3.1.57) becomes

$$\frac{x}{2}T''(t) + f'(t) = 0, \tag{3.1.71}$$

from which we get

$$T(t) = \gamma t + \delta \tag{3.1.72}$$

$$f(t) = \text{constant} = k \tag{3.1.73}$$

where γ, δ and k are constants. Thus, the generators explicitly are

$$\begin{align}
X &= \frac{\gamma}{2}x + k, \tag{3.1.74}\\
T &= \gamma t + \delta, \tag{3.1.75}\\
U &= 0. \tag{3.1.76}
\end{align}$$

This group is applicable to all functions $D(u)$.

If $D(u)$ has the form (3.1.68), then from (3.1.62) we have

$$U = \frac{1}{m}(u + \beta)(2X_x - T'(t)). \tag{3.1.76$_1$}$$

Substituting (3.1.76$_1$) into (3.1.57) we get

$$X_t - D(u)X_{xx} + 4D(u)\frac{1}{m}X_{xx} + 4D'(u)\frac{1}{m}(u + \beta)X_{xx} = 0 \tag{3.1.77}$$

which, on simplification, becomes

$$X_t = -D(u)\left(3 + \frac{4}{m}\right)X_{xx}. \tag{3.1.78}$$

Substituting (3.1.76$_1$) into (3.1.59) and making use of (3.1.78), we get

$$T''(t) = -8D(u)\left(1 + \frac{1}{m}\right)X_{xxx}. \tag{3.1.79}$$

Equations (3.1.78) and (3.1.79) give rise to two distinct cases:

(i) For all constants m we have

$$X_t = X_{xx} = T''(t) = 0 \tag{3.1.80}$$

so that

$$\begin{align}
X &= \lambda x + k \tag{3.1.81}\\
T &= \gamma t + \delta. \tag{3.1.82}
\end{align}$$

Also,

$$U = \frac{1}{m}(u + \beta)(2\lambda - \gamma) \tag{3.1.83}$$

where $\beta, \lambda, \gamma, \delta$, and k are arbitrary constants.

(ii) $m = -4/3$

In this case $X_t = X_{xxx} = T''(t) = 0$, leading to the generators

$$X = \mu x^2 + \lambda x + k \tag{3.1.84}$$
$$T = \gamma t + \delta \tag{3.1.85}$$
$$U = -\frac{3}{4}(u + \beta)(4\mu x + 2\lambda - \gamma) \tag{3.1.86}$$

where $\mu, \lambda, k, \gamma, \delta$, and β are arbitrary constants.

We now consider a special case of (3.1.46) with $D(u) = u^m$, where m is positive. Using the generators (3.1.81) - (3.1.83) from Case (i), the characteristic equations (3.1.18) become

$$\frac{dx}{\lambda x + k} = \frac{dt}{\gamma t + \delta} = \frac{du}{(1/m)(u + \beta)(2\lambda - \gamma)}. \tag{3.1.87}$$

Let $k = 0 = \delta$, and $\gamma = 2\lambda$, then we have

$$\frac{dx}{x} = \frac{dt}{2t} = \frac{du}{0}. \tag{3.1.88}$$

The first equation in (3.1.88) gives the similarity variable

$$\eta = \frac{x}{t^{1/2}}. \tag{3.1.89}$$

Since $u = $ constant in view of the last of (3.1.88), we have the following form of the similarity solution

$$u = \phi(\eta). \tag{3.1.90}$$

Putting (3.1.90) into (3.1.47) with $D(u) = u^m$, we get

$$\frac{d}{d\eta}(\phi^m \phi') + \frac{\eta}{2}\phi' = 0, \tag{3.1.91}$$

which can be written as

$$\frac{d}{d\phi}(\phi^m \phi') + \frac{\eta}{2} = 0. \tag{3.1.92}$$

Putting $u^* = \phi'$ in (3.1.92) and differentiating, we get

$$\frac{d^2}{d\phi^2}(\phi^m u^*) + \frac{1}{2u^*} = 0. \tag{3.1.93}$$

Now put $u^* = v\phi^{1-m/2}$ in (3.1.93) to get the equation

$$\phi^2 \frac{d^2 v}{d\phi^2} + (m+2)\phi \frac{dv}{d\phi} + \frac{m(m+2)}{4} v + \frac{1}{2v} = 0. \tag{3.1.94}$$

We consider special cases of (3.1.94).

a) $m = -2$

Equation (3.1.94) becomes

$$\phi^2 \frac{d^2 v}{d\phi^2} + \frac{1}{2v} = 0. \tag{3.1.95}$$

Now, writing $v = \phi V$, $\phi = 1/\theta$ in (3.1.95) we have

$$\frac{d^2 V}{d\theta^2} + \frac{1}{2V} = 0, \tag{3.1.96}$$

which integrates to yield

$$\left(\frac{dV}{d\theta}\right)^2 + \log V = C_1 \tag{3.1.97}$$

or

$$\theta = \int^V \frac{dV}{\sqrt{C_1 - \log V}} + C_2, \tag{3.1.98}$$

that is,

$$\frac{1}{\phi} = \int^{v/\phi} \frac{dV}{\sqrt{C_1 - \log V}} + C_2. \tag{3.1.99}$$

But with $m = 2$, $v = u\phi^{-3} - \phi'/\phi^3$ so that

$$\frac{1}{\phi} = \int^{\phi'/\phi^3} \frac{dV}{\sqrt{C_1 - \log V}} + C_2. \tag{3.1.100}$$

C_1 and C_2 in the above are constants. Now we put $\psi = 1/\phi^2$ in (3.1.100) and obtain

$$\int^{-\psi'/2} \frac{dV}{\sqrt{C_1 - \log V}} = \psi^{1/2} - C_2. \tag{3.1.101}$$

Therefore, for some function F, we may write $F(\psi') = \psi^{1/2} - C_2$ or $\psi' = f_1(\psi^{1/2} - C_2)$. That is,

$$\eta + C_3 = \int^{1/\phi^2} \frac{d\psi}{f(\psi^{1/2} - C_2)}. \tag{3.1.102}$$

Here, C_3 is another constant. Only two of $C_i (i = 1, 2, 3)$ are arbitrary since Equation (3.1.91) must also be satisfied.

b) $m = -1$

In this case (3.1.94) becomes

$$\phi^2 \frac{d^2 v}{d\phi^2} + \phi \frac{dv}{d\phi} - \frac{v}{4} + \frac{1}{2v} = 0. \tag{3.1.103}$$

Put $\theta = \log \phi$ in (3.1.103) to obtain

$$\frac{d^2 v}{d\theta^2} - \frac{v}{4} + \frac{1}{2v} = 0. \tag{3.1.104}$$

Multiplying (3.1.104) by $2dv/d\theta$ and integrating, we get

$$\left(\frac{dv}{d\theta}\right)^2 - \frac{v^2}{4} + \log v = c_1 \tag{3.1.105}$$

where c_1 is constant of integration. An integration of (3.1.105) gives

$$\log \phi = \int^{\phi'/\phi^{3/2}} \frac{dv}{\sqrt{c_1 + v^2/4 - \log v}} + c_2. \tag{3.1.106}$$

If we write, $\psi = \phi^{-1/2}$ in (3.1.106), then, as for Case (a), we may write it as

$$\psi' = f_2(c_2 + 2\log \psi) \tag{3.1.107}$$

for some function f_2. Integrating (3.1.107) we get

$$\eta + c_3 = \int^{\phi^{-1/2}} \frac{d\psi}{f_2(c_2 + 2\log \psi)}; \tag{3.1.108}$$

$c_i (i = 1, 2, 3)$ in (3.1.105)-(3.1.108) are arbitrary constants. In the above derivation, we have taken the positive sign in the square roots in the first integrals for both Cases (a) and (b). We could also have taken the negative sign in the square roots. The appropriate sign must be decided by reference to a given physical problem.

3.2 Systems of Partial Differential Equations

Logan and Pérez (1980) gave an equivalent treatment for the invariance of a system of PDEs under a group of infinitesimal transformations. For convenience we consider a system of two first-order PDEs in two independent variables x, t and two dependent variables u, v:

$$H_1(x, t, u, v, u_x, u_t, v_x, v_t) = 0$$

$$H_2(x, t, u, v, u_x, u_t, v_x, v_t) \quad = \quad 0. \tag{3.2.1}$$

Consider a one-parameter group of transformations

$$\begin{aligned} \bar{x} &= x + \epsilon X, \quad \bar{t} = t + \epsilon T \\ \bar{u} &= u + \epsilon U, \quad \bar{v} = v + \epsilon V. \end{aligned} \tag{3.2.2}$$

We define the operator

$$\begin{aligned} L &= X \frac{\partial}{\partial x} + T \frac{\partial}{\partial t} + U \frac{\partial}{\partial u} + V \frac{\partial}{\partial v} \\ &\quad + \tilde{U}_x \frac{\partial}{\partial u_x} + \tilde{U}_t \frac{\partial}{\partial u_t} + \tilde{V}_x \frac{\partial}{\partial v_x} + \tilde{V}_t \frac{\partial}{\partial v_t}. \end{aligned} \tag{3.2.3}$$

The system of PDEs (3.2.1) is said to be constantly conformally invariant under the infinitesimal group of transformations (3.2.2) if there exist constants $\alpha_{ij}(i, j = 1, 2)$ such that

$$\begin{aligned} LH_1 &= \alpha_{11}H_1 + \alpha_{12}H_2 \\ LH_2 &= \alpha_{21}H_1 + \alpha_{22}H_2. \end{aligned} \tag{3.2.4}$$

Equating in (3.2.4) the coefficients of like derivative terms to zero, we get a system of linear PDEs to determine the generators X, T, U, and V. We can then proceed as for the case of a single PDE discussed in Section 3.1. In this formulation the equations $H_k = 0(k = 1, 2,)$ need not be substituted after calculating LH_k; the H_k's appear on the right-hand sides of (3.2.4) in linear combinations.

To illustrate the use of infinitesimal trasformations to systems of PDEs, we consider invariance properties of unsteady two-dimensional flow equations in the hydraulic approximation, namely

$$u_x + v_y = 0 \tag{3.2.5a}$$

$$u_t + uu_x + vu_y + gh_x = 0 \tag{3.2.5b}$$

where x and y are Cartesian coordinates measured along and perpendicular to the uniform horizontal bottom, and u and v are the associated velocity components; g is the acceleration due to gravity. The fluid depth $h(x, t)$ is to be determined such that the boundary conditions given below are satisfied.

At the bottom of the channel

$$v = 0 \text{ on } y = 0. \tag{3.2.6}$$

On the free surface we must have

$$v = h_t + uh_x \text{ on } y = h(x, t). \tag{3.2.7}$$

Using the transformations

$$z = \frac{y}{h(x,t)}, \quad w = \frac{Dz}{Dt} = v - z(h_t + uh_x), \tag{3.2.8}$$

Equations (3.2.5a) and (3.2.5b) change to

$$h_t + uh_x + h(u_x + w_z) = 0 \tag{3.2.9}$$

$$u_t + uu_x + wu_z + gh_x = 0 \tag{3.2.10}$$

while the boundary conditions become

$$w = 0 \text{ on } z = 0, 1. \tag{3.2.11}$$

Sachdev and Philip (1986) have used infinitesimal transformations, as formulated by Logan and Pérez (1980), to obtain similarity solutions of the system of Equations (3.2.9) - (3.2.10).

We seek a one-parameter infinitesimal group of transformations which takes the (t, x, z, u, w) - space into itself, and under which the sytem (3.2.9) - (3.2.10) is invariant:

$$\begin{aligned} \bar{t} &= t + \epsilon T, \bar{x} = x + \epsilon X, \bar{z} = z + \epsilon Z \\ \bar{u} &= u + \epsilon U, \bar{w} = w + \epsilon W \end{aligned}$$

where the generators $T, X, Z, U,$ and W are functions of $t, x, z, u,$ and w.

Using Equations (3.1.17) - (3.1.22) of Section 3.1 modified to include three variables t, x, z, the invariance of (3.2.9) gives

$$T = T(t,x,z), X = X(t,x,z), Z = Z(t,x,z) \tag{3.2.12}$$

$$h(U_u - X_x) + Xh_x + Th_t = \alpha_{11}h + \alpha_{12}u \tag{3.2.13}$$

$$h(W_u - Z_x) = \alpha_{12}W \tag{3.2.14}$$

$$-hT_x = \alpha_{12} \tag{3.2.15}$$

$$h(U_w - X_z) = 0 \tag{3.2.16}$$

$$h(W_w - Z_z) + (Xh_x + Th_t) = \alpha_{11}h \tag{3.2.17}$$

$$hT_z = 0 \tag{3.2.18}$$

$$\begin{aligned} Xh_{tx} + Th_{tt} + u(Xh_{xx} + Th_{xt}) \\ +Uh_x + hU_x + hW_z &= \alpha_{11}(h_t + uh_x) + \alpha_{12}gh_x. \end{aligned} \tag{3.2.19}$$

Invariance of (3.2.10) yields

$$U + u(U_u - X_x) - wX_z - X_t = \alpha_{21}h + \alpha_{22}u \tag{3.2.20}$$

$$W + w(U_u - Z_z) - uZ_x - Z_t = \alpha_{22}w \tag{3.2.21}$$

$$U_u - T_t - uT_x - wT_z = \alpha_{22} \tag{3.2.22}$$

$$uU_w = 0 \tag{3.2.23}$$

$$wU_w = \alpha_{21}h \tag{3.2.24}$$

$$U_w = 0 \tag{3.2.25}$$

$$U_t + uU_x + wU_z + g(Xh_{xx} + Th_{xt}) = \alpha_{21}(h_t + uh_x) + \alpha_{22}gh_x. \quad (3.2.26)$$

We now solve the determining differential equations. Differentiating (3.2.14) with respect to w and (3.2.17) with respect to u, we get $\alpha_{12} = W_{uw} = 0$. Equations (3.2.25) and (3.2.24) imply $\alpha_{21} = 0$, since $h \neq 0$. Differentiating (3.2.20) with respect to w and using (3.2.25) and (3.2.12), we get $X = X(x, t)$. Equations (3.2.18), (3.2.15), and (3.2.12) now imply that $T = T(t)$. Further, (3.2.13) gives $U_{uu} = 0$. Equations (3.2.17), (3.2.19), and (3.2.14) give $W_{ww} = W_{zw} = W_{uu} = W_{uw} = 0$. These equations yield

$$W = k(x, t)w + l_1(x, z, t)u + m(x, z, t) \quad (3.2.27)$$

where k, l, and m are functions of their arguments to be determined. Equation (3.2.26) differentiated with respect to w gives $U_z = 0$, which implies that $U = U(x, t, u)$. Equation (3.2.22) now gives

$$U = [\alpha_{22} + T'(t)]u + f(x, t) \quad (3.2.28)$$

where $f(x, t)$ is an arbitrary function. Differentiating (3.2.20) with respect to u and using (3.2.28) we obtain

$$X = [\alpha_{22} + 2T'(t)]x + \beta(t) \quad (3.2.29)$$

where β is an arbitrary function of t. The equations obtained by differentiating (3.2.20) with respect to x and (3.2.26) with respect to u, and use of (3.2.28) and (3.2.29), yield $f = f(t)$ and

$$T = at + b \quad (3.2.30)$$

where a and b are arbitrary constants. Differentiating (3.2.21) with respect to w and using (3.2.27), (3.2.28), and (3.2.30), we get $Z = [k(x, t) + a]z + \gamma(x, t)$ where γ is an arbitrary function of x and t. Invariance of the boundary $z = 0$ implies $\bar{z} = z + \epsilon Z = 0$ when $z = 0$; this implies that $Z = 0$ when $z = 0$ which, in turn, requires that $\gamma(x, t) = 0$. Invariance of the other boundary $z = 1$ implies that $\bar{z} = z + \epsilon Z = 1$ when $z = 1$, requiring that $Z = 0$ when $z = 1$. This implies that $k(x, t) = -a$. Thus, we have $Z = 0$. Equations (3.2.14) and (3.2.27) now give $l_1 = 0$, and (3.2.21) yields $m = 0$. Equation (3.2.20) leads to $f(t) = \beta'(t)$. Thus, the generators of the invariance group for Equations (3.2.9) - (3.2.11) are

$$T = at + b, X = (c + a)x + \beta(t),$$
$$Z = 0, U = cu + \beta'(t), W = -aw \quad (3.2.31)$$

where $a, b, c = \alpha_{22} + a$ are arbitrary constants and $\beta(t)$ is an arbitrary function of t.

Using (3.2.31) in (3.2.13) we get

$$(at + b)h_t + [(c + a)x + \beta(t)]h_x - (\alpha_{11} + a)h = 0. \quad (3.2.32)$$

Differentiating (3.2.32) with respect to x and combining it with the equation obtained by substituting (3.2.31) in (3.2.26), we get

$$(2c - a - \alpha_{11})h_x = \frac{\beta''(t)}{g}. \tag{3.2.33}$$

From (3.2.33) we get the following alternatives:

(i) $2c = a + \alpha_{11}$. In this case,

$$\beta(t) = a_0 t + a_1 \tag{3.2.34}$$

where a_0 and a_1 are arbitrary constants.

(ii) $h = \dfrac{\beta''(t)}{g(2c - a - \alpha_{11})} + \beta_1(t),$

where $\beta_1(t)$ is an arbitrary function of t. But this does not lead to any new solution.

The invariant surface conditions for u and w, respectively, are (see (3.1.17))

$$T\frac{\partial u}{\partial t} + X\frac{\partial u}{\partial x} + Z\frac{\partial u}{\partial z} = U \tag{3.2.35}$$

$$T\frac{\partial w}{\partial t} + X\frac{\partial w}{\partial x} + Z\frac{\partial w}{\partial z} = W. \tag{3.2.36}$$

Case (i) $l = c/a \neq -1$.

The characteristic system corresponding to (3.2.35) becomes

$$\frac{dt}{at + b} = \frac{dx}{(c + a)x + a_0 t + a_1} = \frac{dz}{0} = \frac{du}{cu + a_0}. \tag{3.2.37}$$

Integration of the first two equations gives the similarity variables

$$\xi = \left[x + A + \frac{a_0}{c}\left(t + \frac{B}{l + 1}\right)\right](t + B)^{-l-1}, \quad \eta = z \tag{3.2.38}$$

where $A = \dfrac{a_1}{c + a}, B = \dfrac{b}{a}(a \neq 0)$. Integration of the third equation of the characteristic system (3.2.37) gives the similarity form

$$u = (t + B)^l F(\xi, \eta) - \frac{a_0}{c} \tag{3.2.39}$$

where $F(\xi, \eta)$ is an arbitrary function of ξ and η. In a similar manner, (3.2.36) gives

$$w = (t + B)^{-1}G(\xi, \eta) \tag{3.2.40}$$

where $G(\xi, \eta)$ is another arbitrary function of ξ and η. The characteristic system of (3.2.32) with $\beta(t)$ as given by (3.2.34) and $2c = a + \alpha_{11}$ is

$$\frac{dt}{at+b} = \frac{dx}{(c+a)x + a_0 t + a_1} = \frac{dh}{2ch}$$

from which we get

$$h = (t+B)^{2l} s(\xi) \tag{3.2.41}$$

where s is an arbitrary function of ξ.

By substituting (3.2.39) and (3.2.40) in (3.2.9) and (3.2.10), we obtain a system of PDEs with two independent variables ξ and η (for convenience, we designate ξ, η, F, and G as x, z, u, and w, respectively):

$$u_x + w_z + [u - (l+1)x]\left(\frac{s_x}{s}\right) + 2l = 0 \tag{3.2.42}$$

$$[u - (l+1)x]u_x + wu_z + lu + gs_x = 0. \tag{3.2.43}$$

Again we seek a one-parameter infinitesimal group of transformations

$$\bar{x} = x + \epsilon X, \bar{z} = z + \epsilon Z$$

$$\tag{3.2.44}$$

$$\bar{u} = u + \epsilon U, \bar{w} = w + \epsilon W$$

under which the system (3.2.42) - (3.2.43) is invariant. The generators X, Z, U, and W are functions of x, z, u, and w.

Invariance of (3.2.42) yields

$$X = X(x, z), Z = Z(x, z) \tag{3.2.45}$$

$$U_u - X_x = \beta_{11} + \beta_{12}[u - (l+1)x] \tag{3.2.46}$$

$$W_u - Z_x = \beta_{12} w \tag{3.2.47}$$

$$U_w - X_z = 0 \tag{3.2.48}$$

$$W_w - Z_z = \beta_{11} \tag{3.2.49}$$

$$U_x + W_z + [u - (l+1)x]X\left(\frac{s_x}{s}\right)_x + [U - (l+1)X]\left(\frac{s_x}{s}\right)$$
$$= \beta_{11}[u - (l+1)x]\left(\frac{s_x}{s}\right) + 2l\beta_{11} + \beta_{12}(lu + gs_x). \tag{3.2.50}$$

Invariance of (3.2.43) gives

$$U - (l+1)X + [u-(l+1)x](U_u - X_x) - wX_z = \beta_{21} + \beta_{22}[u-(l+1)x] \tag{3.2.51}$$

$$W + w(U_u - Z_z) - [U - (l+1)x]Z_x = \beta_{22}w \tag{3.2.52}$$

$$[U - (l+1)x]U_w = 0 \tag{3.2.53}$$

$$wU_w = \beta_{21} \tag{3.2.54}$$

$$[u-(l+1)x]U_x+U_z+lU+gXs_{xx} = \beta_{21}[u-(l+1)x]\left(\frac{s_x}{s}\right)+2l\beta_{21}+\beta_{22}(lu+gs_x). \tag{3.2.55}$$

Equations (3.2.47) and (3.2.49) imply that $W_{wu} = \beta_{12} = 0$. Equations (3.2.53) and (3.2.54) yield $U_w = 0$, $U \neq (l+1)x$, $\beta_{21} = 0$. Differentiation of (3.2.46) with respect to u gives $U_{uu} = 0$. Differentiating (3.2.47) with respect to u and (3.2.49), and (3.2.50) with respect to w, we get $W_{uu} = W_{ww} = W_{zw} = 0$. These equations together with $W_{uw} = 0$ imply that

$$W = m(x)w + p(x,z)u + q(x,z) \tag{3.2.56}$$

where m, p and q are arbitrary functions of the indicated arguments. Employing (3.2.56) in (3.2.49) and integrating with respect to z, we get $Z = [m(x) - \beta_{11}]z + f_1(x)$, where $f_1(x)$ is an arbitrary function of x. Invariance of $z = 0$ implies $f_1 = 0$, and invariance of $z = 1$ implies $m(x) = \beta_{11}$, so that

$$Z = 0. \tag{3.2.57}$$

Equations (3.2.47) and (3.2.57) give $W_u = 0$ which, by the use of (3.2.56), implies that $p = 0$. Equation (3.2.55) differentiated with respect to w gives $U_z = 0$ so that $U = U(x,u)$. Equation (3.2.51) differentiated with respect to w yields $X_z = 0$ so that $X = X(x)$. Differentiating (3.2.51) with respect to w and using (3.2.46), we get

$$2U_u = \beta_{22} + X'(x) = 2[\beta_{11} + X'(x)] \tag{3.2.58}$$

from which we get $X = a_1x+b_1$ where $a_1(= \beta_{22}-2\beta_{11})$ and b_1 are arbitrary constants. Therefore, from (3.2.58)

$$U = (\beta_{11} + a_1)u + f_2(x) \tag{3.2.59}$$

where f_2 is an arbitrary function of x. Equations (3.2.56), (3.2.57), (3.2.59), and (3.2.52) now give $q = 0$, so that

$$W = \beta_{11}w. \tag{3.2.60}$$

Equation (3.2.52), with the help of (3.2.59) and (3.2.60), gives $f_2(x) = (l+1)(b_1 - \beta_{11}x)$, and so from (3.2.51)

$$U = (\beta_{11} + a_1)u + (l+1)(b_1 - \beta_{11}x). \tag{3.2.61}$$

Substituting (3.2.58), (3.2.60), and (3.2.61) in (3.2.50), and differentiating the resulting equation with respect to u, we get $(a_1x + b_1)(s_x/s)_x = -a_1(s_x/s)$, from which it follows that

$$s(x) = k_2(a_1x + b_1)^{k_1/a_1} \tag{3.2.62}$$

where k_1 and k_2 are arbitrary constants. Substituting (3.2.62) in (3.2.55) and equating to zero the coefficient of u and the terms independent of u, respectively, we get $\beta_{11} = 0$, provided that $l \neq -1/2$, and

$$l(l+1)b_1 = gk_1k_2(a_1x + b_1)^{k_1/a_1 - 1}(2a_1 - k_1).\qquad(3.2.63)$$

The only meaningful rel   om (3.2.63) is $b_1 = 0$, $k_1 = 2a_1$. Therefore, the infinitesimal generators in terms of the original variables ξ, η, F and G are

$$\begin{aligned}
X &= a_1\xi, Z = 0, U = a_1F, W = 0 &(3.2.64)\\
h &= (t+B)^{2l}s(\xi) = (t+B)^{2l}k_2(a_1\xi)^2\\
&= k(t+B)^{2l}\xi^2\\
&= k\left[x + A + \left(\frac{a_0}{c}\right)\left(t + \frac{B}{l+1}\right)\right]^2 /(t+B)^2 &(3.2.65)
\end{aligned}$$

where $k = k_2a_1^2$. The invariant surface conditions for F and G are (see (3.1.17))

$$X\frac{\partial F}{\partial \xi} + Z\frac{\partial F}{\partial \eta} = U \qquad(3.2.66)$$

$$X\frac{\partial G}{\partial \xi} + Z\frac{\partial G}{\partial \eta} = W. \qquad(3.2.67)$$

The characteristic system for (3.2.66) is

$$\frac{d\xi}{a_1\xi} = \frac{d\eta}{0} = \frac{dF}{a_1F}.$$

The first equation gives the similarity variable as η. Also, we get

$$F = \xi P(\eta) \qquad(3.2.68)$$

where P is an arbitrary function of η. Similarly, from (3.2.67) we get

$$G = Q(\eta) \qquad(3.2.69)$$

where Q is an arbitrary function of η. Substituting (3.2.68) and (3.2.69) in (3.2.42) and (3.2.43) (after changing x, z, u, and w back to ξ, η, F, and G, respectively in (3.2.42) and (3.2.43)) and using (3.2.65) for h, we get the system of ODEs

$$3P + Q' - 2 = 0 \qquad(3.2.70)$$

$$QP' - P + P^2 + 2gk = 0. \qquad(3.2.71)$$

where prime denotes differentiation with respect to η. Denoting gk by $1/(4\alpha^2)$ in (3.2.71) we get

$$Q = \frac{P - P^2 - 1/2\alpha^2}{P'}. \qquad(3.2.72)$$

Substituting (3.2.72) in (3.2.70) and writing $P = S/2\alpha$, the latter becomes

$$(S^2 - 2\alpha S + 2)S'' + (S - 2\alpha)S'^2 = 0. \tag{3.2.73}$$

Interchanging the independent and dependent variables in (3.2.73) we have

$$(S^2 - 2\alpha S + 2)\frac{d^2\eta}{dS^2} + (2\alpha - S)\frac{d\eta}{dS} = 0$$

or

$$(S - \lambda_1)(S - \lambda_2)\frac{d^2\eta}{dU^2} + (2\alpha - S)\frac{d\eta}{dS} = 0 \tag{3.2.74}$$

where λ_1 and λ_2 are roots of the quadratic

$$\lambda^2 - 2\alpha\lambda + 2 = 0, \tag{3.2.75}$$

assumed to be real, requiring $\alpha^2 > 2$. We also assume that $F \neq (l+1)\xi$. This is the condition for the absence of critical levels in the flow. Making the substitution $\tau = (S - \lambda_1)/(\lambda_2 - \lambda_1)$ in (3.2.74) we get the hypergeometric equation

$$\tau(1-\tau)\frac{d^2\eta}{d\tau^2} + \left(\frac{2}{\lambda_1^2 - 2} + \tau\right)\frac{d\eta}{d\tau} = 0. \tag{3.2.76}$$

The solution of (3.2.76) can be expressed as

$$\eta = A_1 + B_1\beta\left(\frac{4 - \lambda_1^2}{2 - \lambda_1^2}, \frac{2 - 2\lambda_1^2}{2 - \lambda_1^2}, \frac{S - \lambda_1}{2\lambda_1^{-1} - \lambda_1}\right) \tag{3.2.77}$$

where A_1 and B_1 are arbitrary constants. β is the incomplete beta function. Equation (3.2.72) may be written as

$$Q = \frac{S^2 - 2\alpha S + 2}{2\alpha S'} = \frac{(S - \lambda_1)(S - \lambda_2)}{2\alpha S'}$$

so that the boundary conditions become $\eta = 0$ at $S = \lambda_1$ and $\eta = 1$ at $S = \lambda_2 = 2/\lambda_1$. Using these boundary conditions in (3.2.77) we get the solution as

$$\eta = \frac{\beta\left(\frac{4-\lambda_1^2}{2-\lambda_1^2}, \frac{2-2\lambda_1^2}{2-\lambda_1^2}, \frac{S-\lambda_1}{2\lambda_1^{-1}-\lambda_1}\right)}{\beta\left(\frac{4-\lambda_1^2}{2-\lambda_1^2}, \frac{2-2\lambda_1^2}{2-\lambda_1^2}, 1\right)}. \tag{3.2.78}$$

We now consider the case $l = c/a = -1$. The characteristic system for (3.2.35) in this case is

$$\frac{dt}{at+b} = \frac{dx}{a_0 t + a_1} = \frac{dz}{0} = \frac{du}{-au + a_0}. \tag{3.2.79}$$

Integrating the first two of these equations we have the similarity variables

$$\xi = x - A_0 t - (A_1 - A_0 B)\ln(t + B) \tag{3.2.80}$$

$$\eta = z \qquad\qquad (3.2.81)$$

where $A_0 = a_0/a$, $A_1 = a_1/a$ and $B = b/a$. The last two of the characteristic relations (3.2.79) integrate to give

$$u = A_0 + (t + B)^{-1}F(\xi, \eta) \qquad\qquad (3.2.82)$$

where F is an arbitrary function of ξ and η. Similarly, from the characteristic system for (3.2.36) we get

$$w = (t + B)^{-1}G(\xi, \eta) \qquad\qquad (3.2.83)$$

where G is another arbitrary function of ξ and η. The characteristic system for (3.2.32) now is

$$\frac{dt}{at + b} = \frac{dx}{a_0 t + a_1} = \frac{dh}{-2ah}. \qquad\qquad (3.2.84)$$

The general solution of (3.2.32), therefore, is

$$h = (t + B)^{-2}s(\xi) \qquad\qquad (3.2.85)$$

where s is an arbitrary function of ξ. Substituting (3.2.82), (3.2.83), and (3.2.85) in (3.2.9) and (3.2.10), we get

$$F_\xi + G_\eta + (F - A_2)\left(\frac{s_\xi}{s}\right) - 2 = 0 \qquad\qquad (3.2.86)$$

$$(F - A_2)F_\xi + GF_\eta + gs_\xi = 0 \qquad\qquad (3.2.87)$$

where $A_2 = A_1 - A_0 B$. The determining differential equations for the transformation group (3.2.44) which leave (3.2.86) and (3.2.87) invariant can be obtained from (3.2.45) - (3.2.55) by replacing $(l + 1)x$ by A_2 and l by -1. Proceeding as in the above case, we get $A_2 = 0$ and

$$\xi = x - A_0 t. \qquad\qquad (3.2.88)$$

The generators in the present case are

$$X = a_1 \xi + b_1, Z = 0, U = a_1 F, W = 0. \qquad\qquad (3.2.89)$$

We also get

$$\begin{aligned} h &= \frac{k_2(a_1\xi + b_1)^2}{(t + B)^2} \\ &= \frac{k(x + B_1 - A_0 t)^2}{(t + B)^2} \end{aligned} \qquad\qquad (3.2.90)$$

where $B_1 = b_1/a_1$. The characteristic system for the invariant surface condition (3.2.66) is

$$\frac{d\xi}{a_1\xi + b_1} = \frac{d\eta}{0} = \frac{dF}{a_1 F}. \qquad\qquad (3.2.91)$$

From the system (3.2.91) we get η as the similarity variable and

$$F = (\xi + B_1)P(\eta) \tag{3.2.92}$$

where $P(\eta)$ is an arbitrary function of η. In a similar manner, we get from (3.2.67)

$$G = Q(\eta) \tag{3.2.93}$$

where $Q(\eta)$ is another arbitrary function of η. Substituting (3.2.92) and (3.2.93) into (3.2.86) and (3.2.87), we get the ODEs

$$3P + Q' - 2 = 0$$
$$P^2 + QP' - P + 2gk = 0,$$

which are the same as (3.2.70) and (3.2.71) and, hence, may be solved in the form (3.2.78). The solutions of Freeman (1972) and Sachdev (1980) are thus recovered and generalised.

3.3 Self-Similar Solutions of the Second Kind - Viscous Gravity Currents

We conclude our analysis of self-similar solutions by discussion of a special class of self-similar solutions: "self-similar solutions of the second kind." The first kind of self-similar solutions are those for which the exponent in the similarity variable is determined by dimensional arguments alone; a famous example of this kind of solution is the Taylor-Sedov self-similar solution describing the point explosion in a uniform medium. An excellent discussion of these two classes of solutions in gasdynamic context is given by Zel'dovich and Raizer (1967). We quote from Zel'dovich's foreword to Barenblatt's book on intermediate asymptotics, giving the characterising property of self-similar solutions of second kind: "We shall reserve the name solution of the second kind for the large and evergrowing class of solutions for which the exponents are found in the process of solving the problem, analogous to the determination of eigenvalues for linear equations. For this case, conservation laws and dimensional considerations prove to be insufficient."

A more recent review of this topic is due to Peletier (1998), who discusses mainly nonlinear diffusion equations with special reference to turbulent outbursts.

Here we content ourselves with the discussion of one nonlinear diffusion equation which occurs in many applications, namely

$$\frac{\partial h}{\partial t} = x^{-n} \frac{\partial}{\partial x}\left(x^n h^m \frac{\partial h}{\partial x}\right). \tag{3.3.1}$$

Equation (3.3.1) may be interpreted as a nonlinear heat conduction equation if h is taken to be the temperature and $\lambda = \lambda_0 h^m$ as the coefficient of heat conduction: $n = 0, 1$ for plane and axisymmetric geometry, respectively. We shall, however, follow the work of Gratton and Minotti (1990) who consider (3.3.1) as a model for the spreading of viscous gravity currents over a rigid horizontal surface. Actually, we can rewrite (3.3.1) for $m = 3$ as the system

$$h^2 h_x + v = 0 \tag{3.3.2}$$

$$h_t + (vh)_x + n\left(\frac{vh}{x}\right) = 0 \tag{3.3.3}$$

where, in the gravity current context, $h(x,t)$ denotes the thickness of the current and $v(x,t)$ is the average horizontal velocity. A derivation of (3.3.2) - (3.3.3) may be found in Smith (1969). The basic assumptions in this derivation are that the viscous gravity flow on a rigid horizontal surface is slow and the motion is essentially horizontal so that the pressure is purely hydrostatic $\left(\dfrac{\partial p}{\partial z} = -\rho g\right)$, the inertial effects are negligible, and the length of the current is much larger than its depth.

Since Equations (3.3.2) - (3.3.3) do not contain any constant dimensional parameter and involve only quantities having the dimensions of length [L] or time [T] or combinations thereof, the dependent variables h and v can be expressed as

$$h = (x^2 t^{-1} Z)^{1/3}, \quad v = xt^{-1}V \tag{3.3.4}$$

where Z and V are dimensionless functions of x, t and the constant parameters of the problem, which arise from the initial and boundary conditions. Substituting (3.3.4) into (3.3.2) - (3.3.3), we get

$$x\frac{\partial Z}{dx} + 2Z + 3V = 0 \tag{3.3.5}$$

$$t\frac{\partial Z}{\partial t} + 3xZ\frac{\partial V}{\partial x} + xV\frac{\partial Z}{\partial x} + (5 + 3n)VZ - Z = 0. \tag{3.3.6}$$

If the boundary/initial conditions are such that they involve two (or more) constant parameters with independent dimensions, enabling forming of two combinations $\hat{l}$ and $\hat{t}$ of them such that $[\hat{l}] = L, [\hat{t}] = T$, then Z and V will in general depend on two dimensionless independent variables $x/\hat{l}, t/\hat{t}$; they may also depend on some other dimensionless parameters $\pi_1, \pi_2, \ldots$, etc. In this case, the problem is non-self-similar, and no reduction to ODEs is possible.

On the other hand, suppose that the problem depends only on one parameter b with independent dimensions, which without loss of generality may be written as

$$[b] = LT^{-\delta}$$

where δ is a numerical constant. In this case there will be a single dimensionless combination of x, t, and b, which may be written as

$$\zeta = x/bt^{\delta}. \tag{3.3.7}$$

This is the similarity variable and the functions Z and V depend only on ζ : $Z = Z(\zeta)$ and $V = V(\zeta)$. Equations (3.3.5) - (3.3.6) reduce to the ODEs

$$\zeta Z' + 2Z + 3V = 0 \tag{3.3.8}$$

$$3\zeta Z V' - \zeta Z'(\delta - V) + (5 + 3n)VZ - Z = 0 \tag{3.3.9}$$

where prime denotes derivative with respect to ζ. Eliminating ζ from the system (3.3.8) - (3.3.9), we can write it in the (V, Z) plane as

$$\frac{dV}{dZ} = \frac{Z(2\delta - 1) + 3(1 + n)VZ + 3(\delta - V)V}{3Z(2Z + 3V)}. \tag{3.3.10}$$

The variation of V and Z with ζ is obtained with the help of (3.3.8), written as

$$\frac{d}{dZ}(\ln|\zeta|) = -\frac{1}{2Z + 3V}. \tag{3.3.11}$$

Typically, for self-similar solutions of the second kind, the analysis reduces to the discussion of an ODE in the phase plane, Equation (3.3.10) in the present case, and a connecting Equation (3.3.11) which relates the dependent variables with the independent variable ζ (See Sachdev (1991)). A curve in the (V, Z) (phase) plane represents an integral curve. If it passes through some particular points representing the boundary conditions, it represents the solution of the corresponding physical problem. Each such piece represents the flow in a certain domain of the independent variables. The discussion of all the singular points of (3.3.10) and solutions in their neighborhoods would indicate what kind of self-similar problems may be solved. One of the earliest and most exhaustive discussions of the phase plane for solutions of the self-similar kind in the context of one-dimensional spherical and cylindrical flows, with or without shocks, may be found in the classical work of Courant and Friedrichs (1948).

Equation (3.3.10) has six singular points. We discuss each of them separately.

(i) Point $O(Z_0 = 0, V_0 = 0)$. This point for $\delta \neq 0$ represents points at infinity, namely $\zeta = \infty$.

If $Z > 0$, the point O is a node, and as it is approached all integral curves (except $Z = 0$) converge to the curve

$$V = -\frac{2\delta - 1}{3\delta}Z\left[1 - \frac{(5 + 3n)\delta - 4}{3\delta^2}Z + \cdots\right]. \tag{3.3.12}$$

If $Z < 0$, O is a saddle and only a single curve given by (3.3.12) reaches O. For both $Z > 0$ and $Z < 0$, the asymptotic solution near O for $\delta \neq 0, 1/2$ (see (3.3.12)) may be found to be

$$
\begin{aligned}
Z &= K'\zeta^{-1/\delta} \\
h &= (K')^{1/3}(b/3)^{1/3\delta}x^{(2\delta-1)/3\delta} \\
v &= -K'\frac{2\delta-1}{3\delta}b^{1/\delta}x^{(\delta-1)/\delta}.
\end{aligned}
\tag{3.3.13}
$$

K and K' here and in the following are arbitrary constants. It may be observed that h and v in (3.3.13) do not depend on t as $x \to \infty$.

For the special case $\delta = 1/2$ and $Z > 0$, the integral curve near O may be found to be

$$
V = K \exp(-1/4Z),
\tag{3.3.14}
$$

with the corresponding asymptotic formulae

$$
Z = K'\zeta^{-2}, h = (3b^2K')^{1/3}, v = (Kx/t)\exp(-\zeta^2/4K').
\tag{3.3.15}
$$

For $Z < 0$, only the curve corresponding to the trivial solution $V = 0$ arrives at O.

For the other special case $\delta = 0$, the point O is a saddle. Ignoring the trivial solution $Z = 0$, the integral curves near O are described by

$$
V = \pm \left(\frac{1}{9}KZ^{-2/3} - \frac{2}{15}Z \right)^{1/2},
\tag{3.3.16}
$$

and the only curves that pass through O are

$$
V = \pm \left(-\frac{2Z}{15} \right)^{1/2}
\tag{3.3.17}
$$

corresponding to $K = 0$. For $n = 0$, (3.3.17) is an exact solution of (3.3.10).

(ii) Point A $(Z_A = 0, V_A = \delta)$ is a saddle.

If $\delta = 0$, the point A coalesces with O; therefore, only the case $\delta \neq 0$ may be considered. Again ignoring the trivial solution $Z = 0$, we obtain the other approximate solution near this point, namely

$$
V = \delta + \frac{(5+3n)\delta - 1}{12\delta}Z.
\tag{3.3.18}
$$

The point A represents an advancing front of a gravity viscous current with the locus $x_f = \zeta_f bt^\delta$, where ζ_f is a constant. Introducing the notation $\eta = x/x_f = \zeta/\zeta_f$, the solution near the front may be found to be

$$Z = 3\delta(1 - \eta) \tag{3.3.19}$$

$$h = \zeta_f^{2/3}(9b^2\delta)^{1/3}t^{(2\delta-1)/3}(1 - \eta)^{1/3}$$
$$\times \left[1 - \frac{(1 + 3n)\delta - 1}{24\delta}(1 - \eta) + \cdots\right] \tag{3.3.20}$$

$$v = \delta x_f t^{-1}\eta\left[1 + \frac{(5 + 3n)\delta - 1}{4\delta}(1 - \eta) + \cdots\right]. \tag{3.3.21}$$

For $Z > 0$, the integral curves in the present case correspond to currents produced by sources whose flux depends on time according to a power law. Here, too, there are special exact solutions of (3.3.10):

a) $n = 0$, $\delta = 1$

$$Z = 3V(V - 1). \tag{3.3.22}$$

This curve represents a current whose profile moves with constant velocity without change in shape.

(b) $n = 0, \delta = 1/8$

In this case, (3.3.18) with the above values of n and δ is an exact solution of (3.3.10) and represents a current that is drained from the origin and that has a front moving away.

(iii) Point B ($Z_B = -3/2(5 + 3n)$, $V_B = 1/(5 + 3n)$ is a node for $\delta \leq \delta_-$ and for $\delta \geq \delta_+$ where $\delta_\pm = 13/10 \pm (6/5)^{1/2}$ if $n = 0$ and $\delta_\pm = 1 \pm \sqrt{3}/2$ if $n = 1$. With $Z^* = Z - Z_B$, $V^* = V - V_B$, the approximate representation of the integral curves in the neighbourhood of B is

$$(V^* - \gamma_+Z^*)^\gamma(V^* - \gamma_-Z^*) = K \tag{3.3.23}$$

$$\gamma_\pm = \frac{1}{18}(1 + 3n) - \frac{1}{9}(\delta \pm \Delta)(5 + 3n) \tag{3.3.24}$$

$$\gamma = (\Delta + \Gamma)/(\Delta - \Gamma) \tag{3.3.25}$$

$$\Delta = [(\delta - \delta_+)(\delta - \delta_-)]^{1/2}, \quad \Gamma = \delta - (13 + 3n)/2(5 + 3n). \tag{3.3.26}$$

Near point B, we have

$$Z^*\zeta^{2+3\gamma_\pm} = K'. \tag{3.3.27}$$

Since $2 + 3\gamma_\pm$ is positive for $\delta \leq \delta_-$ and negative for $\delta \geq \delta_+$, point B corresponds to $\zeta - +\infty$ in the former, and to $\zeta = 0$ in the latter. The asymptotic behaviour of h and v near B for $t < 0$ is given by

$$h = -\left(\frac{9}{10 + 6n}\frac{x^2}{t}\right)^{1/3} \tag{3.3.28}$$

$$v = \frac{1}{5 + 3n}xt^{-1}. \tag{3.3.29}$$

For $\delta_- < \delta < \delta_0$ and $\delta_0 < \delta < \delta_+$ where $\delta_0 = 13/10$ for $n = 0$ and $\delta_0 = 1$ for $n = 1$, B is a focus; the integral curves spiral counterclockwise as ζ tends to infinity in the first case (stable focus), and away from B as ζ increases starting from zero at B in the second case (unstable spiral). The phase variables V and Z near B have an oscillatory behaviour that, as may be verified, the physical variables themselves do not exhibit.

We also note that $Z = Z_B$, $V = V_B$ is an exact solution of (3.3.8) - (3.3.9) for any n and δ, so that (3.3.28) - (3.3.29) are exact solutions of (3.3.2)-(3.3.3). This special solution, represented by a single point in the phase plane, describes a current with a fixed front at $x = 0$, as does (3.3.17).

(iv) Point C ($Z_C = 0, V_C = \infty$) is a node. The integral curves near C are given by $Z^{1/3}V = K$, where K is constant. C represents a point of the fluid at a finite distance $x_f = \zeta_f b t^\delta$, $\zeta_f = $ constant, from the origin. With $\eta = x/x_f = \zeta/\zeta_f$, the asymptotic solution near C is

$$Z = (4K)^{3/4}(1 - \eta)^{3/4} \tag{3.3.30}$$

$$h = \zeta_f^{1/3}(3b^2)^{1/3}\left(\frac{4K}{3}\right)^{1/4} t^{(2\delta-1)/3}(1 - \eta)^{1/4}\left[1 - \frac{13}{24}(1 - \eta)\right] \tag{3.3.31}$$

$$v = \zeta_f\frac{Kb}{3}\left(\frac{3}{4K}\right)^{1/4} t^{\delta-1}\eta(1 - \eta)^{-1/4}. \tag{3.3.32}$$

As $\eta \to 1$ so that C is approached, we have $h \to 0$, $v \to \infty$ such that $(2\pi x)^n hv$ is finite.

(v) Point D ($Z_D = \infty, V_D = (1 - 2\delta)/3(1 + n)$) is a saddle. The integral curves in the neighbourhood of this point are

$$Z^{-(3+n)/2}(ZV^* - \gamma_0) - K, \quad \gamma_0 = \frac{(2\delta - 1)[(5 + 3n)\delta - 1]}{9(3 + n)(1 + n)^2} \tag{3.3.33}$$

where $V^* = V - V_D$. In this case, the only curve which reaches D from points in the finite (Z, V) plane is $V^* = \gamma_0 Z^{-1}$. If we move along this curve, D represents the origin $x = 0$, and the following asymptotic formulae hold there:

$$Z = K'\zeta^{-2} \tag{3.3.34}$$

$$h = (3K'b^2)^{1/3}t^{(2\delta-1)/3} \tag{3.3.35}$$

$$v = -\left[\frac{(2\delta - 1)}{3(1 + n)}\right]x/t. \tag{3.3.36}$$

This solution represents spreading due to gravity with no mass inflow at the origin.

(vi) Point E ($Z_E = \infty, V_E = \infty$) is a saddle node. It represents the origin $x = 0$. Equation (3.3.10) in the present case approximates to

$$\frac{dY}{dW} = \frac{Y(2W + 3Y)}{W[(1 + n)W - Y]}, \quad Y = Z^{-1}, W = V^{-1}. \tag{3.3.37}$$

For $n = 0$, (3.3.37) integrates to yield

$$W^3 Y^{-4}(Y + \frac{1}{4}W)^5 = K, \tag{3.3.38}$$

where K is a constant. It has the following approximate representation near the point E:

$$V = -\frac{1}{4}Z \ (K = 0) \tag{3.3.39}$$

$$V = \pm 4^{-5/8} K^{-1/8} Z^{1/2} \ (K \neq 0). \tag{3.3.40}$$

For the curve (3.3.39), the asymptotic formulae near E are

$$Z = K'\zeta^{-5/4}$$
$$h = (3K'b^{5/4})^{1/3}x^{1/4}t^{(5\delta/4-1)/3} \tag{3.3.41}$$
$$v = -(K'b^{5/4}/4)x^{-1/4}t^{5\delta/4-1}. \tag{3.3.42}$$

This solution represents a current with an outflow at the origin. Corresponding to (3.3.40), we have the following approximate behaviour near E:

$$Z = K'\zeta^{-2} \tag{3.3.43}$$
$$h = (3K'b^2)^{1/3}t^{(2\delta-1)/3} \tag{3.3.44}$$
$$v = \pm 4^{-5/8} K^{-1/8}(K')^{1/2}bt^{\delta-1}. \tag{3.3.45}$$

This solution describes currents with inflow or outflow at the origin corresponding to $+$ or $-$ sign in (3.3.45), respectively. For $n = 1$, Equation (3.3.37) exactly integrates to give

$$YW^3 \exp(2W/Y) = K \tag{3.3.46}$$

where K is constant of integration. We may check from (3.3.46) that the integral curves that reach E are given approximately by

$$V = Z/\ln(Z^2). \tag{3.3.47}$$

The following asymptotic formulae hold in the neighbourhood of the point E:

$$Z = K'\zeta^{-2}|\ln \zeta|^{3/4} \tag{3.3.48}$$
$$h = (3K'b^2)^{1/3}t^{(2\delta-1)/3}|\ln \zeta|^{1/4} \tag{3.3.49}$$
$$v = \left(\frac{1}{4}K'b^2\right)x^{-1}t^{2\delta-1}|\ln \zeta|^{-1/4}. \tag{3.3.50}$$

These describe a current with inflow at the origin. No integral curves can enter E via second and fourth quadrants.

Gratton and Minotti (1990) illustrate, with the help of several figures, different flows in the $V - Z$ plane as obtained by the numerical integration of (3.3.10). They also list various types of trajectories that connect pairs of singular points with short descriptions of the currents they represent. Their general conclusion regarding the physical validity of these solutions, particularly in the lubrication theory, is that these solutions are in good agreement with the experiments as supported by the work of Huppert (1982).

To check whether a given boundary value problem can be solved in the phase plane, one must identify, by inspection of the boundary or initial conditions, the parameter b that determines the self-similar variable ζ and the exponent δ defining it, and then select the appropriate integral curve among the (n, δ) family, that is, the curve whose asymptotic behaviour corresponds to the boundary condition at hand. If the boundary or initial conditions do not determine the parameter b, the self-similar exponent δ must be found by other methods. In this case, we say the self-similar solution is of the second kind (see Sachdev (1991)).

Here, we give a specific example which has a close analogy to the classical problem of converging shock waves first considered by Guderley (1942). We consider a pool of fluid outside a circular wall; inside the wall there is no fluid. Suppose the wall is suddenly removed so that the fluid rushes in, with a convergent front whose radius decreases to zero as the fluid moves in. We consider the asymptotic behaviour when the converging fluid is close to the center. During this stage the characteristic length is too small compared to any other length parameter, say the initial radius of the collapsing wall. Initial conditions also do not provide any characteristic parameters in this flow regime. Therefore, even if the flow is self-similar, the parameter δ cannot be found from initial or boundary conditions, a situation typical of the second kind of self-similar flows.

Since we are concerned with an advancing front, the solution must be represented by a trajectory leaving the singular point A. Besides, $t < 0$ before the collapse time $t = 0$; therefore, the integral curve must lie in the half-plane $Z < 0$. In this domain, the trajectory leaving A can go to B, O, or C. The trajectories from A to B or from A to C represent flows tending to infinity and zero, respectively, and, therefore, must be ruled out. The trajectory joining A and O has the required property that it reaches the center with a finite (non-zero) velocity v at a finite distance h behind the front, as $t \to 0$. The value of $\delta = \delta_c$ for which the trajectory from A reaches O must be found by numerical integration of (3.3.10), starting with the local solution near A. The latter solution is

$$h = \left(-9\delta_c \frac{x_f^2}{t} \right)^{1/3} (1 - \eta)^{1/3} \left[1 - \frac{4\delta_c - 1}{24\delta_c} (1 - \eta) + \cdots \right] \qquad (3.3.51)$$

$$v = -\frac{\delta_c x_f}{t} \eta \left[1 + \frac{8\delta_c - 1}{4\delta_c}(1 - \eta) + \cdots \right] \tag{3.3.52}$$

where $\eta = x/x_f$ and $x_f = K(-t)^{0.762}$. The exponent $\delta_c = 0.762$ is found by integration of (3.3.10) from A so as to reach the point O by suitable interpolation, etc. It is clear that the front has infinite velocity as it reaches the center.

In the present case, the similarity exponent is found by solving a non-linear eigenvalue problem, and not by dimensional considerations.

Gratton and Minotti (1990) solve several other physical problems — mostly self-similar solutions of the first kind — by either solving (3.3.10) numerically, or by considering its special exact solutions.

There are two cases for which it is possible to solve (3.3.2) - (3.3.3) explicitly.

(a) $n = 0, \delta = 0$ in (3.3.8) - (3.3.9)

This case with plane symmetry occurs when b has the dimension of length so that $b = l$ and $\zeta = \xi/l$. If we write $Z(\zeta) = W(\zeta)\zeta^{-2}$, Equation (3.3.8) becomes simply

$$V = -\frac{1}{3\zeta}\frac{dW}{d\zeta}. \tag{3.3.53}$$

Equation (3.3.9) then becomes

$$\frac{d^2 W}{d\zeta^2} + 1 + \frac{1}{3W}\left(\frac{dW}{d\zeta}\right)^2 = 0. \tag{3.3.54}$$

Equation (3.3.54) is autonomous and has the first integral

$$\frac{1}{2}\left(\frac{dW}{d\zeta}\right)^2 = -\frac{3}{5}W + \frac{1}{2}KW^{-2/3} \tag{3.3.55}$$

where K is the constant of integration. The case $K = 0$ yields the exact solution (3.3.17). For $K \neq 0$, (3.3.55) can be integrated to yield

$$\zeta = \zeta_0 \pm \frac{3}{|K|^{1/2}}\left(\frac{5}{6}K\right)^{4/5} \int_{w(\zeta_0)}^{w(\zeta)} \frac{|w|^3 dw}{[\operatorname{sgn}(K)(1 - w^5)]^{1/2}} \tag{3.3.56}$$

where $w = (6/5K)^{1/5}W^{1/3}$ and $K(1 - w^5) \geq 0$.

The integral in (3.3.56) must be evaluated numerically. Upon inversion of (3.3.56) one may write $w = w(\zeta)$ and, hence, the solution.

(b) Travelling wave solutions

The basic system (3.3.2)-(3.3.3) for $n = 0$ is translation-invariant and so admits progressive or travelling wave solutions

$$h = h(\xi), v = v(\xi), \xi = ct - x \qquad (3.3.57)$$

where c is constant. Putting (3.3.57) into (3.3.2) - (3.3.3), we get

$$v = h^2 h' \qquad (3.3.58)$$

$$(h^3 h')' - ch' = 0 \qquad (3.3.59)$$

where prime denotes derivative with respect to ξ. Equation (3.3.59) immediately integrates to give

$$h^3 h' - ch = K \qquad (3.3.60)$$

where K is constant of integration. For the special case $K = 0$, the solution of (3.3.60) and (3.3.58) is

$$h = [(9c)(\xi - \xi_0)]^{1/3}, v = c, \xi_0 = \text{ constant.} \qquad (3.3.61)$$

For $K \neq 0$, we consider the cases $K < 0$ and $K > 0$ separately.

(i) $K < 0$. We let

$$K = -ch_0, h = h_0 H(\phi), \phi = c\xi/h_0^3. \qquad (3.3.62)$$

Equation (3.3.60) changes to

$$\frac{d\phi}{dH} = \frac{H^3}{H - 1}, \qquad (3.3.63)$$

which integrates differently depending upon whether $H \geq 1$ or $0 \leq H \leq 1$. For the former we have

$$\phi = \frac{1}{3}(H - 1)^3 + \frac{3}{2}(H - 1)^2 + 3(H - 1) + \ln(H - 1) - \frac{29}{6} \qquad (3.3.64)$$

where the integration constant has been chosen such that $H(\phi = 0) = 2$. The profile (3.3.64) represents a gravity current moving with constant speed c with no change in shape.

For $0 \leq H \leq 1$, the solution of (3.3.63) is

$$\phi = -\frac{1}{3}(1 - H)^3 + \frac{3}{2}(1 - H)^2 - 3(1 - H) + \ln(1 - H) + \frac{11}{6} \qquad (3.3.65)$$

where we have chosen $H = 0$ when $\phi = 0$. This profile has a well-defined front at $\phi = 0$.

(ii) $K > 0$. In this case we let $K = ch_0$ in (3.3.62). Equation (3.3.60) then reduces to

$$\frac{d\phi}{dH} = \frac{H^3}{H+1} \qquad (3.3.66)$$

and integrates to yield

$$\phi = \frac{1}{3}(1+H)^3 - \frac{3}{2}(1+H)^2 + 3(1+H) - \ln(1+H) - \frac{11}{6}. \qquad (3.3.67)$$

Here, $H(\phi = 0) = 0$. The solution (3.3.67) is interpreted as arising out of some piston motion.

Gratton and Minotti (1990) also consider steady flows of the basic system (3.3.2)-(3.3.3) and interpret them appropriately.

PART B

The Direct Similarity Approach

3.4   Introduction

The method using the group of infinitesimal transformations to identify the similarity form of the solution has been much in vogue in the last three decades. Now, computer programs are available to lead directly to the form, thus mitigating the considerable effort involved in the procedure. A more direct and intuitive approach to finding similarity solutions, requiring no knowledge of the group invariance property of PDEs, was proposed by Clarkson and Kruskal (1989). This approach is a direct extension of the intuitive argument which was adopted by scientists before the group theoretic method became popular. It is simply assumed, with reference to a single PDE in two independent variables, for example, that the solution is expressible in the form $u(x, t) = \alpha(x, t) + \beta(x, t) f(\eta)$, where $\eta = \eta(x, t)$ is the similarity variable. This expression is substituted into the given PDE. It is then required that the PDE reduces to an ODE in f with η as the independent variable. This leads to an overdetermined system of PDEs for the coefficient functions $\Gamma_i(\eta)$ appearing in the reduced ODE, and α, β, and η. This overdetermined system is easily solved with the help of some preliminary remarks. The functions α, β, and η as well as the coefficients $\Gamma_i(\eta)$ are thus determined. It has been shown that this approach leads essentially to the same class of ODEs that the method of infinitesimal transformation does. However, the direct approach is more transparent and requires less effort to arrive at the final results.

3.5   A Nonlinear Heat Equation in Three Dimensions

We discuss in this section the nonlinear heat equation in Cartesian coordinates

$$
\begin{aligned}
u_t &= \left(1 + \frac{\beta}{(1 + u_x)^2}\right) u_{xx} + u_{yy} + u_{zz} \\[2mm]
&= \Delta u - \left(\frac{\beta}{1 + u_x}\right)_x
\end{aligned}
\tag{3.5.1}
$$

and its analogue

$$
\begin{aligned}
u_t &= \frac{1}{r}\left(1 + \frac{\beta}{(1 + ru_r)^2}\right)(ru_r)_r + \frac{1}{r^2}u_{\theta\theta} + u_{zz} \\
&= \Delta u - \frac{1}{r}\left(\frac{\beta}{1 + ru_r}\right)_r
\end{aligned}
\tag{3.5.2}
$$

in cylindrical coordinates, where $\beta > 0$ is the nonlinearity parameter; it is assumed that $1 + u_x > 0$. This equation was originally derived by Stikker (1970) to describe the conduction of heat in steel coils during the batch annealing process, and was rederived by Willms (1995).

Our purpose in discussing (3.5.1) is to show that sometimes a direct separation of variables does not lead to a sensible result, and an additive separation term may instead yield a solution to some initial/boundary value problems. This example thus motivates the ansatz in the so-called direct method of finding similarity solution presented in the following sections. Following Willms (1995), we write

$$
u(x, y, z, t) = X(x)V(y, z, t)
\tag{3.5.3}
$$

in (3.5.1) to obtain

$$
XV_t = VX''\left(1 + \frac{\beta}{(1 + VX')^2}\right) + X(V_{yy} + V_{zz})
\tag{3.5.4}
$$

or

$$
\frac{V_t - V_{yy} - V_{zz}}{V} = \frac{X''}{X}\left(1 + \frac{\beta}{(1 + VX')^2}\right).
\tag{3.5.5}
$$

It is clear that V and X do not completely separate, but one possible solution is

$$
X'' = 0 = V_t - V_{yy} - V_{zz}
\tag{3.5.6}
$$

yielding

$$
X(x) = Ax + B
\tag{3.5.7}
$$

where A and B are arbitrary constants. The solutions governed by (3.5.6) are not very interesting since they also solve the linear heat equation, the case $\beta = 0$ in (3.5.1). We may now seek additive separability

$$
u(x, y, z, t) = X(x) + V(y, z, t).
\tag{3.5.8}
$$

The form (3.5.8) is again a rather special case of the so-called direct method (see Section 3.6). Substitution of (3.5.8) into (3.5.1) gives

$$
V_t = X'' + V_{yy} + V_{zz} - \beta\frac{\partial}{\partial x}\left(\frac{1}{1 + X'}\right)
$$

or

$$V_t - V_{yy} - V_{zz} = \frac{d}{dx}\left(1 + X' - \frac{\beta}{1 + X'}\right). \tag{3.5.9}$$

Writing the constant of separation as 4λ, (3.5.9) separates into

$$V_t - V_{yy} - V_{zz} = 4\lambda \tag{3.5.10}$$

$$1 + X' - \frac{\beta}{1 + X'} = 4(\lambda x + A) \tag{3.5.11}$$

where A is constant of integration. Equation (3.5.10) is an inhomogeneous heat equation in two space variables and may be treated by standard methods (see Friedman (1964)) given a set of boundary and initial conditions. Equation (3.5.11) has the two branches

$$1 + X'_\pm = 2\lambda x + 2A \pm \sqrt{4(\lambda x + A)^2 + \beta} \tag{3.5.12}$$

which, on integration, yields

$$X_\pm = \lambda x^2 + (2A - 1)x \pm \int \sqrt{4(\lambda x + A)^2 + \beta}\, dx + B \tag{3.5.13}$$

where B is constant of integration. It may be observed that, since $\beta > 0$, $1 + X'$ in (3.5.12) is either always positive or always negative. In (3.5.1) it is also assumed that $1 + u_x > 0$; we conclude that X_+ is the only meaningful solution. We also assume that $\lambda \neq 0$, since otherwise we again have only the solution relating to the heat equation. Carrying out the integration in (3.5.13) we have

$$\begin{aligned} X_\pm &= \lambda x^2 + (2A - 1)x + B \pm \frac{1}{2\lambda}\Big[(\lambda x + A)\sqrt{4(\lambda x + A)^2 + \beta} \\ &\quad + \frac{\beta}{2}\ln\Big|2(\lambda x + A) + \sqrt{4(\lambda x + A)^2 + \beta}\Big|\Big]. \end{aligned} \tag{3.5.14}$$

The exact solutions are therefore given by

$$u_\pm(x, y, z, t) = X_\pm(x) + V(y, z, t) \tag{3.5.15}$$

where $V(y, z, t)$ is any solution of (3.5.10). It may be shown that the following asymptotic behaviours of $1 + X'_\pm$ hold:

$$\lim_{x \to 0}\left(1 + X'_\pm\right) = 2A \pm \sqrt{4A^2 + \beta}$$

$$\lim_{x \to \infty}\left(1 + X'_+\right) = \begin{cases} \infty & \text{if } \lambda > 0 \\ 0 & \text{if } \lambda < 0 \end{cases}$$

$$\lim_{x \to \infty}\left(1 + X'_-\right) = \begin{cases} 0 & \text{if } \lambda > 0 \\ -\infty & \text{if } \lambda < 0 \end{cases} \tag{3.5.16}$$

$$\lim_{x \to -\infty} (1 + X'_+) = \begin{cases} 0 & \text{if } \lambda > 0 \\ \infty & \text{if } \lambda < 0 \end{cases}$$

$$\lim_{x \to -\infty} (1 + X'_-) = \begin{cases} -\infty & \text{if } \lambda > 0 \\ 0 & \text{if } \lambda < 0. \end{cases}$$

The analysis of (3.5.2) is entirely analogous to that for (3.5.1) and the reader may refer to Willms (1995) for details. Willms (1995) has also indicated what kind of initial/boundary conditions for the additively separable solutions of (3.5.2) (and similarly of (3.5.1)) may be imposed.

It is quite likely that Equation (3.5.1) (as also (3.5.2)) has more general solutions of the form

$$u = A(x, y, z, t) + B(x, y, z, t)U(\zeta(x, y, z, t)) \qquad (3.5.17)$$

where $\zeta(x, y, z, t)$ is the similarity variable.

3.6   Similarity Solution of Burgers Equation by the Direct Method

Sachdev and his collaborators (See Sachdev (1987), (1991)) have studied similarity solutions of a class of GBEs:

$$u_t + uu_x + \frac{ju}{2t} = u_{xx} \qquad (3.6.1)$$

$$u_t + u^2 u_x + \frac{ju}{2t} = u_{xx} \qquad (3.6.2)$$

$$u_t + u^2 u_x = u_{xx} \qquad (3.6.3)$$

$$u_t + uu_x + f(x, t) = g(t)u_{xx} \qquad (3.6.4)$$

$$u_t + u^\beta u_x + f(t)u^\alpha = g(t)u_{xx} \qquad (3.6.5)$$

Each of the equations simulates some physical situation (Sachdev (1987)). We derived ODE forms of these equations via similarity assumption with a view to finding single hump solutions, themselves governed by a special class of ODEs we denominated as Euler-Painlevé transcendents. Here we show how the entire class of similarity solutions may be obtained by the direct approach, extending the ones known by intuitive arguments.

We begin with Burgers equation itself, that is

$$u_t + uu_x = u_{xx} \qquad (3.6.6)$$

wherein we let $\delta = 2$ to make the coefficient of u_{xx} unity in the original equation (see Section 6.2).

Let
$$u(x,t) = A(x,t) + B(x,t)U(z(x,t)), \quad B(x,t) \neq 0 \qquad (3.6.7)$$
where $z = z(x,t)$ is the similarity variable. Putting (3.6.7) into (3.6.6), we get

$$A_{xx} - A_t - AA_x + (B_{xx} - B_t - AB_x - BA_x)U - BB_xU^2$$
$$- B^2 z_x UU' + [2B_x z_x + Bz_{xx} - B(z_t + Az_x)]U' + Bz_x^2 U'' = 0. \quad (3.6.8)$$

For (3.6.8) to reduce to an ODE in U as a function of the similarity variable z alone, we introduce functions $\Gamma_n(z)$, $n = 1,2,\ldots5$ such that (3.6.8) becomes

$$\Gamma_1(z) + \Gamma_2(z)U + \Gamma_3(z)U^2 + \Gamma_4(z)UU' + \Gamma_5(z)U' + U'' = 0. \qquad (3.6.9)$$

Comparing (3.6.8) and (3.6.9), we get

$$
\begin{aligned}
A_{xx} - A_t - AA_x &= Bz_x^2\Gamma_1(z) & (3.6.10) \\
B_{xx} - B_t - AB_x - BA_x &= Bz_x^2\Gamma_2(z) & (3.6.11) \\
-B_x &= z_x^2\Gamma_3(z) & (3.6.12) \\
-B &= z_x\Gamma_4(z) & (3.6.13) \\
2B_x z_x + Bz_{xx} - Bz_t - ABz_x &= Bz_x^2\Gamma_5(z). & (3.6.14)
\end{aligned}
$$

The following remarks help simplify the system (3.6.9) - (3.6.14).

(i) If $A(x,t)$ has the form $A(x,t) = \hat{A}(x,t)+B(x,t)\Gamma(z)$, we put $\Gamma(z) \equiv 0$.

(ii) If $B(x,t)$ is found to have the form $B(x,t) = \hat{B}(x,t)\Gamma(z)$, then we may choose $\Gamma(z) \equiv 1$.

(iii) If $z(x,t)$ is determined from the implicit relation $f(z) = \hat{z}(x,t)$ where $f(z)$ is an invertible function, then we may simply put $f(z) \equiv z$.

The following cases arise.

(a) In the first instance we may assume that $z_{xx} = 0$ so that z is a linear function of x. Let $\Gamma_3(z) = \Lambda_3'(z)$; we may easily integrate (3.6.12) and obtain $B(x,t) = -z_x\Lambda_3(z)$. Remark (ii) then permits us to write

$$B(x,t) = -z_x. \qquad (3.6.15)$$

Since $z_{xx} = 0$, we have $B_x = 0$ from (3.6.15), so that $B = B(t)$. It is easy to check that (3.6.13) and (3.6.11), with the help of Remarks (iii) and (i), give $\Gamma_4(z) = 1$, $\Gamma_2(z) = 0$ so that

$$
\begin{aligned}
z(x,t) &= -xB(t) + D(t) & (3.6.16) \\
A(x,t) &= -x[B'(t)/B(t)] + K(t) & (3.6.17)
\end{aligned}
$$

where the functions $D(t)$ and $K(t)$ must be determined. On using (3.6.16) - (3.6.17) in (3.6.10) and (3.6.14), we get

$$\beta B^2 + D' - KB = 0 \tag{3.6.18}$$

$$[(B^{-1}B')' - (B^{-1}B')^2]x - K' + B^{-1}B'K = B^3\Gamma_1(z) \tag{3.6.19}$$

provided we choose $\Gamma_5(z) = \beta$, a constant. The accent in (3.6.18) and (3.6.19) denotes derivative with respect to t. Since LHS of (3.6.19) is linear in x, and z is given by (3.6.16), we must have $\Gamma_1(z) = \alpha z$, where α is a constant.

Equating coefficients of x and x^0 on both sides of (3.6.19) we get

$$BB'' - 2B'^2 + \alpha B^6 = 0 \tag{3.6.20}$$

$$BK' - KB' + \alpha DB^3 = 0. \tag{3.6.21}$$

Equations (3.6.20)-(3.6.21), being highly nonlinear, are difficult to solve generally. However, a special solution is easily obtained as

$$B(t) = a(bt + b_0)^{-1/2}, \quad D(t) = 0 \quad \text{and} \quad K(t) = a\beta(bt + b_0)^{-1/2} \tag{3.6.22}$$

where a, b and b_0 are arbitrary constants. All the functions $A(x,t), B(x,t)$ and $z(x,t)$, are now fully determined (see (3.6.16), (3.6.17), and (3.6.22)) and the similarity form (3.6.7) is therefore found to be

$$u(x,t) = \frac{b}{2}x(bt + b_0)^{-1} + a(bt + b_0)^{-1/2}[\beta + U(z)] \tag{3.6.23}$$

where

$$z(x,t) = -a(bt + b_0)^{-1/2}x. \tag{3.6.24}$$

Putting $\Gamma_n, n = 1, 2, 3, 4, 5$ thus obtained, namely $\Gamma_1 = \alpha z$, $\Gamma_2 = 0, \Gamma_3 = 0, \Gamma_4 = 1, \Gamma_5 = \beta$ into (3.6.9), we get

$$U'' + (\beta + U)U' - \frac{b^2}{4a^4}z = 0. \tag{3.6.25}$$

We may rewrite (3.6.23) as

$$u(x,t) = (bt + b_0)^{-1/2}F(z) \tag{3.6.26}$$

where

$$F(z) = a[\beta + U(z)] - (b/2a)z. \tag{3.6.27}$$

Equation (3.6.25) now becomes

$$F'' + (1/a)FF' + (b/2a^2)(zF)' = 0. \tag{3.6.28}$$

(b) Here we solve (3.6.10)-(3.6.14) in a different manner. Making use of Remark (ii), we may put $\Gamma_4(z) = 1$ in (3.6.13); it then reduces to

$$B(x,t) = -z_x. \tag{3.6.29}$$

Further, we may put $\Gamma_3(z) = \Lambda_3'(z)$ in (3.6.12), use (3.6.29), and integrate with respect to x. We obtain

$$\Lambda_3(z) = K(t) + \log B(x,t) \tag{3.6.30}$$

where $K(t)$ is the function of integration. Now we make use of Remark (iii) and let $\Lambda_3(z) = z$. Thus, (3.6.30) becomes

$$z = K(t) + \log B(x,t). \tag{3.6.31}$$

Eliminating z from (3.6.29) and (3.6.31), we have

$$B_x + B^2 = 0 \tag{3.6.32}$$

which integrates to give

$$B(x,t) = b_0[b_0 x + D(t)]^{-1} \tag{3.6.33}$$

where b_0 is an arbitrary constant; the function $D(t)$ remains to be determined. If we insert (3.6.31) and (3.6.33) into (3.6.14) and use Remark (i), we find that we may put $\Gamma_5 = 0$, and the function A from (3.6.14) then is

$$A(x,t) = -3b_0[b_0 x + D(t)]^{-1} + b_0^{-1}[b_0 x + D(t)]K' - D'. \tag{3.6.34}$$

On substituting $B(x,t)$ and $A(x,t)$ from (3.6.33) and (3.6.34) into (3.6.11) and (3.6.10), we have $\Gamma_2(z) = -4$ and

$$\Gamma_1(z) = 3 - \frac{K'' + K'^2}{b_0^4[b_0 x + D(t)]^{-4}} + \frac{D''}{b_0^4[b_0 x + D(t)]^{-3}}. \tag{3.6.35}$$

In view of the expression (3.6.31) for z, Equation (3.6.35) may be satisfied provided that

$$\Gamma_1(z) = 3 \tag{3.6.36}$$

$$K'' + K'^2 = 0 \tag{3.6.37}$$

$$D'' = 0. \tag{3.6.38}$$

Equations (3.6.37) and (3.6.38) integrate to give

$$K(t) = b + \log(b_1 t + a) \tag{3.6.39}$$

$$D(t) = ct + d \tag{3.6.40}$$

where a, b, c, d, and b_1 are arbitrary constants. Equations (3.6.31), (3.6.33), (3.6.34), (3.6.39), and (3.6.40) give the similarity form of the solution (3.6.7) as

$$
\begin{aligned}
u(x,t) &= -\frac{c}{b_0} + (b_1/b_0)\frac{b_0 x + ct + d}{b_1 t + a} \\
&\quad + b_0(b_0 x + ct + d)^{-1}[U(z) - 3] \qquad (3.6.41) \\
z(x,t) &= b + \log(b_1 t + a) + \log b_0(b_0 x + ct + d)^{-1}. \quad (3.6.42)
\end{aligned}
$$

Substituting for $\Gamma_n(z), n = 1, \ldots 5$, thus found, namely 3, -4, 1, 1, 0, respectively, into (3.6.9), we get the following equation for the similarity function $U(z)$:

$$
3 - 4U + U^2 + UU' + U'' = 0. \qquad (3.6.43)
$$

(c) Sometimes a (known) special solution guides the choice of dependent and independent variables so that the direct similarity method when applied to the new form of PDE generalizes the existing solution. For example, suppose we choose

$$
\begin{aligned}
w(\xi, \eta) &= u(x,t) + 1 & (3.6.44) \\
\xi(x,t) &= x + t, & (3.6.45) \\
\eta &= t^{1/2} & (3.6.46)
\end{aligned}
$$

as the new dependent and independent variables (ξ is the moving coordinate). Then Equation (3.6.6) becomes

$$
\frac{1}{2\eta} w_\eta + w w_\xi = w_{\xi\xi}. \qquad (3.6.47)
$$

We substitute (3.6.7) into (3.6.47), with w, ξ and η replaced by u, x, and t, respectively. Proceeding in the manner described above, Equation (3.6.47) changes to

$$
\Gamma_1(z) + \Gamma_2(z)U + \Gamma_3(z)U^2 + \Gamma_4(z)U' + \Gamma_5(z)UU' + U'' = 0 \qquad (3.6.48)
$$

where, as before, the functions $\Gamma_i(z), i = 1, 2, 3, 4, 5$ are introduced such that

$$
\begin{aligned}
(1/2t)A_t + AA_x - A_{xx} &= -Bz_x^2\Gamma_1(z) & (3.6.49) \\
(1/2t)B_t + AB_x + BA_x - B_{xx} &= -Bz_x^2\Gamma_2(z) & (3.6.50) \\
B_x &= -z_x^2\Gamma_3(z) & (3.6.51) \\
B &= -z_x\Gamma_4(z) & (3.6.52) \\
(1/2t)Bz_t + ABz_x - 2B_x z_x - Bz_{xx} &= -Bz_x^2\Gamma_5(z). & (3.6.53)
\end{aligned}
$$

Here, again, if we assume that $z_{xx} = 0$, then it is easy to check that (3.6.51), (3.6.53), and (3.6.50) give, on the use of Remarks (ii), (iii), and (i), respectively, $\Gamma_3(z) = 0$, $\Gamma_5(z) = 1$ and $\Gamma_2(z) = 0$; therefore, $B(x,t) = B(t)$ and

$$B(t) = -z_x \tag{3.6.54}$$

$$z(x,t) = -xB(t) \tag{3.6.55}$$

$$A(x,t) = -\frac{x}{2t}\frac{B'}{B} + D(t). \tag{3.6.56}$$

On substituting (3.6.55) into (3.6.52), we get $\Gamma_4(z) = D(t) = 0$. Therefore, (3.6.56) reduces to

$$A(x,t) = -\frac{x}{2t}\frac{B'}{B}. \tag{3.6.57}$$

Putting (3.6.57) into (3.6.49) we get

$$\Gamma_1(z) = -\frac{x}{4t^3}B^{-4}B' + \frac{x}{4t^2}B^{-5}(BB'' - 2B'^2). \tag{3.6.58}$$

Since the right-hand side of (3.6.58) is linear in x, we must have $\Gamma_1(z) = \alpha z$, where α is a constant, and so (3.6.58) reduces to

$$BB'' - 2B'^2 - t^{-1}BB' = -4\alpha t^2 B^6. \tag{3.6.59}$$

A special solution of (3.6.59) is

$$B(t) = -\alpha_1 t^{-1}, \quad 4\alpha\alpha_1^4 + 1 = 0. \tag{3.6.60}$$

Equation (3.6.57) now gives $A = x/2t^2$. Now, reverting to the variables w, ξ, and η of Equation (3.6.47), we have the similarity solution as

$$w(\xi,\eta) = \frac{\xi}{2\eta^2} + \frac{\alpha_1}{\eta}U(z), \quad z(\xi,\eta) = \frac{\alpha_1\xi}{\eta}. \tag{3.6.61}$$

Equation (3.6.48), with $\Gamma_1 = \alpha z$, $\Gamma_2 = 0$, $\Gamma_3 = 0$, $\Gamma_4 = 0$, and $\Gamma_5 = 1$, reduces to

$$U'' + UU' - \frac{1}{4\alpha_1^4}z = 0. \tag{3.6.62}$$

If we write $f(z) = z/(2\alpha) - 1 - \alpha_1 U(z)$, then (3.6.62) becomes

$$f'' - \alpha_1^{-1}ff' + (1/2\alpha_1^2)(zf)' = 0. \tag{3.6.63}$$

(d) Another interesting solution of Burgers equation obtained via the Cole-Hopf transformation has one of the forms

$$
\begin{aligned}
u &= \frac{x}{t} - \frac{2}{t} \tanh \frac{x}{t} \\[4pt]
u &= \frac{x}{t} + \frac{2}{t} \tan \frac{x}{t} \\[4pt]
u &= \frac{x}{t} - \frac{2}{t} \coth \frac{x}{t} \\[4pt]
u &= \frac{x}{t} - \frac{2}{t} \cot \frac{x}{t}.
\end{aligned}
\tag{3.6.64}
$$

To generalize (3.6.64), we let $z_{xx} = 0$ in (3.6.7) so that z is linear in x. Set $\Gamma_3(z) = \Lambda_3'(z)$ in (3.6.12) and integrate with respect to x to obtain $B = -z_x \Lambda_3(z)$ where the function of integration has been chosen to be zero. Using Remark (ii), we may put $\Lambda_3(z) = 1$ and thus have

$$
B = -z_x. \tag{3.6.65}
$$

Remarks (iii) and (i) used in Equations (3.6.13) and (3.6.14), respectively, give $\Gamma_4(z) = 1$ and $\Gamma_5(z) = 0$. Since $\Gamma_3 = \Lambda_3' = 0$, (3.6.12) shows that $B = B(t)$. From (3.6.13) and (3.6.14) we get

$$
z(x,t) = -xB(t) + D(t) \tag{3.6.66}
$$

$$
A(x,t) = -x\frac{B'(t)}{B(t)} + \frac{D'(t)}{B(t)} \tag{3.6.67}
$$

where the function $D(t)$ remains to be determined. Equations (3.6.11), (3.6.66), and (3.6.67) imply that $\Gamma_2(z) = 0$, and so (3.6.11) becomes

$$
B'(t) + B(t)A_x = 0 \tag{3.6.68}
$$

implying that

$$
A(x,t) = A_0(t)x. \tag{3.6.69}
$$

Equation (3.6.10) now becomes

$$
B^3 \Gamma_1(z) = -x(A_0' + A_0^2). \tag{3.6.70}
$$

We may assume $\Gamma_1(z) = 0$ so that $A_0 = (t + c)^{-1}$ where c is an arbitrary constant. From (3.6.69), we have

$$
A(x,t) = x(t + c)^{-1} \tag{3.6.71}
$$

and hence from (3.6.68)

$$
B(t) = b(t + c)^{-1} \tag{3.6.72}
$$

where b is an arbitrary constant. On use of (3.6.71) and (3.6.72), (3.6.67) gives $D(t) = d$, a constant. The similarity form of the solution of (3.6.6), therefore, is

$$u(x,t) = (t+c)^{-1}[x + bU(z)], \qquad (3.6.73)$$

$$z(x,t) = -bx(t+c)^{-1} + d. \qquad (3.6.74)$$

The ODE (3.6.9) in this case is

$$U'' + UU' = 0. \qquad (3.6.75)$$

Equation (3.6.75) has solutions of the form tan, cot, tanh, coth, and, therefore, (3.6.73)-(3.6.74) generalize (3.6.64). The cases considered so far cover and generalize the entire catalogue of similarity forms of solutions of Burgers equation prepared by Benton and Platzman (1972). Mayil Vaganan (1994) has applied the direct similarity analysis to several generalized Burgers equations. We give below a summary of his results.

1.
$$u_t + uu_x + \frac{ju}{2t} = u_{xx}, \quad j > 0 \qquad (3.6.76)$$

$$u(x,t) = \frac{x}{2t} + \frac{\alpha d_0}{b_0} t^{\alpha - 1/2} + b_0 t^{-1/2} U(z) \qquad (3.6.77)$$

$$z = -b_0 x t^{-1/2} + d_0 t^{\alpha} \qquad (3.6.78)$$

$$U'' + UU' - \lambda U + az = 0, \qquad (3.6.79)$$

where α, d_0, b_0, and a are constants; α satisfies the quadratic

$$\alpha^2 + \frac{j\alpha}{2} + \frac{j-1}{4} = 0. \qquad (3.6.80)$$

2.
$$u_t + u^2 u_x + \frac{ju}{2t} = u_{xx} \qquad (3.6.81)$$

$$u(x,t) = t^{-1/4} U(z) \qquad (3.6.82)$$

$$z = xt^{-1/2} \qquad (3.6.83)$$

$$U'' - U^2 U' + 2zU' + 2\left(\frac{1}{2} - j\right)U - 0 \qquad (3.6.84)$$

Observe the diminution of symmetries because of the cubic nonlinearity in (3.6.81) in contrast to those for the quadratic case in (3.6.76).

3.
$$u_t + u^2 u_x = u_{xx} \qquad (3.6.85)$$

$$u(x,t) = t^{-1/4}U(z), \tag{3.6.86}$$

$$z = xt^{-1/2} \tag{3.6.87}$$

$$U'' - U^2 U' + \frac{1}{2}zU' + \frac{1}{4}U = 0 \tag{3.6.88}$$

4. $$u_t + uu_x + f(x,t) = g(t)u_{xx} \tag{3.6.89}$$

$$u(x,t) = -xB'(t)/B(t) + K(t) + B(t)U(z) \tag{3.6.90}$$

$$z = -xB(t)/g(t) + D(t) \tag{3.6.91}$$

where
$$gD' + aB^2 D - KB = 0. \tag{3.6.92}$$

The function $g(t)$ is prescribed; Equation (3.6.92) relates the functions $B(t)$, $D(t)$, and $K(t)$. a is an arbitrary constant. The functions $g(t)$ and $f(x,t)$ appearing in the given PDE satisfy the equations

$$g'(t) = aB^2 \tag{3.6.93}$$

$$f(x,t) = \left[\left(\frac{B'}{B}\right)' - \left(\frac{B'}{B}\right)^2 \right]x - K'$$

$$+ \frac{KB'}{B} - g^{-1}B^3 F(z) \tag{3.6.94}$$

where $F(z)$ depends on x and t through the similarity variable $z = -xB(t)/g(t) + D(t)$. The reduced ODE for $U(z)$ is

$$U'' + UU' + azU' + F(z) = 0. \tag{3.6.95}$$

5. $$u_t + u^\beta u_x + f(t)u^\alpha = g(t)u_{xx} \tag{3.6.96}$$

where α and β are real.

Equation (3.6.96) must first be simplified to have integral powers of the dependent variable and its derivatives. So if we write

$$u = v^{-1/\beta}, \tag{3.6.97}$$

we get

$$g(t)\left[vv_{xx} - \frac{\alpha+1}{\alpha-1}v_x^2\right] = vv_t + v_x - \frac{\alpha-1}{2}f(t) \tag{3.6.98}$$

provided that

$$\beta = \frac{\alpha-1}{2}. \tag{3.6.99}$$

α and β in (3.6.96) must satisfy (3.6.99). Applying the direct similarity method, we get the similarity form for $v(x,t)$:

$$v(x,t) = t^{1-a}H(z) \tag{3.6.100}$$
$$z = (a/b)xt^{-a} \tag{3.6.101}$$

provided that

$$f(t) = \frac{2\lambda}{\alpha-1}t^{1-2a} \tag{3.6.102}$$

$$g(t) = t^{2a-1}. \tag{3.6.103}$$

The ODE for the function $H(z)$ in (3.6.100) is

$$HH'' - \frac{\alpha+1}{\alpha-1}H'^2 - \frac{(1-a)}{a^2}b^2H^2 + \frac{b^2}{a}zHH' - \frac{b}{a}H' + \frac{\lambda b^2}{a^2} = 0. \tag{3.6.104}$$

In the above, a, b, and λ are arbitrary constants.

3.7 Exact Free Surface Flows for Shallow-Water Equations via the Direct Similarity Approach

Now we extend the approach of Clarkson and Kruskal (1989) to systems of PDEs involving three independent variables. Here we take up the system of nonlinear PDEs which governs flows generated by large amplitude, long gravity waves as they propagate over a horizontal bed into a region where the flow is steady but sheared in a vertical direction. The interesting feature of this study is that the wave speed depends on the local height of the wave, which itself must be found as part of the solution. Freeman (1972) and Sachdev (1980) found simple wave and time-dependent self-similar solutions by an intuitive argument. Here the forms of these solutions will be determined (and to some extent generalised) by the direct approach. Moreover, the method of solving the resulting ODEs is quite different from that of Freeman (1972) and Sachdev (1980). We closely follow the work of Sachdev and Mayil Vaganan (1994). The governing equations of motion are

$$u_x + v_y = 0 \tag{3.7.1}$$

$$u_t + uu_x + vu_y + \frac{1}{\rho}p_x = 0 \tag{3.7.2}$$

$$p_y + g\rho = 0 \tag{3.7.3}$$

where x is horizontal distance in the direction of wave propagation and y is the vertical distance measured from the horizontal bottom. The (incompressible) fluid has density ρ and pressure p; g denotes acceleration due to gravity.

The flow is bounded below by the horizontal bottom $y = 0$ and above by the free surface $y = h(x, t)$. The boundary conditions, therefore, are

$$v = 0 \quad \text{on} \quad y = 0 \tag{3.7.4}$$

$$v = h_t + uh_x \quad \text{on} \quad y = h(x, t). \tag{3.7.5}$$

On the free surface, the pressure is assumed to be constant, equal to p_0, say. The hydrostatic equation (3.7.3) therefore integrates to yield

$$p = p_0 + g\rho(h - y). \tag{3.7.6}$$

Elimination of p from (3.7.2) with the help of (3.7.6) yields

$$u_t + uu_x + vu_y + gh_x = 0. \tag{3.7.7}$$

Equations (3.7.1) and (3.7.7) must be solved subject to BCs (3.7.4) and (3.7.5), and the free surface $y = h(x, t)$ must be found as part of the solution. We have already incorporated the BC $p = p_0$ on the free surface.

It is convenient to introduce the so-called σ-variables

$$z \;=\; \frac{y}{h} \tag{3.7.8}$$

$$w(x, z, t) \;=\; \frac{1}{h}[v - z(h_t + uh_x)] \tag{3.7.9}$$

into the given PDEs and BC so that (3.7.1) and (3.7.7) become

$$h_t + uh_x + h(u_x + w_z) \;=\; 0 \tag{3.7.10}$$

$$u_t + uu_x + wu_z + gh_x \;=\; 0. \tag{3.7.11}$$

The BCs (3.7.4) and (3.7.5) become homogeneous:

$$w = 0 \quad \text{on} \quad z = 0, 1. \tag{3.7.12}$$

This is the major advantage accruing from the introduction of the σ-variables.

Freeman (1972) introduced simple waves on shear flows by an intuitive argument. We shall derive their form by the direct similarity approach. We let

$$u \;=\; A(z) + B(z)F(\xi, \eta), \quad B(z) \neq 0 \tag{3.7.13}$$

$$w \;=\; I(x, z, t) + J(x, z, t)G(\xi, \eta), J(x, z, t) \neq 0 \tag{3.7.14}$$

$$\xi \;=\; \xi(x, t) \tag{3.7.15}$$

$$\eta \;=\; \eta(z) \tag{3.7.16}$$

and substitute these in (3.7.10) and (3.7.11). We get

$$(h_t + Ah_x + hI_z) + Bh_x F + hJ_z G + hB\xi_x F_\xi + hJ\eta' G_\eta \ = \ 0 \quad (3.7.17)$$
$$(IA_z + gh_x) + IB_z F + JA_z G + JB_z FG + B(\xi_t + A\xi_x)F_\xi$$
$$+BI\eta' F_\eta + B^2 \xi_x FF_\xi + BJ\eta' GF_\eta \ = \ 0. \quad (3.7.18)$$

For (3.7.17) - (3.7.18) to reduce to PDEs in two independent variables ξ and η for the determination of the function $F(\xi, \eta)$ and $G(\xi, \eta)$, the coefficients of F and G and their derivatives must be functions of ξ and η alone. We should therefore be able to write (3.7.17) and (3.7.18) as

$$\Gamma_1 + \Gamma_2 F + \Gamma_3 G + \Gamma_4 F_\xi + G_\eta = 0 \tag{3.7.19}$$
$$\Gamma_5 + \Gamma_6 F + \Gamma_7 G + \Gamma_8 FG + \Gamma_9 F_\xi + \Gamma_{10} F_\eta + \Gamma_4 FF_\xi + GF_\eta = 0 \tag{3.7.20}$$

where the functions $\Gamma_n = \Gamma_n(\xi, \eta), n = 1, 2, \ldots 10$ are defined by comparison of (3.7.19)-(3.7.20) with (3.7.17)-(3.7.18):

$$h_t + Ah_x + hI_z = hJ\eta'\Gamma_1 \tag{3.7.21}$$
$$Bh_x = hJ\eta'\Gamma_2 \tag{3.7.22}$$
$$J_z = J\eta'\Gamma_3 \tag{3.7.23}$$
$$B\xi_x = J\eta'\Gamma_4 \tag{3.7.24}$$
$$BA_z + gh_x = BJ\eta'\Gamma_5 \tag{3.7.25}$$
$$IB_z = BJ\eta'\Gamma_6 \tag{3.7.26}$$
$$A_z = B\eta'\Gamma_7 \tag{3.7.27}$$
$$B_z = B\eta'\Gamma_8 \tag{3.7.28}$$
$$\xi_t + A\xi_x = J\eta'\Gamma_9 \tag{3.7.29}$$
$$I = J\Gamma_{10}. \tag{3.7.30}$$

Equations (3.7.21) - (3.7.30) may be solved for $A(z)$, $B(z)$, $I(x, z, t)$, $J(x, z, t)$, $\xi(x, t)$, $\eta(z)$, and $\Gamma_n(\xi, \eta), n = 1, 2, \ldots 10$ in the light of the following remarks.

Remark 1

If the function $A(z)(I(x, z, t))$ is found to have the form $A(z) = \hat{A}(z) + B(z) \times \Gamma(\xi, \eta)(I(x, z, t) = \hat{I}(x, z, t) + J(x, z, t)\Omega(\xi, \eta))$, we may set $\Gamma(\xi, \eta) \equiv 0$ $(\Omega(\xi, \eta) \equiv 0)$.

Remark 2

If $B(z)(J(x, z, t))$ is found to have the form $B(z) = \hat{B}(z)\Gamma(\xi, \eta)(J(x, z, t) = \hat{J}(x, z, t)\Omega(\xi, \eta))$, then we may set $\Gamma(\xi, \eta) \equiv 1(\Omega(\xi, \eta) \equiv 1)$.

Remark 3

If $\xi(x,t)(\eta(z))$ is to be determined from $f(\xi) \equiv \hat{\xi}(x,t)(g(\eta) \equiv \hat{\eta}(z))$, where $f(\xi)(g(\eta))$ is an invertible function, then we may simply take $f(\xi) \equiv \xi \; (g(\eta) \equiv \eta)$.

Now we shall attempt to solve the system (3.7.21) - (3.7.30) in some convenient order. If in (3.7.23) we set $\Gamma_3(\xi,\eta) = \Omega_3'(\eta)/\Omega_3(\eta)$ and integrate with respect to z, we get $J = \hat{J}(x,t)\Omega_3(\eta)$ where $\hat{J}(x,t) > 0$ is the function of integration. We use Remark 2 to set $\Omega_3(\eta) = 1$ so that $J = J(x,t)$. In the same way, (3.7.28) gives $\Gamma_8(\xi,\eta) = 0$ and $B = B_0$, a constant. If we put $\Gamma_7(\xi,\eta) = \Omega_7'(\eta)$ in (3.7.27) and integrate with respect to z, we get $A = A_0 + B_0\Omega_7(\eta)$ where A_0 is another constant. In view of Remark 1, we may set $\Omega_7(\eta) = 0$ so that $A = A_0$. We may similarly deduce from (3.7.30) that $\Gamma_{10}(\xi,\eta) = 0$ and, therefore, $I(x,z,t) = 0$. Without loss of generality we choose $A_0 = 0$ and $B_0 = 1$. Therefore, we arrive at the determination

$$A = 0, \quad B = 1, \quad I = 0 \quad \text{and} \quad J = J(x,t). \qquad (3.7.31)$$

Since we have shown that $I(x,z,t) = 0$, (3.7.26) gives $\Gamma_6 = 0$. In view of (3.7.31), (3.7.24) separates to give $J(x,t) = \xi_x$ and $\eta'\Gamma_4 = 1$, if the separation constant is chosen to be 1. Again, if we let $\Gamma_4 = \Omega_4'(\eta)$ in $\eta'\Gamma_4 = 1$ and integrate with respect to z, we get $\Omega_4(\eta) = z$; here we have put the constant of integration equal to zero. Using Remark 3, we can choose $\Omega_4(\eta) \equiv \eta$ and so $\eta = z$. In a similar manner, (3.7.25) yields $\Gamma_5 = g$ and $\xi(x,t) = h(x,t)$. Thus, we find that

$$\xi(x,t) = h(x,t), \quad \eta(z) = z \quad \text{and} \quad J(x,t) = h_x(x,t). \qquad (3.7.32)$$

Using (3.7.31) and (3.7.32), (3.7.22) reduces to $\Gamma_2 = 1/h$, and (3.7.21) and (3.7.29) each reduce to

$$h_t + c(h)h_x = 0 \qquad (3.7.33)$$

provided that $h\Gamma_1 = \Gamma_9 = -c(h)$. Thus, the solution of the system (3.7.10)-(3.7.11) has the form

$$u \;=\; F(h,z) \qquad (3.7.34)$$

$$w \;=\; h_x G(h,z). \qquad (3.7.35)$$

It is easy to check from (3.7.33) and (3.7.34) that $u(x,z,t)$ satisfies the one-dimensional wave equation

$$u_t + c(h)u_x = 0 \qquad (3.7.36)$$

proving analytically the existence of simple wave form of the solution. Introducing $\Gamma_n(z)$, $n = 1,2,\ldots 10$ thus found into (3.7.19) - (3.7.20), we get the PDEs governing F and G:

$$h(F_h + G_z) + F - c = 0 \tag{3.7.37}$$

$$(F - c)F_h + GF_z + g = 0. \tag{3.7.38}$$

The boundary conditions (3.7.12) (see (3.7.35)) reduce to

$$G(h, 0) = 0 \quad \text{and} \quad G(h, 1) = 0. \tag{3.7.39}$$

The system (3.7.37) - (3.7.38) in two independent variables may be reduced in the usual manner

$$F(h, z) = M(h, z) + N(h, z)P(\eta), \quad N(h, z) \neq 0 \tag{3.7.40}$$

$$G(h, z) = U(h, z) + V(h, z)Q(\eta), \quad V(h, z) \neq 0 \tag{3.7.41}$$

where $\eta = \eta(z)$ is the new independent variable. Since we have now dealt with several examples in two independent variables, we skip the details (see Mayil Vaganan (1994)) and give the final results. It is found that

$$\begin{aligned} M(h, z) &= (2n_1 - \alpha_0)h^{1/2} \\ N(h) &= h^{1/2} \\ V(h) &= \frac{1}{2}h^{-1/2} \\ c(h) &= (3n_1 - 2\alpha_0)h^{1/2} \end{aligned} \tag{3.7.42}$$

where n_1 and α_0 are arbitrary constants; the form of $h = h(x, t)$ remains arbitrary. We discuss a special case of (3.7.40)-(3.7.42) which conforms to that of Freeman (1972) and Sachdev and Philip (1986) through a simple scaling. Here we have

$$\begin{aligned} c(h) &= 2\alpha(gh)^{1/2}, \quad F(h, z) = (gh)^{1/2}P(z) \\ G(h, z) &= \frac{1}{2}g^{1/2}h^{-1/2}Q(z) \end{aligned} \tag{3.7.43}$$

so that the system (3.7.37)-(3.7.38) reduces to

$$Q_z + 3P - 4\alpha = 0 \tag{3.7.44}$$

$$QP_z + P^2 - 2\alpha P + 2 = 0. \tag{3.7.45}$$

The boundary conditions (3.7.39), in view of the last of (3.7.43), become

$$Q(0) = 0, \quad Q(1) = 0. \tag{3.7.46}$$

Eliminating $Q(z)$ from (3.7.44) and (3.7.45), we have

$$(P^2 - 2\alpha P + 2)\frac{d^2 P}{dz^2} + (P - 2\alpha)\left(\frac{dP}{dz}\right)^2 = 0. \tag{3.7.47}$$

Interchanging the dependent and independent variables in (3.7.47), we have

$$(P-a)(P-b)\frac{d^2z}{dP^2} + (2\alpha - P)\frac{dz}{dP} = 0 \tag{3.7.48}$$

where a and b are roots of $P^2 - 2\alpha P + 2 = 0$. For a and b to be real we must have $\alpha^2 > 2$; this is the condition for the absence of critical levels in the flow (Freeman (1972)). In view of (3.7.45) and (3.7.46), the conditions on P are

$$P(0) = a, \qquad P(1) = b. \tag{3.7.49}$$

Introducing the variable

$$\tau = \frac{P-a}{b-a} \tag{3.7.50}$$

in (3.7.48) and using the formula $2\alpha = a + b = a + 2/a$, we arrive at the hypergeometric equation satisfied by $z(\tau)$:

$$\tau(1-\tau)\frac{d^2z}{d\tau^2} + \left(\frac{2}{a^2-2} + \tau\right)\frac{dz}{d\tau} = 0. \tag{3.7.51}$$

It is clear from (3.7.49) and (3.7.50) that $\tau = 0$ at $z = 0$ and $\tau = 1$ at $z = 1$. The solution of (3.7.51) satisfying these BCs is

$$\begin{aligned}
z = \frac{\beta(p,q;\tau)}{\beta(p,q;1)} &= \tau^p \frac{{}_2F_1(p,1-q,1+p;\tau)}{{}_2F_1(p,1-q,1+p;1)} \\
&= \sum_{k=0}^{\infty} \lambda_k \tau^{p+k}
\end{aligned} \tag{3.7.52}$$

where

$$\lambda_k = \frac{\Gamma(1+p)\Gamma(k+p)\Gamma(1+k-q)}{k!\,{}_2F_1(p,1-q,1+p;1)\Gamma(p)\Gamma(1-q)\Gamma(1+k+p)}, k \geq 0. \tag{3.7.53}$$

Here, $\beta(p,q;\tau)$ is the incomplete beta function with $p = (a-2b+2)/(a-b)$ and $q - (2a - b - 2)/(a - b)$.

The analysis so far follows Sachdev and Philip (1986). Here, we proceed further in a manner which permits generalisation of the solution expressed in terms of the incomplete beta function.

The (implicit) incomplete beta function solution (3.7.52)-(3.7.53) may be inverted by using the method described by Keener (1988). We let

$$\tau = \sum_{n=1}^{\infty} \sigma_n Z^n, \qquad z = Z^P. \tag{3.7.54}$$

Now substitute (3.7.54) in (3.7.52) and equate coefficients of Z^k for $k \geq p$ on both sides. We get

$$\lambda_0 \sigma_1^p - 1 = 0, \quad \sum_{k=0}^{j} \lambda_{j-k}\mu_{j-k,k} = 0, \quad j \geq 1 \tag{3.7.55}$$

where

$$\mu_{j0} = \sigma_1^{j+p}$$

$$\mu_{jk} = \frac{1}{k\sigma_1}\sum_{l=1}^{k}(l(j+p)-k+l)\sigma_{l+1}\mu_{j,k-l}, \quad k \geq 1. \quad (3.7.56)$$

The algebraic system (3.7.55) - (3.7.56) may be solved to express $\sigma_j, j \geq 1$ in terms of $\lambda_k, k = 0, 1, \ldots j - 1$. Since $\tau = (P - a)/(b - a)$, (3.7.54) may be rewritten in terms of $P(z)$:

$$P(z) = a + (b - a)\sum_{n=1}^{\infty}\sigma_n z^{n/p}. \quad (3.7.57)$$

We may also obtain an equation for Q alone from (3.7.44)-(3.7.45). We have

$$3QQ_{zz} - Q_z^2 + 2\alpha Q_z + 8\alpha^2 - 18 = 0. \quad (3.7.58)$$

The solution of (3.7.58) may be obtained by inserting (3.7.57) in (3.7.44), integrating and putting the constant of integration equal to zero: We obtain

$$Q(z) = z\left[4\alpha - 3a - 3(b - a)\sum_{n=1}^{\infty}\frac{p\sigma_n}{n+p}z^{n/p}\right]. \quad (3.7.59)$$

We may now write the explicit solution of shallow-water equations (3.7.10) - (3.7.11) satisfying the BC (3.7.12) using (3.7.34), (3.7.35), (3.7.40), (3.7.41), (3.7.57), and (3.7.59):

$$u(x, z, t) = (gh)^{1/2}\left[a + (b - a)\sum_{n=1}^{\infty}\sigma_n z^{n/p}\right] \quad (3.7.60)$$

$$w(x, z, t) = \frac{1}{2}h_x(g/h)^{1/2}z\left[4\alpha - 3a - 3(b - a)\sum_{n=1}^{\infty}\frac{p\sigma_n}{n+p}z^{n/p}\right]. \quad (3.7.61)$$

The solution (3.7.60) -(3.7.61) of the original PDE system is the simple wave solution, arising from its reduction to the ODEs (3.7.44) and (3.7.45) and their solution subject to BCs (3.7.46). It is possible to generate more general solutions of the "reduced" PDE system (3.7.37)-(3.7.38), satisfying (3.7.39). That becomes possible by generalising the form (3.7.60)-(3.7.61). We write the solution of (3.7.37) - (3.7.38) as

$$F(h, z) = \sum_{n=0}^{\infty}F_n(h)z^{nr} \quad (3.7.62)$$

$$G(h, z) = z(1 - z^r)\sum_{n=0}^{\infty}G_n(h)z^{nr} \quad (3.7.63)$$

where r is a (real) positive number. The series form (3.7.63) is written so that the boundary conditions (3.7.39) on G at $z = 0$ and $z = 1$ are automatically satisfied. We rewrite (3.7.63) for convenience of subsequent calculations as

$$G(h, z) = \sum_{n=0}^{\infty} K_n(h) z^{nr+1} \qquad (3.7.64)$$

where

$$\begin{aligned} K_0(h) &= G_0(h), \\ K_n(h) &= G_n(h) - G_{n-1}(h), \qquad n \geq 1 \end{aligned} \qquad (3.7.65)$$

or

$$G_n(h) = \sum_{k=0}^{n} K_k(h), \qquad n \geq 0. \qquad (3.7.66)$$

Substituting (3.7.62) and (3.7.64) into (3.7.37) - (3.7.38) and equating coefficients of $z^{nr}, n \geq 0$ to zero, we get

$$hK_0 + hF_0' + F_0 - c = 0 \qquad (3.7.67)$$

$$(nr + 1)hK_n(h) + hF_n'(h) + F_n(h) = 0, n \geq 1 \quad (3.7.68)$$

$$(F_0 - c)F_0' + g = 0 \qquad (3.7.69)$$

$$\sum_{k=0}^{n} F_k(h)[F_{n-k}'(h) + krK_{n-k}(h)] - cF_n'(h) = 0, \quad n \geq 1. (3.7.70)$$

Equations (3.7.69) - (3.7.70) are ODEs while (3.7.67) - (3.7.68) are algebraic relations which together give the solution for $F_n(h)$ and $K_n(h)$, $n \geq 0$. The general solution can be found to be

$$c(h) = c_0 h^{1/2} \qquad (3.7.71)$$

$$F_n(h) = a_n h^{1/2+n(\lambda-1/2)} \qquad (3.7.72)$$

$$K_n(h) = \frac{1}{1+nr} \left(c_n - \left[\frac{3}{2} + n \left(\lambda - \frac{1}{2} \right) \right] a_n \right) h^{-1/2+n(\lambda-1/2)} \qquad (3.7.73)$$

where c_0 and λ are arbitrary constants; $c_n = 0$, $n \geq 1$, while $a_n, n \geq 0$ are to be found from the following algebraic equations:

$$a_0(a_0 - c_0) + 2g = 0 \qquad (3.7.74)$$

$$\left(c_0 \left[\frac{1}{2} + n \left(\lambda - r - \frac{1}{2} \right) \right] - a_0 \left\{ 1 + n \left(\lambda - \frac{3}{2}r - \frac{1}{2} \right) \right\} \right) a_n$$

$$= \sum_{k=1}^{n-1} a_k \left\{ a_{n-k} \left[\frac{1}{2} + (n - k) \left(\lambda - \frac{1}{2} \right) \right] \right.$$

$$-\frac{kr}{1+(n-k)r}\left[\frac{3}{2}+(n-k)\left(\lambda-\frac{1}{2}\right)\right]a_{n-k}\Bigg\}, \quad n \geq 1. \quad (3.7.75)$$

For $n = 1$, (3.7.75) gives

$$r = \left(a_0\left(\lambda+\frac{1}{2}\right)-\lambda c_0\right)\left(\frac{3}{2}a_0-c_0\right)^{-1}, \quad (3.7.76)$$

requiring, however, that a_1 is a (nonzero) arbitrary constant. The functions $G_n(h)$, $n \geq 0$ are obtained from (3.7.65), (3.7.66), and (3.7.73) as

$$G_n(h) = \sum_{k=0}^{n}\frac{1}{1+kr}\left(c_k-\left[\frac{3}{2}+k\left(\lambda-\frac{1}{2}\right)\right]a_k\right)h^{-1/2+k\left(\lambda-\frac{1}{2}\right)}, \quad n \geq 0. \quad (3.7.77)$$

Thus, a new class of progressive wave solutions of the shallow-water equations (3.7.10)-(3.7.11) subject to the boundary conditions (3.7.12) is found to be

$$u(x,z,t) = h^{1/2}\sum_{n=0}^{\infty}a_n h^{n(\lambda-1/2)}z^{nr} \quad (3.7.78)$$

$$w(x,z,t) = h^{-1/2}h_x z(1-z^r)\times$$
$$\sum_{n=0}^{\infty}\left[\sum_{k=0}^{n}\frac{c_k-[3/2+k(\lambda-1/2)]a_k}{1+kr}h^{k(\lambda-1/2)}\right]z^{nr}$$
$$= h^{-1/2}h_x z\sum_{n=0}^{\infty}\frac{c_n-[3/2+n(\lambda-1/2)]a_n}{1+nr}h^{n(\lambda-1/2)}z^{nr}.$$
$$(3.7.79)$$

It is not difficult to check that the solution (3.7.60) - (3.7.61) is embedded in the more general solution (3.7.78) - (3.7.79) and may be recovered from the latter by putting $\lambda = 1/2$ and appropriately choosing the constant a_1.

The present approach has been much exploited for several other solutions, including the time-dependent similarity solutions, in the work of Sachdev and Mayil Vaganan (1994). Several limiting cases leading to singular solutions have also been identified. The entire gamut of analysis for the free surface flows for shallow-water equations has been extended to the more complicated situation when the medium is compressible and barotropic. The work of Sachdev and Philip (1988) showed that for this medium, the neat beta function form of the solution does not exist. However, the present general approach is readily adapted to the study of simple waves and time-dependent self-similar flows for the compressible medium (Sachdev and Mayil Vaganan (1995)).

3.8   Multipronged Approach to Exact Solutions of Nonlinear PDEs – an Example from Gasdynamics

We give below a formulation of the set of non-isentropic one-dimensional gasdynamic equations due to Ustinov (1982), (1984), (1986), which arises from one of its conservative forms and admits exact treatment as well as comparison of different approachs to exact solutions. We shall seek intermediate integrals which generalise Riemann invariants to non-isentropic flows, obtain exact solutions by the direct approach, and compare them with those found by the group theoretic approach; we also use some intuitive ideas for obtaining the solution (Sachdev, Dowerah, Mayil Vaganan, and Philip (1997)).

We begin with one-dimensional adiabatic motion of an ideal gas. The governing PDEs are

$$\rho_t + (\rho u)_r = 0 \tag{3.8.1}$$

$$u_t + u u_r + \frac{1}{\rho} p_r = 0 \tag{3.8.2}$$

$$(p\rho^{-\gamma})_t + u(p\rho^{-\gamma})_r = 0 \tag{3.8.3}$$

where ρ, u, and p are density, particle velocity, and pressure at a point r and time t; γ is the ratio of specific heats. First, we recast the system (3.8.1) - (3.8.3), by making use of the following conservation laws arising from them:

$$(\rho u)_t + (p + \rho u^2)_r = 0 \tag{3.8.4}$$

$$\left(\frac{p}{\gamma - 1} + \frac{1}{2}\rho u^2 \right)_t + \left(\frac{\gamma}{\gamma - 1} pu + \frac{\rho u^3}{2} \right)_r = 0. \tag{3.8.5}$$

Equation (3.8.1) is already in a conservation form. Equations (3.8.1) and (3.8.4) are equivalent to the differential relations

$$d\tau = \rho dr - \rho u dt = m dy \tag{3.8.6}$$

$$d\xi = \rho u dr - (p + \rho u^2) dt = -m a_0 dx, \quad \text{say,} \tag{3.8.7}$$

where m and a_0 are positive constants, having dimensions ML^{-2} and LT^{-2}, respectively. Using (3.8.6) and (3.8.7) and the equation of state

$$f(\tau) = \frac{p^{1/\gamma}}{(\gamma - 1)\rho}, \tag{3.8.8}$$

we can replace (3.8.5) by the total differential

$$d\eta = u d\xi + \left(fv - \frac{1}{2}u^2 \right) d\tau = m a_0^2 dz, \quad v = p^{(\gamma-1)/\gamma}. \tag{3.8.9}$$

It is clear from (3.8.6), (3.8.7), and (3.8.9) that x, y, and z are dimensionless variables. We manipulated these equations to obtain a single equation for z as a function of x and y. From (3.8.9), we get

$$u = -a_0 z_x \tag{3.8.10}$$

$$\frac{1}{\gamma - 1}\frac{p}{\rho} - \frac{1}{2}u^2 = a_0^2 z_y. \tag{3.8.11}$$

On elimination of u from (3.8.10) and (3.8.11), we get

$$\frac{p}{\rho} = a_0^2(\gamma - 1)\left[z_y + \frac{1}{2}(z_x)^2\right]. \tag{3.8.12}$$

Differentiating (3.8.12) with respect to x and y, respectively, we have

$$\frac{1}{\rho}p_x - \frac{p}{\rho^2}\rho_x = a_0^2(\gamma - 1)(z_{xy} + z_x z_{xx}) \tag{3.8.13}$$

$$\frac{1}{\rho}p_y - \frac{p}{\rho^2}\rho_y = a_0^2(\gamma - 1)(z_{yy} + z_x z_{xy}). \tag{3.8.14}$$

Similarly, from (3.8.8) we have

$$\frac{1}{\rho}\rho_x = \frac{1}{\gamma p}p_x \tag{3.8.15}$$

$$\frac{1}{\rho}\rho_y = \frac{1}{\gamma p}p_y - \frac{m}{f}\frac{df}{d\tau} \tag{3.8.16}$$

(see (3.8.6)). On inserting (3.8.15) into (3.8.13) and (3.8.16) into (3.8.14), respectively, we eliminate ρ_x and ρ_y:

$$\frac{1}{\rho}p_x = \gamma a_0^2(z_{xy} + z_x z_{xx}) \tag{3.8.17}$$

$$\frac{1}{\rho}p_y = \gamma a_0^2(z_{yy} + z_x z_{xy}) - \frac{m\gamma}{\gamma - 1}\frac{1}{f}\frac{df}{d\tau}\frac{p}{\rho}. \tag{3.8.18}$$

From (3.8.6) and (3.8.7) we have

$$dt = \frac{ma_0}{p}dx + \frac{mu}{p}dy \tag{3.8.19}$$

so that

$$a_0 p_y = up_x - pu_x. \tag{3.8.20}$$

On using (3.8.10) in (3.8.20), we get

$$p_y + z_x p_x - pz_{xx} = 0. \tag{3.8.21}$$

Finally, using (3.8.12), (3.8.17), and (3.8.18) in (3.8.21), we arrive at a single PDE for z:

$$z_{yy} + 2z_x z_{xy} + \left[\frac{\gamma+1}{2\gamma}(z_x)^2 - \frac{\gamma-1}{\gamma}z_y \right] z_{xx} - \phi(y)\left[z_y + \frac{1}{2}(z_x)^2 \right] = 0$$

$$\tag{3.8.22}$$

$$\phi(y) = \frac{m}{f}\frac{df}{d\tau}. \tag{3.8.23}$$

The Rankine-Hugoniot (RH) relations across a normal plane shock are

$$u = \frac{2U}{\gamma+1} - \frac{2\gamma p_0}{(\gamma+1)\rho_0 U} \tag{3.8.24}$$

$$p = \frac{2\rho_0 U^2}{\gamma+1} - \frac{\gamma-1}{(\gamma+1)}p_0 \tag{3.8.25}$$

$$\rho = \frac{(\gamma+1)\rho_0^2 U^2}{2\gamma p_0 + (\gamma-1)\rho_0 U^2} \tag{3.8.26}$$

where U is the shock velocity and the subscript zero denotes conditions immediately ahead of the shock. The shock locus is defined by

$$\frac{dr}{dt} = U. \tag{3.8.27}$$

We must now transform the RH conditions (3.8.24) - (3.8.26) in terms of the variables x, y, and z. Let the shock locus be given by $y = y_0(x)$. On inserting (3.8.24) - (3.8.27) into the differential relations (3.8.6), (3.8.7), and (3.8.9), we get

$$U\rho_0 dt = mdy_0 \tag{3.8.28}$$

$$p_u dt = ma_0 dx \tag{3.8.29}$$

$$\frac{p_0 U}{\gamma-1}dt = ma_0^2 dz. \tag{3.8.30}$$

From (3.8.28) and (3.8.29) we get

$$\frac{dy_0}{dx} = \frac{\gamma U}{a_0} \tag{3.8.31}$$

where we have used the relation $a_0^2 = \frac{\gamma p_0}{\rho_0}$. Dividing (3.8.30) by (3.8.29) and integrating with respect to x and using the condition $z = 0$ at $y = 0$, we get

$$z(x,t)|_{y=y_0(x)} = \frac{1}{\gamma(\gamma-1)}y_0(x). \tag{3.8.32}$$

Equations (3.8.10), (3.8.24) and (3.8.31) give

$$z_x|_{y=y_0(x)} = 2\frac{\gamma^2 - [y_0'(x)]^2}{\gamma(\gamma+1)y_0'(x)}. \tag{3.8.33}$$

We thus have relations (3.8.31)-(3.8.33) holding along the shock $y = y_0(x)$. As is usual for this class of problems, the shock locus itself must be found as part of the solution.

The other boundary condition arises from the piston motion which drives the flow. Let the velocity of the gas at the piston be denoted by $V(t)$. Therefore, we have the relation

$$m\frac{dV}{dt} = -p, \quad V(0) = u_0 \tag{3.8.34}$$

where u_0 is the initial velocity of the gas. Integrating (3.8.34) and using (3.8.6) - (3.8.7) at $y = 0$, we obtain

$$z(x,0) = \frac{1}{2}x^2 - \frac{u_0}{a_0}x. \tag{3.8.35}$$

The corresponding strong shock condition is

$$z(0,y) = 0, \quad y \geq 0. \tag{3.8.36}$$

Thus, we must solve the nonlinear PDE (3.8.22) subject to the two conditions (3.8.32)-(3.8.33) at the shock $y = y_0(x)$, and (3.8.35) at the piston $y = 0$. The third condition helps determine the shock locus $y = y_0(x)$. The problem is now sensibly posed. The present derivation is due to Ustinov (1982, 1984).

We may observe that the sound speed $a^2 = \dfrac{\gamma p}{\rho}$ follows easily from (3.8.12),

$$a^2 = \gamma(\gamma-1)a_0^2\left(z_y + \frac{1}{2}z_x^2\right), \tag{3.8.37}$$

while the density follows from $a^2 = \dfrac{\gamma p}{\rho}$ and (3.8.8):

$$\rho - \frac{a^{2/(\gamma-1)}}{\gamma^{1/(\gamma-1)}(\gamma-1)^{\gamma/(\gamma-1)}}[f(y)]^{-\gamma/(\gamma-1)}. \tag{3.8.38}$$

We first attempt to find intermediate integrals for (3.8.22), even though the form of the latter does not immediately suggest their existence. We look for first integrals

$$q = I(x,y,z,p) \tag{3.8.39}$$

where, in the usual notation, $p = z_x$ and $q = z_y$. Differentiating (3.8.39) to obtain z_{xy} and z_{yy} and putting the latter into (3.8.22), we have

$$I_y + qI_z \quad + \quad (I_x + pI_z)(2p + I_p) - \phi(y)\left(q + \frac{1}{2}p^2\right)$$

$$+ \left[I_p^2 + 2pI_p - n\left(q + \frac{1}{2}p^2\right) + p^2\right]z_{xx} = 0, \quad n = \frac{\gamma - 1}{\gamma}.$$

$$(3.8.40)$$

Therefore, the necessary conditions for (3.8.39) to be an intermediate integral of (3.8.22) are

$$I_y + qI_z + (I_x + pI_z)(2p + I_p) - \phi(y)\left(q + \frac{1}{2}p^2\right) \;=\; 0 \quad (3.8.41)$$

$$I_p^2 + 2pI_p - n\left(q + \frac{1}{2}p^2\right) + p^2 \;=\; 0. \quad (3.8.42)$$

Solving (3.8.42) for I_p and integrating with respect to p, we get

$$q \equiv I = \left[F(x, y, z) \pm \frac{\sqrt{n}}{2}p\right]^2 - \frac{1}{2}p^2 \qquad (3.8.43)$$

where $F(x, y, z)$ is the function of integration. Using (3.8.10) and (3.8.37) for expressing p and q in terms of u and a, we can write (3.8.43) as

$$u \pm \frac{2a}{\gamma - 1} = \frac{2a_0}{\sqrt{n}}F(x, y, z). \qquad (3.8.44)$$

Equations (3.8.44) generalise the definition of Riemann invariants for non-isentropic flows. These reduce to the usual Riemann invariants for isentropic flows when $F(x, y, z)$ in (3.8.44) is constant. To obtain the unknown function $F(x, y, z)$ we substitute (3.8.43) into (3.8.41) and equate coefficients of different powers of p to zero. We obtain

$$F_y \pm \sqrt{n}FF_x - \frac{1}{2}\phi(y)F \;=\; 0 \qquad (3.8.45)$$

$$\left(\frac{n}{2} + 1\right)F_x \mp \frac{\sqrt{n}}{4}\phi(y) \;=\; 0 \qquad (3.8.46)$$

$$F_z \;=\; 0. \qquad (3.8.47)$$

Integrating (3.8.46) with respect to x we have

$$F(x, y) = \pm \frac{n_0}{2}x\phi(y) + \frac{1}{2}K(y), \quad n_0 = \frac{\sqrt{n}}{(n + 2)} \qquad (3.8.48)$$

where $K(y)$ is a function of integration. Putting $F(x, y)$ into (3.8.45) and equating coefficients of different powers of x to zero, etc., we get

$$\frac{d\phi}{dy} + l_0\phi^2 = 0, \tag{3.8.49}$$

$$\frac{dK}{dy} + l_0\phi K = 0 \tag{3.8.50}$$

where

$$l_0 = -\frac{1}{n+2}.$$

The solution of the system (3.8.49)-(3.8.50) is

$$\phi(y) = (\alpha_0 + l_0 y)^{-1} \tag{3.8.51}$$

$$K(y) = \beta_0(\alpha_0 + l_0 y)^{-1} \tag{3.8.52}$$

where α_0 and β_0 are arbitrary constants. With (3.8.48), (3.8.51), and (3.8.52), the intermediate integrals (3.8.43) become

$$q = \frac{1}{4}\left[\frac{\beta_0 \pm n_0 x}{\alpha_0 + l_0 y} \pm \sqrt{np}\right]^2 - \frac{1}{2}p^2 \tag{3.8.53}$$

or, in physical variables, the Riemann invariants are

$$u \pm \frac{2a}{\gamma - 1} = \frac{a_0}{\sqrt{n}}\frac{\beta_0 + n_0 x}{\alpha_0 + l_0 y}. \tag{3.8.54}$$

We write the first integral (3.8.53) with positive sign more explicitly as

$$l_0 z_Y + \alpha_1 z_X^2 - \beta_1\frac{X}{Y}z_X - \frac{X^2}{4Y^2} = 0 \tag{3.8.55}$$

where $X = \beta_0 + n_0 x$, $Y = \alpha_0 + l_0 y$, $\alpha_1 = [n(2 - n)/(4(n + 2)^2)]$ and $\beta_1 = n/2(n + 2)$.

Following Charpit's method, the characteristics of (3.8.55) are written as

$$\frac{dX}{ds} = \beta_1 Y^{-1}X - 2\alpha_1 p$$

$$\frac{dY}{ds} = -l_0$$

$$\frac{dz}{ds} = \beta_1 Y^{-1}Xp - 2\alpha_1 p^2 - l_0 q \tag{3.8.56}$$

$$\frac{dp}{ds} = -\frac{1}{2}Y^{-2}X - \beta_1 Y^{-1}p$$

$$\frac{dq}{ds} = \frac{1}{2}Y^{-3}X^2 + \beta_1 Y^{-2}Xp$$

where the parameter s is measured along the characteristics. A solution of the system (3.8.56) is

$$X(s) = \frac{\gamma+1}{3\gamma-1}c_0 - \frac{\gamma-1}{3\gamma-1}c_1 S$$

$$Y(s) = S$$

$$z(s) = \frac{(\gamma-1)^2}{2\gamma(3\gamma-1)}c_0 c_1 \log S - \frac{\gamma-1}{3\gamma-1}c_1^2 S + \frac{\gamma+1}{2(3\gamma-1)}c_0^2 S^{-1} + c_4$$

$$p(s) = c_0 S^{-1} + c_1 \tag{3.8.57}$$

$$q(s) = \frac{\gamma-1}{3\gamma-1}c_0 c_1 S^{-1} - \frac{\gamma+1}{2(3\gamma-1)}c_0^2 S^{-2}$$

$$S = \left(\frac{\gamma}{3\gamma-1}s + c_3\right)$$

where c_0, c_1, c_3 and c_4 are arbitrary constants. Eliminating s from $X(s)$, $Y(s)$, and $p(s)\left(=\dfrac{\partial z}{\partial X}\right)$, we get

$$\frac{\gamma-1}{2\gamma}Y\frac{\partial z}{\partial X} + \frac{3\gamma-1}{2\gamma}X - c_1 = 0. \tag{3.8.58}$$

Solving for $\dfrac{\partial z}{\partial X}$ and $\dfrac{\partial z}{\partial Y}$ from (3.8.55) and (3.8.58), putting them into $dz = pdx + qdy$, and integrating, we get the solution

$$z(x,y) = \left(\alpha_0 - \frac{\gamma}{(3\gamma-1)}y\right)^{-1}\left[-\frac{\gamma}{2(3\gamma-1)}x^2\right.$$

$$+ \left(c_1\frac{\sqrt{\gamma(\gamma-1)}}{3\gamma-1} - \beta_0\sqrt{\frac{\gamma}{\gamma-1}}\right)x + c_1\beta_0 - \frac{3\gamma-1}{2(\gamma-1)}\beta_0^2$$

$$\left. - \frac{\gamma^2-1}{4\gamma(3\gamma-1)}c_1^2\right] \tag{3.8.59}$$

where we have changed X and Y back to x and y via the relations $X = \beta_0 + n_0 x$, $Y = \alpha_0 + l_0 y$ (see below (3.8.55)).

The sound speed is obtained from (3.8.37) and (3.8.59) and, hence, ρ from (3.8.23), (3.8.38), and (3.8.51):

$$a^2 = \gamma(\gamma-1)a_0^2\left[\frac{c_1(\gamma-1)}{4(3\gamma-1)}\right]^2\left(\alpha_0 - \frac{\gamma}{3\gamma-1}y\right)^{-2}$$

$$\rho = \frac{a^{2/(\gamma-1)}}{\gamma^{1/(\gamma-1)}(\gamma-1)^{\gamma/(\gamma-1)}}\left(\alpha_0 - \frac{\gamma}{3\gamma-1}y\right)^{(3\gamma-1)/(\gamma-1)}. \tag{3.8.60}$$

We may also seek the solution of the first-order PDE (3.8.55) in the similarity form

$$z(x,y) = Y^{m_1} g(\xi), \quad \xi = XY^{m_2}. \tag{3.8.61}$$

It is easily checked that for this purpose we must have $m_1 = 1$, $m_2 = -1$ and then $g(\xi)$ satisfies the first-order ODE

$$g'^2 + \frac{2(n+2)}{n}\xi g' + \frac{4(n+2)^2}{n(n-2)}\left(\frac{1}{n+2}g + \frac{1}{4}\xi^2\right) = 0. \tag{3.8.62}$$

The quadratic (3.8.62) in g' may be solved and, hence, g found by integration. However, a simple solution of this equation is

$$g(\xi) = c_4^2 \pm \left[-\frac{4(n+2)}{n(n-2)}\right]^{1/2} c_4\xi - \frac{n+2}{2(n-2)}\xi^2 \tag{3.8.63}$$

where c_4 is a constant. Again expressing X and Y in terms of x and y etc., we may write this solution of (3.8.22) as

$$
\begin{aligned}
z(x,y) &= \frac{\gamma(\gamma-1)}{2(\gamma+1)(3\gamma-1)}\left(\alpha_0 - \frac{\gamma}{(3\gamma-1)}y\right)^{-1}x^2 \\
&+ \left[\pm\frac{2c_4\gamma}{\sqrt{(\gamma+1)(3\gamma-1)}} + \beta_0\frac{\sqrt{\gamma(\gamma-1)}}{\gamma+1}\right. \\
&\quad \left.\left(\alpha_0 - \frac{\gamma}{(3\gamma-1)}y\right)^{-1}\right]x \\
&+ c_4^2\left(\alpha_0 - \frac{\gamma}{3\gamma-1}y\right) + \frac{3\gamma-1}{2(\gamma+1)}\beta_0^2\left(\alpha_0 - \frac{\gamma}{(3\gamma-1)}y\right)^{-1} \\
&\pm c_4\beta_0\sqrt{\frac{4\gamma(3\gamma-1)}{\gamma^2-1}}.
\end{aligned} \tag{3.8.64}
$$

The sound speed and density of the gas, etc. for this case may be found as for (3.8.59).

Equation (3.8.22) provides a good opportunity to learn about the effectiveness of the direct method of Clarkson and Kruskal (1989). We let

$$z(x,y) = \alpha(x,y) + \beta(x,y)H(\eta(x,y)), \quad \beta(x,y) \neq 0. \tag{3.8.65}$$

Substituting (3.8.65) into (3.8.22) and requiring the resulting equation to be an ODE for $H(\eta)$, we write it as

$$
\begin{aligned}
&(H'^2 + \Gamma_1 HH' + \Gamma_2 H' + \Gamma_3 H^2 + \Gamma_4 H + \Gamma_5)H'' + \\
&+ \Gamma_6 H'^3 + (\Gamma_7 H + \Gamma_8)H'^2 + (\Gamma_9 H^2 + \Gamma_{10}H + \Gamma_{11})H' \\
&+ \Gamma_{12}H^3 + \Gamma_{13}H^2 + \Gamma_{14}H + \Gamma_{15} = 0
\end{aligned} \tag{3.8.66}
$$

where the functions $\Gamma_n = \Gamma_n(\eta)$ are introduced in the following manner:

$$2\beta_x \;=\; \beta\Gamma_1 \qquad (3.8.67)$$

$$2\omega\alpha_x\eta_x + (2-n)\eta_y \;=\; \omega\beta\eta_x^2\Gamma_2 \qquad (3.8.68)$$

$$\beta_x^2 \;=\; \beta^2\eta_x^2\Gamma_3 \qquad (3.8.69)$$

$$2\omega\alpha_x\beta_x\eta_x + 2\beta_x\eta_y - n\beta_y\eta_x \;=\; \omega\beta^2\eta_x^3\Gamma_4 \qquad (3.8.70)$$

$$\omega\alpha_x^2\eta_x^2 + 2\alpha_x\eta_x\eta_y - n\alpha_y\eta_x^2 + \eta_y^2 \;=\; \omega\beta^2\eta_x^4\Gamma_5 \qquad (3.8.71)$$

$$\beta\eta_{xx} + 2\beta_x\eta_x \;=\; \beta\eta_x^2\Gamma_6 \qquad (3.8.72)$$

$$2\beta\beta_x\eta_{xx} + 4\beta_x^2\eta_x + \beta\beta_{xx}\eta_x \;=\; \beta^2\eta_x^3\Gamma_7 \qquad (3.8.73)$$

$$4\omega\beta\alpha_x\eta_x\eta_{xx} - 2n\beta\eta_y\eta_{xx} + 4\beta\eta_x\eta_{xy}$$
$$+8\omega\alpha_x\beta_x\eta_x - 4(n-1)\beta_x\eta_x\eta_y + 4\beta_y\eta_x^2$$
$$+2\omega\beta\alpha_{xx}\eta_x^2 - \phi(y)\beta\eta_x^2 \;=\; 2\omega\beta^2\eta_x^4\Gamma_8 \qquad (3.8.74)$$

$$\beta_x(\beta\beta_x\eta_{xx} + 2\beta_x^2\eta_x + 2\beta\beta_{xx}\eta_x) \;=\; \beta^3\eta_x^3\Gamma_9 \qquad (3.8.75)$$

$$4\omega\beta\alpha_x\beta_x\eta_{xx} - 2n\beta\beta_y\eta_{xx} + 4\beta\beta_x\eta_{xy}$$
$$+8\omega\alpha_x\beta_x\eta_x + 4\omega\beta\alpha_x\beta_{xx}\eta_x + 4\beta_x\eta_y$$
$$-4(n-1)\beta_x\beta_y\eta_x + 4\omega\beta\alpha_{xx}\beta_x\eta_x$$
$$-2\phi(y)\beta\beta_x\eta_x + 4\beta\beta_{xy}\eta_x - 2n\beta\beta_{xx}\eta_y \;=\; 2\omega\beta^3\eta_x^4\Gamma_{10} \qquad (3.8.76)$$

$$2\omega\beta\alpha_x^2\eta_{xx} + 2n\beta\alpha_y\eta_{xx} + 2\beta\eta_{yy}$$
$$+4\beta\alpha_x\eta_{xy} + 4\omega\alpha_x^2\beta_x\eta_x + 4\alpha_x\beta_x\eta_y$$
$$+4\alpha_x\beta_y\eta_x + 4\omega\beta\alpha_x\alpha_{xx}\eta_x - 4n\alpha_y\beta_x\eta_x$$
$$+4\beta_y\eta_y + 4\beta\alpha_{xy}\eta_x - 2n\beta\alpha_{xx}\eta_y$$
$$-2\phi(y)\beta(\alpha_x\eta_x + \eta_y) \;=\; 2\omega\beta^3\eta_x^4\Gamma_{11} \qquad (3.8.77)$$

$$\beta_x^2\beta_{xx} \;=\; \beta^3\eta_x^4\Gamma_{12} \qquad (3.8.78)$$

$$4\omega\alpha_x\beta_x\beta_{xx} + 2\omega\alpha_{xx}\beta_x^2 + 4\beta_x\beta_{xy}$$
$$-2n\beta_y\beta_{xx} - \phi(y)\beta_x^2 \;=\; 2\omega\beta^3\eta_x^4\Gamma_{13} \qquad (3.8.79)$$

$$2\omega\alpha_x\beta_{xx} + 4\omega\alpha_x\alpha_{xx}\beta_x + 4\alpha_x\beta_{xy}$$
$$-2n\alpha_y\beta_{xx} + 4\beta_x\alpha_{xy} - 2n\alpha_{xx}\beta_y$$
$$+2\beta_{yy} - 2\phi(y)(\alpha_x\beta_x + \beta_y) \;=\; 2\omega\beta^3\eta_x^4\Gamma_{14} \qquad (3.8.80)$$

$$2\alpha_{yy} + 4\alpha_x\alpha_{xy} + 2\omega\alpha_x^2\alpha_{xx}$$
$$-2n\alpha_y\alpha_{xx} - \phi(y)(\alpha_x^2 + 2\alpha_y) \;=\; 2\omega\beta^3\eta_x^4\Gamma_{15} \qquad (3.8.81)$$

where $\omega = (\gamma + 1)/(2\gamma) = 1 - n/2$.

The following remarks help to solve the system (3.8.67) - (3.8.81) for $\alpha(x,y)$, $\beta(x,y)$, $\eta(x,y)$, and $\Gamma_n(\eta)$, $n = 1, 2, \ldots 15$.

Remark 1

If $\alpha(x,y)$ has the form $\alpha(x,y) = \hat{\alpha}(x,y) + \beta(x,y)\Omega(\eta)$, we may choose $\Omega = 0$.

Remark 2

If $\beta(x,y)$ is found to have the form $\beta(x,y) = \hat{\beta}(x,y)\Omega(\eta)$, we may put $\Omega(\eta) = 1$.

Remark 3

If $\eta(x,y)$ is determined from the equation $f(\eta) = \hat{\eta}(x,y)$, where $f(\eta)$ is any invertible function, then we may put $f(\eta) = \eta$ without any loss of generality.

Setting $\Gamma_1(\eta) = 2\Omega_1'(\eta)/\Omega_1(\eta)$ in (3.8.67) and integrating with respect to η, we get $\beta = \hat{\beta}(y)\Omega_1(\eta)$ where $\hat{\beta}(y)$ is a function of integration. Using Remark 2, we put $\Omega_1(\eta) = 1$ so that

$$\beta = \beta(y) \tag{3.8.82}$$

where we have dropped the hat from β. Putting $\Gamma_6(\eta) = -\dfrac{\Omega_6''(\eta)}{\Omega_6'(\eta)}$ in (3.8.72) and integrating twice with respect to x, we get $\Omega_6(\eta) = x\theta(y) + \psi(y)$ where $\theta(y)$ and $\psi(y)$ are functions of integration. In view of Remark 3, we put $\Omega_6(\eta) = \eta$ and arrive at the similarity variable

$$\eta = x\theta(y) + \psi(y). \tag{3.8.83}$$

Putting $\Gamma_2(\eta) = 2\Omega_2'(\eta)$ in (3.8.68) and integrating with respect to x, we get $\alpha = \beta\Omega_2 - (x/2\theta)(x\theta' + 2\psi') + \lambda(y)$ where we have used (3.8.83). $\lambda(y)$ is a function of integration. Using Remark 1, we may put $\Omega_2(\eta) = 0$. Thus, we have

$$\alpha(x,y) = -\frac{x}{2\theta(y)}[x\theta'(y) + 2\psi'(y)] + \lambda(y). \tag{3.8.84}$$

It is easily shown from (3.8.67)-(3.8.69), (3.8.72), (3.8.73), (3.8.75), (3.8.76), (3.8.78), (3.8.79), and (3.8.82) that

$$\Gamma_n(\eta) = 0, \quad n = 1, 2, 3, 6, 7, 9, 10, 12, 13. \tag{3.8.85}$$

In view of (3.8.82)-(3.8.84), (3.8.70) requires that $\Gamma_4(\eta) = k$, a constant, and then it becomes

$$\beta' = -\omega_0 k\theta^2\beta^2, \quad \omega_0 = \omega/n. \tag{3.8.86}$$

On insertion of (3.8.83) and (3.8.84), (3.8.71) becomes

$$(x\theta' + \psi')^2 - \theta(x^2\theta'' + 2x\psi'') + \theta'(x^2\theta' + 2x\psi') + 2\theta^2\lambda' = -2\omega_0\beta^2\theta^4\Gamma_5(\eta). \tag{3.8.87}$$

Since LHS of (3.8.87) is quadratic in x, $\Gamma_5(\eta)$ must have the form $\Gamma_5(\eta) = a\eta^2 + b\eta + c$, where a, b, and c are constants. With this choice, (3.8.87) may be satisfied provided

$$\theta\theta'' - 2\theta'^2 = 2a\omega_0\beta^2\theta^6 \tag{3.8.88}$$

$$\psi'' - 2\theta^{-1}\theta'\psi' = \omega_0\beta^2\theta^4(2a\psi + b) \tag{3.8.89}$$

$$2\theta^2\lambda' + \psi'^2 = -2\omega_0\beta^2\theta^4(a\psi^2 + b\psi + c). \tag{3.8.90}$$

Again, in view of (3.8.82) - (3.8.84), (3.8.74) requires that $\Gamma_8(\eta) = l$, a constant, and then it becomes

$$\phi(y) = 2(2 - \omega)\theta^{-1}\theta' - 2\omega(2k_0 + l)\theta^2\beta, \quad k_0 = k/n. \tag{3.8.91}$$

After substitution of (3.8.88) and (3.8.89) into (3.8.77) and (3.8.81), we find that

$$\Gamma_{11}(\eta) = (-1/n)(2a\eta + b) \tag{3.8.92}$$

$$\Gamma_{15}(\eta) = -2\omega_0(2k/n + l)(a\eta^2 + b\eta) + d \tag{3.8.93}$$

and

$$\begin{aligned}\lambda'' - 2\theta^{-1}\theta'\lambda' &= [2k\omega_0^2(a\psi^2 + b\psi) + 2\omega\omega_0(2k_0 + l)c + \omega d]\theta^4\beta^3 \\ &\quad -2\omega_0\beta^2\theta^2\psi'(2a\psi + b)\end{aligned} \tag{3.8.94}$$

where d is an arbitrary constant. Further, Equation (3.8.80) requires that $\Gamma_{14} = h$, a constant, and then it reduces to

$$\beta'' - 2\theta^{-1}\theta'\beta' = [h\omega + (2\omega\omega_0 k)(2k/n + l)]\theta^4\beta^3. \tag{3.8.95}$$

We now have six equations (3.8.86), (3.8.88), (3.8.89), (3.8.90), (3.8.94), and (3.8.95) for the determination of four functions $\theta(y), \beta(y), \psi(y)$, and $\lambda(y)$. If we choose the constants h and d suitably, this overdetermined system can be made determinate. If we let $h = (-2k\omega_0)(k/n + l)$, Equation (3.8.95) can be obtained from (3.8.86) by differentiation. Similarly, with $d = (-2\omega_0)(k/n + l)c$, (3.8.94) can found by differentiating (3.8.90) with respect to y and using (3.8.86) and (3.8.89). Thus, the solution (3.8.65) of (3.8.22) may be written as

$$z(x, y) = \lambda(y) - \frac{x}{2\theta(y)}[x\theta'(y) + 2\psi'(y)] + \beta(y)H(\eta) \tag{3.8.96}$$

$$\eta(x, y) = x\theta(y) + \psi(y) \tag{3.8.97}$$

where the functions $\beta(y), \theta(y), \psi(y)$, and $\lambda(y)$ are governed by the system (3.8.86), (3.8.88) - (3.8.90). The function $H(\eta)$ is governed by (3.8.66) with $\Gamma_n(\eta), n = 1, 2, \ldots 15$, found above, appropriately substituted into it:

$$(H'^2 + kH + a\eta^2 + b\eta + c)H'' + lH'^2 - \frac{1}{n}(2a\eta + b)H'$$

$$-2kw_0(k_0 + l)H - 2w_0(k_0 + l)(a\eta^2 + b\eta + c) = 0. \quad (3.8.98)$$

Equation (3.8.98) seems difficult to solve generally. However, a special solution may be easily found if we assume that

$$H(\eta) = A\eta^2 + B\eta + C \quad (3.8.99)$$

and substitute into (3.8.98), here, A, B, and C are constants. Then the coefficients of η^2, η, η^0, etc. put equal to zero give

$$4A^3 + (2l + k)A^2 + w(a - k_0^2 - lk_0)A - aw_0(k_0 + l) \quad = \quad 0 \quad (3.8.100)$$

$$[4A^2 + (2l + k)A - a/n - kk_0w_0 - klw_0]B$$

$$+b(1 - 1/n)A - bw_0(k_0 + l) \quad = \quad 0 \quad (3.8.101)$$

$$2k(A - k_0w_0 - lw_0)C + (2A + l)B^2 - (b/n)B$$

$$+2cA - 2w_0(k_0 + l)c \quad = \quad 0. \quad (3.8.102)$$

The algebraic system (3.8.100) - (3.8.102) will be solved for those cases for which the system of ODEs (3.8.86), (3.8.88), (3.8.89), and (3.8.90) for $\beta(y), \theta(y), \psi(y)$, and $\lambda(y)$ can be solved. The following simple cases may be noted.

1. The following power law solutions of the ODEs are easily found:

$$\theta(y) \quad = \quad m_0(m_1y + m_2)^{n_1},$$

$$\beta(y) \quad = \quad \frac{nm_1(2n_1 + 1)}{k\omega m_0^2}(m_1y + m_2)^{-2n_1-1}$$

$$\psi(y) \quad = \quad -b/2a,$$

$$\lambda(y) \quad = \quad -\frac{(b^2 - 4ac)nm_1(2n_1 + 1)}{4a\omega k^2 m_0^2}(m_1y + m_2)^{-2n_1-1} \quad (3.8.103)$$

where m_0, m_1, and m_2 are arbitrary constants and n_1 satisfies the quadratic $n_1^2 + n_1 + 2an/(8an + \omega k^2) = 0$. When (3.8.103) is inserted into (3.8.96) and (3.8.97), we get the similarity form of the solution

$$z(x,y) = -\frac{(b^2 - 4ac)nm_1(2n_1 + 1)}{4awk^2m_0^2}(m_1y + m_2)^{-2n_1-1}$$

$$-\frac{1}{2}n_1m_1(m_1y + m_2)^{-1}x^2$$

$$+\frac{nm_1(2n_1 + 1)}{kwm_0^2}(m_1y + m_2)^{-2n_1-1}H(\eta) \qquad (3.8.104)$$

$$\eta(x,y) = m_0(m_1y + m_2)^{n_1}x - b/(2a). \qquad (3.8.105)$$

Equations (3.8.104), (3.8.105), and (3.8.99) yield

$$z(x,y) = m_1\left[-\frac{n_1}{2} + \frac{n}{k\omega}(1 + 2n_1)m_0^2A\right](m_1y + m_2)^{-1}x^2$$

$$+\frac{n}{k\omega}(1 + 2n_1)m_0^2\left[-\frac{b}{a}A + B\right](m_1y + m_2)^{-n_1-1}x$$

$$+\left[\frac{b^2A}{4a^2} - \frac{bB}{a} + C - \frac{b^2 - 4ac}{4ak}\right]$$

$$\times\frac{n}{k\omega}(1 + 2n_1)m_0^2m_1(m_1y + m_2)^{-2n_1-1}. \qquad (3.8.106)$$

The function

$$\phi(y) = 2m_1[n_1(2 - \omega) - k_0^{-1}(2k_0 + l)(1 + 2n_1)](m_1y + m_2)^{-1}. \qquad (3.8.107)$$

2. If we let $b = c = 0$ in (3.8.86) and (3.8.88) - (3.8.90), we can find an exponential type of solution

$$\theta(y) = m_0e^{-m_1y}$$

$$\beta(y) = -\frac{2nm_1}{\omega km_0^2}e^{2m_1y} \qquad (3.8.108)$$

$$\psi(y) = d_0e^{-m_1y}$$

$$\lambda(y) = \lambda_0 - \frac{d_0^2m_1^2}{2m_0^2}\left(1 + \frac{8an}{\omega k^2}\right)y$$

where m_0, m_1, d_0, and λ_0 are arbitrary constants. Substituting (3.8.108) into (3.8.96) and (3.8.97), we get the similarity form

$$z(x,y) = \lambda_0 - \frac{d_0^2m_1^2}{2m_0^2}\left(1 + \frac{8an}{\omega k^2}\right)y + \frac{1}{2}m_1^2x^2$$

$$+\frac{m_1d_0}{m_0}x - \frac{2nm_1}{\omega km_0^2}e^{2m_1y}H(\eta)$$

$$\eta(x,y) = (m_0x + d_0)e^{-m_1y}. \qquad (3.8.109)$$

If we use the form (3.8.99) for $H(\eta)$, we have the explicit solution

$$
\begin{aligned}
z(x,y) = {} & \lambda_0 - \frac{2nm_1 d_0^2 A}{\omega k m_0^2} - \frac{d_0^2 m_1^2}{2m_0^2}\left(1 + \frac{8an}{\omega k^2}\right) y - \frac{2nm_1 d_0}{\omega k m_0^2} \\
& \times (B + Ce^{m_1 y})e^{m_1 y} + \frac{m_1}{m_0}\left[d_0 - \frac{2n}{\omega k}(2d_0 A + Be^{m_1 y})\right] x \\
& + m_1\left(\frac{1}{2} - \frac{2nA}{\omega k}\right) x^2.
\end{aligned}
\tag{3.8.110}
$$

Here the function $\varphi(y) = 2m_1(\omega + lk_0^{-1})$.

Since the details are entirely similar, we summarise the results for some other sets of parameters a, b, c, and k appearing in (3.8.86) and (3.8.88) - (3.8.90) (see Sachdev et al. (1997) for details).

3. $k = 0$, $b^2 = 4ac$

$$
\begin{aligned}
z(x,y) = {} & m_1\left(\frac{1}{4} + A\sqrt{\frac{n}{8a\omega}}\right)(m_1 y + m_2)^{-1} x^2 \\
& + m_0 m_1\left(-\frac{bA}{a} + B\right)\sqrt{\frac{n}{8a\omega}}(m_1 y + m_2)^{-1/2} x \\
& + \lambda_0 + m_0^2 m_1\left(\frac{b^2 A}{4a^2} - \frac{bB}{a} + C\right)\sqrt{\frac{n}{8a\omega}}
\end{aligned}
\tag{3.8.111}
$$

$$
\phi(y) = -m_1\left[2 - \omega + (2k_0 + l)\sqrt{\frac{n\omega}{2a}}\right](m_1 y + m_2)^{-1}
$$

4. $k = b = c = 0$

$$
\begin{aligned}
z(x,y) = {} & m_1\left(\frac{1}{4} - \frac{l}{2}\sqrt{\frac{n}{8a\omega}}\right)(m_1 y + m_2)^{-1} x^2 \\
& + \left(d_0 m_0 m_1\left(\frac{1}{4} - l\sqrt{\frac{n}{8a\omega}}\right)(m_1 y + m_2)^{-1}\right) x \\
& + d_0\left(\frac{1}{4} m_0^2 m_1 - \frac{l d_0}{2}\right)(m_1 y + m_2)^{-1} + C
\end{aligned}
\tag{3.8.112}
$$

where C is an arbitrary constant.

$$
\phi(y) = -m_1\left(2 - \omega + (2k_0 + l)\sqrt{\frac{n\omega}{2a}}\right)(m_1 y + m_2)^{-1}
\tag{3.8.113}
$$

5. $a = b = c = 0$

$$
\begin{aligned}
z(x,y) = {} & m_1\left(\frac{1}{2} - \frac{nA}{\omega k}\right)(m_1 y + m_2)^{-1} x^2 - \frac{nm_1 B}{\omega k m_0} x \\
& - \frac{nm_1 C}{\omega k m_0^2}(m_1 y + m_2)
\end{aligned}
$$

where A satisfies the quadratic

$$4A^2 + (2l + k)A - k\omega_0(k/n + l) = 0$$

and

$$C = -(1/2k)\left[(2l + A)/\left\{A - \omega_0\left(\frac{k}{n} + l\right)\right\}\right]B^2;$$

B is an arbitrary constant.

$$\phi(y) = 2m_1[\omega + k_0^{-1}l](m_1y + m_2)^{-1} \qquad (3.8.114)$$

6. $a = 0$.

   In this case, two distinct possibilities arise

(i) $z(x,y) = \dfrac{nm_1 A}{\omega k}(m_1y + m_2)^{-1}x^2$

$\qquad + \left[\dfrac{nm_1}{\omega k m_0}\left(\dfrac{b}{k} + B - \dfrac{2bnA}{\omega k^2}\log(m_1y + m_2) + 2A(d_0y + d_1)\right)\right.$

$\qquad \left. - \dfrac{d_0}{m_0}(m_1y + m_2)\right](m_1y + m_2)^{-1}x$

$\qquad + \dfrac{nm_1}{\omega k m_0^2}\left[\dfrac{1}{k}\left(c - \dfrac{nb^2}{2\omega k^2}\right) + \dfrac{b}{k}(d_0y + d_1)\right.$

$\qquad - \dfrac{nb^2}{\omega k^3 m_0^2}\log(m_1y + m_2) + \dfrac{n^2b^2A}{\omega^2 k^4}[\log(m_1y + m_2)]^2$

$\qquad + A(d_0y + d_1)^2 - \dfrac{2nbA}{\omega k^2}(d_0y + d_1)\log(m_1y + m_2)$

$\qquad \left. - \dfrac{nbB}{\omega k^3}\log(m_1y + m_2) + B(d_0y + d_1) + C\right](m_1y + m_2)^{-1}$

$\qquad - \dfrac{d_0^2}{2m_0^2}y + \lambda_0 \qquad\qquad\qquad (3.8.115)$

$\phi(y) = -(2nm_1/k)(2k_0 + l)(m_1y + m_2)^{-1} \qquad (3.8.116)$

(ii) $z(x,y) = m_1\left(\dfrac{1}{2} - \dfrac{nA}{\omega k^2}\right)(m_1y + m_2)^{-1}x^2$

$\qquad + \left[-\dfrac{nm_1}{\omega k^2 m_0}\left(b(m_1y + m_2)^{-1} + 2d_1A + B\right.\right.$

$\qquad \left.+ \dfrac{2nbA}{\omega k^2}\log(m_1y + m_2)\right) + \dfrac{d_0}{m_0}\left(\dfrac{2nA}{\omega k^2} - 1\right)(m_1y + m_2)^{-1}\right]x$

$$+\frac{d_0^2}{m_1 m_0^2}\left(\frac{1}{2} - \frac{nA}{\omega k^2}\right)(m_1 y + m_2)^{-1}$$

$$+\frac{nm_1}{\omega k^2 m_0^2}\left(\frac{nb^2}{2\omega k^2} - bd_1 - c + d_1^2 A + d_1 B + C\right)(m_1 y + m_2)$$

$$+\frac{bn^2}{\omega^2 k^4 m_0^2}\Big(2d_0 A \log(m_1 y + m_2) - m_1(b + 2d_1 A + B)$$

$$\times (m_1 y + m_2)\log(m_1 y + m_2)$$

$$-\frac{bnm_1 A}{\omega k^2}(m_1 y + m_2)[\log(m_1 y + m_2)]^2\Big)$$

$$+\frac{nd_0}{\omega k^2 m_0^2}(2d_1 A + B) + \lambda_0 \tag{3.8.117}$$

$$\phi(y) = 2m_1(\omega + k_0^{-1} l)(m_1 y + m_2)^{-1} \tag{3.8.118}$$

7. $a = b = c = k = 0$

$$z = C + \frac{1}{2}(m_1 - l d_1 m_0^2 (m_1 y + m_2)^{-1})(m_1 y + m_2)^{-1} x^2$$

$$+ \left[\frac{d_0 m_1}{m_0} + m_0 d_1(-l d_0 (m_1 y + m_2)^{-1} + B)\right](m_1 y + m_2)^{-1} x$$

$$+ \left[\frac{m_1 d_0^2}{2m_0^2} + d_0 d_1\left(-\frac{l d_0}{2}(m_1 y + m_2)^{-1} + B\right)\right](m_1 y + m_2)^{-1}$$

$$\tag{3.8.119}$$

where B and C are arbitrary constants.

$$\phi(y) = -2m_1(2 - \omega)(m_1 y + m_2)^{-1} - 2\omega d_1 m_0^2 (2k_0 + l)(m_1 y + m_2)^{-2} \tag{3.8.120}$$

Solutions Quadratic in x

The boundary condition on the piston (3.8.35) suggests that we may seek solutions of (3.8.22) in the form

$$z(x, y) = F_0(y) + G(y)x + \frac{1}{2}H(y)x^2. \tag{3.8.121}$$

Substituting (3.8.121) into (3.8.22) and equating coefficients of x^2, x, x^0, etc. to zero, we get the following coupled system of nonlinear ODEs for

$F_0(y), G(y)$ and $H(y)$:

$$H'' + [(4 - n)H - \phi]H' + (2 - n)H^3 - \phi H^2 = 0 \qquad (3.8.122)$$

$$G'' + [(2 - n)H - \phi]G' + [(2 - n)H^2 - \phi H + 2H']G = 0 \qquad (3.8.123)$$

$$F_0'' - (nH + \phi)F_0' - \frac{1}{2}[\phi - (2 - n)H]G^2 + 2GG' = 0. \qquad (3.8.124)$$

If we choose $\phi = 4A_0 H$, where A_0 is a constant, Equation (3.8.122) is easily solved:

$$H = (C_0 + B_0 y)^{-1}, \quad B_0 = \omega - 2A_0 \qquad (3.8.125)$$

where C_0 is a constant.

Equations (3.8.123) and (3.8.124) then yield

$$G(y) = D_0(C_0 + B_0 y)^{m_1}, \quad m_1 = -1 - 2B_0^{-1} \qquad (3.8.126)$$

$$F_0(y) = \frac{1}{6}D_0^2(C_0 + B_0 y)^{2m_1 + 1} \qquad (3.8.127)$$

where D_0 is another arbitrary constant. The solution in the present case can be written as

$$
\begin{aligned}
z(x, y) = {} & (1/6)D_0^2(C_0 + B_0 y)^{2m_1 + 1} + D_0(C_0 + B_0 y)^{m_1} x \\
& + (1/2)(C_0 + B_0 y)^{-1} x^2.
\end{aligned} \qquad (3.8.128)
$$

The speed of sound and the gas density are now given by

$$
\begin{aligned}
a^2 = {} & \gamma(\gamma - 1)a_0^2\{(1/2)D_0^2[1 + (1/3)(2n_2 + 1)B_0](C_0 + B_0 y)^{2n_2} \\
& + D_0(n_2 B_0 + 1)(C_0 + B_0 y)^{n_2 - 1} x \\
& + (1/2)(1 - B_0)(C_0 + B_0 y)^{-2} x^2\}
\end{aligned} \qquad (3.8.129)
$$

$$\rho = \frac{a^{2/(\gamma - 1)}}{\gamma^{1/(\gamma - 1)}(\gamma - 1)^{\gamma/(\gamma - 1)}}(C_0 + B_0 y)^{4\gamma A_0/(\gamma - 1)B_0} \qquad (3.8.130)$$

(see (3.8.37) and (3.8.38)).

Solutions by Equation – Splitting

In a rather ad hoc manner we split (3.8.22) into two equations

$$\omega z_x^2 - n z_y = 0 \qquad (3.8.131)$$

$$z_{yy} + 2z_x z_{xy} - \phi(y)\left(z_y + \frac{1}{2}z_x^2\right) = 0. \qquad (3.8.132)$$

We recall that $\omega = \dfrac{\gamma+1}{2\gamma}$, $n = \dfrac{\gamma-1}{\gamma}$, and $\phi(y) = \dfrac{1}{f}\dfrac{df}{dy}$ (see (3.8.23)). We manipulate (3.8.131) and (3.8.132) to enable us to integrate. We thus get

$$z_{yy} + 2z_x z_{xy} - 2\omega_1 \varphi(y) z_y = 0, \quad \omega_1 = 1/2\omega \qquad (3.8.133)$$

which may be written as

$$\frac{\partial}{\partial y}\left(z_x^2 - \frac{1}{\omega_0}z_y\right) = 2\omega_1 \frac{f'(y)}{f(y)} z_y - \left(1 + \frac{1}{\omega_0}\right) z_{yy} \qquad (3.8.134)$$

where ω_0 is an arbitrary constant. Taking (3.8.131) into account, (3.8.134) reduces to

$$2\omega_1 f^{-1} f' z_y - \left(1 + \frac{1}{\omega_0}\right) z_{yy} = 0 \qquad (3.8.135)$$

which, on integration twice with respect to y, gives

$$z = v(x)\int^y f^{\alpha_2}(t)dt + w(x), \quad \alpha_2 = 2\gamma/(3\gamma-1) \qquad (3.8.136)$$

where $v(x)$ and $w(x)$ are functions of integration. We substitute (3.8.136) into (3.8.131) to find that

$$v(x) = (b_0 x + b_2)^2, \quad w(x) = b_0 b_1 x^2 + 2b_1 b_2 x + b_3. \qquad (3.8.137)$$

The function $f(y)$ is obtained by using (3.8.131), (3.8.136), and (3.8.137):

$$f(y) = \left(\frac{1}{2b_1\sqrt{\omega_0}} - 2\sqrt{\omega_0}Ay\right)^{(-3\gamma+1)/\gamma} \qquad (3.8.138)$$

where A is an arbitrary constant. The solution of (3.8.22) may now be written as

$$\begin{aligned}
z(x,y) &= \frac{1}{2A\sqrt{\omega_0}}(b_0 x + b_2)^2 \left(\frac{1}{2b_1\sqrt{\omega_0}} - 2\sqrt{\omega_0}Ay\right)^{-1}\\
&\quad + b_0 b_1 x^2 + 2b_1 b_2 x + b_3.
\end{aligned} \qquad (3.8.139)$$

The equation of state (3.8.8) in the present case becomes

$$p = (\gamma-1)^\gamma \rho^\gamma \left(\frac{1}{2b_1\sqrt{\omega_0}} - 2\omega_0^{1/2}Ay\right)^{-3\gamma+1}. \qquad (3.8.140)$$

An alternative splitting of (3.8.22) leads to

$$n z_{xx} + \phi(y) = 0 \qquad (3.8.141)$$

$$z_{yy} + 2z_x z_{xy} - \frac{1}{n}\phi(y)z_x^2 = 0. \qquad (3.8.142)$$

Equation (3.8.141) integrates immediately to give

$$z(x,y) = -\frac{1}{n}\left[\frac{1}{2}\phi(y)x^2 + V(y)x + W(y)\right].$$

(3.8.143)

To get the functions $\phi(y), V(y)$, and $W(y)$, we substitute (3.8.143) into (3.8.142) and equate coefficients of x^2, x, x^0 to zero. We get

$$n^2\phi'' - 4n\phi\phi' + 2\phi^3 = 0$$

(3.8.144)

$$n^2V'' - 2n\frac{d}{dy}(\phi V') + 2\phi^2V = 0$$

(3.8.145)

$$n^2W'' - 2nVV' + \phi V^2 = 0.$$

(3.8.146)

The system (3.8.144) - (3.8.146) integrates consecutively to give

$$\phi(y) = -\frac{\gamma-1}{\gamma}\frac{\beta}{\alpha+\beta y}$$

$$V(y) = b_4(\alpha+\beta y)^{-1} + b_5$$

(3.8.147)

$$W(y) = \frac{\gamma}{\gamma-1}\left[b_6 y + b_7 + \frac{b_5^2}{\beta}(\alpha+\beta y)[-1 + \log(\alpha+\beta y)]\right.$$

$$\left. -\frac{1}{2\beta}b_4^2(\alpha+\beta y)^{-1}\right]$$

(3.8.148)

where b_4, b_5, b_6, and b_7 are arbitrary constants. The solution (3.8.143) of (3.8.22) can now be written as

$$z(x,y) = -\frac{1}{n}\left\{-\frac{\gamma-1}{2\gamma}\frac{\beta}{\alpha+\beta y}x^2 + (b_4(\alpha+\beta y)^{-1} + b_5)x\right.$$

$$+\frac{\gamma}{\gamma-1}\left[(b_6 y + b_7) + \frac{b_5^2}{\beta}(\alpha + \beta y)[-1 + \log(\alpha+\beta y)]\right.$$

$$\left.\left. -\frac{1}{2\beta}b_4^2(\alpha+\beta y)^{-1}\right]\right\}.$$

(3.8.149)

The equation of state in the present case becomes

$$p = (\gamma-1)^\gamma(\alpha+\beta y)^{1-\gamma}\rho^\gamma.$$

(3.8.150)

The main point of finding the whole gamut of solutions for the complicated nonlinear PDE (3.8.22) is to show that even though it seems to have no apparent symmetries, it admits considerable analysis – by Lie group transformations, direct similarity approach, method of intermediate integrals, and the ad hoc approach of equation-splitting. The nonlinear ODEs (3.8.122) - (3.8.124) governing the quadratic form (3.8.121) of the solution has not been solved generally; it seems difficult to handle analytically. But even its special cases give rise to a variety of exact solutions of (3.8.22).

Sachdev et al. (1997) also demonstrated the effectiveness of the direct similarity approach which recovers all the solutions obtained by the classical Lie group approach. More recently it has been shown that the direct approach and the general group theoretic approach do not always lead to the same results (Ludlow, Clarkson and Bassom (1999)).

Thus, one must explore all methods of solving a nonlinear problem analytically before resorting to numerical solution and relying upon the computer-generated numbers alone. However, the latter may give a clue to what kind of exact or asymptotic solutions one may look for.

Here we have not treated specific shocked solutions, but it is not difficult to generalise those found earlier by Ustinov (1986). All his solutions are contained in the large families of solutions reported here.

Chapter 4

Exact Travelling Wave Solutions

4.1 Travelling Wave Solutions

1. A wave form which describes waves of permanent shape, that is, waves that do not change with time, is often written as

$$u(x,t) = f(x - ct) \tag{4.1.1}$$

where $c > 0$ is a constant. This profile or wave has form $f(x)$ at $t = 0$ and simply (and bodily) propagates as it is, with speed c to the right. $f(x + ct)$ describes a similar wave moving to the left with speed c. It is easy to see that (4.1.1) is a solution of the first-order linear PDE (often called advection equation)

$$u_t + cu_x = 0. \tag{4.1.2}$$

As we shall see later, in more complicated situations c may depend on u. Then (4.1.2) describes the well-known phenomenon of shock formation from a smooth initial profile (see Equation (2.2.10)). The function f is usually required to hold for all time t, and satisfy some finite boundary conditions at $x = +\infty$ and $x = -\infty$:

$$f(+\infty, t) = c_1, \quad f(-\infty, t) = c_2, \tag{4.1.3}$$

where c_1 and c_2 are finite constants. Not all PDEs may have solutions of the form (4.1.1) satisfying (4.1.3). The simplest example of this kind is the diffusion equation

$$u_t = Du_{xx} \tag{4.1.4}$$

where D is a constant. If we write $u = f(\eta), \eta = x - ct$, then (4.1.4) reduces to

$$-cf'(\eta) - Df''(\eta) = 0, \quad -\infty < \eta < \infty. \tag{4.1.5}$$

Equation (4.1.5) easily integrates to give

$$f(\eta) = a + b \exp\left(-\frac{c\eta}{D}\right) \tag{4.1.6}$$

where a and b are arbitrary constants.

To satisfy (4.1.3) we must have $b = 0$. Therefore, there is no nonconstant travelling wave solution of the diffusion equation (4.1.4): simple diffusion is unable to transmit information of the present kind over long distances. As we shall show later, if nonlinear convection and/or reaction kinetics are coupled with diffusion, meaningful travelling wave solutions do exist.

We shall discuss in the sequel travelling or simple waves for hyperbolic and diffusive systems. For historical reasons, we first discuss the simple wave solutions in the context of one-dimensional gasdynamics, which we shall generalize later to systems of both homogeneous and inhomogeneous hyperbolic systems.

4.2 Simple Waves in 1-D Gasdynamics

The equations governing compressible flows in one dimension are

$$\rho_t + u\rho_x + \rho u_x = 0 \tag{4.2.1}$$

$$u_t + uu_x + \frac{1}{\rho}p_x = 0 \tag{4.2.2}$$

$$S_t + uS_x = 0 \tag{4.2.3}$$

where ρ, u, p, and S are density, particle velocity, pressure, and entropy at a position x and time t. If we assume that no shocks are present and the gas is homogeneous at some initial time such that $S = $ constant, then (4.2.3) implies that S is constant at all later times and we have to deal with (4.2.1) and (4.2.2) along with the relation $c^2 = \left(\dfrac{\partial p}{\partial \rho}\right)_S$, giving the square of the speed of sound. Equation (4.2.2) can then be written as

$$u_t + uu_x + \frac{c^2}{\rho}\rho_x = 0. \tag{4.2.4}$$

Now we use an argument due originally to Earnshaw (1858). We look for solutions of (4.2.1) and (4.2.4) in the form $p = p(\rho)$, $u = V(\rho)$; this system then reduces to

$$\rho_t + (V + \rho V')\rho_x = 0 \tag{4.2.5}$$

$$(\rho_t + V\rho_x)V' + \frac{c^2}{\rho}\rho_x = 0. \tag{4.2.6}$$

Equations (4.2.5)-(4.2.6) are consistent only if

$$V' = \pm\frac{c}{\rho}. \tag{4.2.7}$$

Equations (4.2.5)-(4.2.6) are consistent only if

$$V' = \pm \frac{c}{\rho}. \tag{4.2.7}$$

In view of (4.2.7), (4.2.5) becomes

$$\rho_t + (V \pm c)\rho_x = 0. \tag{4.2.8}$$

On using the definition $c^2(\rho) = \gamma p/\rho = k\rho^{\gamma-1}$ in (4.2.7), where k is a constant, we get

$$u = V(\rho) = \int_{\rho_0}^{\rho} \frac{c(\rho)}{\rho} d\rho = \frac{2}{\gamma - 1}[c(\rho) - c_0]. \tag{4.2.9}$$

Using (4.2.9) and the relation between c and ρ above, we arrive at an exact equation in u alone governing simple wave motions in isentropic 1-D gasdynamic flows:

$$u_t + \left(c_0 + \frac{\gamma + 1}{2}u\right)u_x = 0. \tag{4.2.10}$$

Equation (4.2.10) can be reduced to the form $U_t + UU_x = 0$ by a simple change of variables (see Section 2.2 for a discussion of (4.2.10)). The solution of (4.2.10) by the method of characteristics has the implicit form

$$x = t\left(c_0 + \frac{\gamma + 1}{2}u\right) + f(u) = t(c + u) + f(u) \tag{4.2.11}$$

where $f(u)$ is an arbitrary function. A similar solution may be found if we choose the lower sign in (4.2.7) and (4.2.8). For a given value of u, we may write (4.2.11) as $x = At + B$ so that u is constant along this line and, hence, in this sense (4.2.11) represents a travelling wave. We have (using (4.2.7) and (4.2.8)) two types of waves moving with speeds $c - u$ and $c + u$, respectively. Using thermodynamic relations for an isentropic flow and (4.2.9), we have the corresponding results for other physical variables:

$$c = c_0 \pm \frac{1}{2}(\gamma - 1)u \tag{4.2.12}$$

$$\rho = \rho_0 \left(1 \pm \frac{1}{2}\frac{(\gamma - 1)u}{c_0}\right)^{2/(\gamma-1)} \tag{4.2.13}$$

$$p = p_0 \left(1 \pm \frac{1}{2}\frac{(\gamma - 1)u}{c_0}\right)^{2\gamma/(\gamma-1)}. \tag{4.2.14}$$

If it is possible to invert the function $f(u)$, we may write (4.2.11) in a more transparent form:

$$u = F\left[x - t\left(c_0 + \frac{(\gamma + 1)}{2}u\right)\right]. \tag{4.2.15}$$

The solution (4.2.15) shows that for the nonlinear PDE (4.2.10), the profile has a wave speed which itself depends on u and is responsible for the phenomenon of wave breaking and shock formation. In the special case $f(u) = 0$ in (4.2.11), u (and hence c, ρ, and p) are functions of $(x/c_0 t)$ and we have a centered simple wave. One could directly find this solution from the original system (4.2.2) and (4.2.4) by assuming that all the functions u, p, and ρ depend on the combination $(x/c_0 t)$ only, that is, seek the self-similar solution with $x/c_0 t$ as the similarity variable. It is again not difficult to see that the travelling wave solutions of the form

$$u = f(\zeta - \lambda\tau + c), \qquad (4.2.16)$$

where ζ and τ are variables and λ and c are arbitrary constants, can be put in a self-similar form by the change of variables

$$\zeta = \ln x, \quad \tau = \ln t, \qquad (4.2.17)$$

and renaming c as $- \ln A$. Then (4.2.16) becomes

$$u = f\left(\ln \frac{x}{At^\lambda}\right) = F\left(\frac{x}{At^\lambda}\right), \qquad (4.2.18)$$

the form $F\left(\dfrac{x}{At}\right)$ of a centered simple wave with $\lambda = 1$ being a special case (see Barenblatt (1979)).

Several physical problems governed by the nonlinear PDE

$$u_t + uu_x = 0, \qquad (4.2.19)$$

or its generalization $u_t + u^n u_x = 0$, may have solutions expressible as travelling waves or in a self-similar form. Here we consider a simple IVP:

$$u(x, 0) = 0 \text{ if } x < 0; \quad u(x, 0) = 1 \text{ if } x > 0.$$

For this Riemann problem one may easily draw the characteristic diagram. As the characteristic relations

$$\frac{dt}{1} = \frac{dx}{u} = \frac{du}{0} \qquad (4.2.20)$$

shows, u is constant along the characteristic $\dfrac{dx}{dt} = u$. It is clear that the data $u = 1$ is carried into the region $x > t$ along the characteristics $x = t + x_0$ issuing from various points x_0. Similarly, the value $u = 0$ is carried along the lines $x = x_0$ into the region $x_0 < 0$. There is a region $0 < x < t$ which remains void of characteristics. This void can be filled by drawing characteristics, all passing through 0, in the form of a centered fan: $x = ct, 0 < c < 1$. Each of these characteristics carries the value $u = c$ along it. Thus the solution may be made up as

$$\begin{aligned}
u(x,t) &= 0 \text{ if } x < 0 \\
u(x,t) &= \frac{x}{t} \text{ if } 0 < x/t < 1 \\
u(x,t) &= 1 \text{ if } x > t.
\end{aligned} \qquad (4.2.21)$$

The solution (4.2.21) is self-similar, depending on x/t only, and describes a centered expansion fan or a rarefaction wave, the fan spreading out from the point $x = 0$.

Other solutions of (4.2.19) and its generalisations describing flows with or without shocks may be found in Section 2.5.

4.3 Elementary Nonlinear Diffusive Travelling Waves

We observed in Section 4.1 that the linear diffusion equation (4.1.4) therein does not support a travelling wave satisfying the end conditions (4.1.3). We shall now add various nonlinear physical terms to that equation to see how the situation changes. First, if we put in a nonlinear convection term (see (4.2.19)), we get the celebrated Burgers equation

$$u_t + u u_x = D u_{xx} \qquad (4.3.1)$$

which is the simplest equation describing the balance between nonlinear convection and linear diffusion (see Sachdev (1987) for a formal derivation of (4.3.1) in the gas dynamic context); the diffusion coefficient D is small but finite. We seek what are referred to as shock wave solutions of (4.3.1), assuming values u_1, u_2 as $x \to +\infty$ and $-\infty$, respectively. Introducing the travelling wave form $u = u(x - Ut) \equiv u(\xi)$ into (4.3.1), we have

$$-U u_\xi + u u_\xi = D u_{\xi\xi}. \qquad (4.3.2)$$

Equation (4.3.2) must satisfy the end conditions $u \to u_1, u_2$ as $\xi \to +\infty, -\infty$, respectively. Integrating this equation once we have

$$\frac{1}{2} u^2 - U u + C = D u_\xi \qquad (4.3.3)$$

where C is the constant of integration. Imposing the conditions at $\xi = \pm\infty$, we get

$$\frac{1}{2} u_1^2 - U u_1 = \frac{1}{2} u_2^2 - U u_2 = -C. \qquad (4.3.4)$$

Equation (4.3.4) yields the constants

$$U = \frac{1}{2}(u_1 + u_2), \quad C = \frac{1}{2} u_1 u_2. \qquad (4.3.5)$$

Putting (4.3.5) into (4.3.3), we get

$$(u - u_1)(u_2 - u) = -2Du_\xi \tag{4.3.6}$$

which, on integration, gives

$$\frac{\xi}{2D} = \frac{1}{u_2 - u_1} \ln \frac{u_2 - u}{u - u_1}. \tag{4.3.7}$$

On solving (4.3.7) for u, we have

$$u = u_1 + \frac{u_2 - u_1}{1 + \exp\left[\frac{u_2 - u_1}{2D}(x - Ut)\right]}, \quad U = \frac{u_1 + u_2}{2}. \tag{4.3.8}$$

Equation (4.3.8) describes the structure of a uniformly moving shock with end conditions u_1 and u_2, its speed being the average of these values: $u = \frac{u_1 + u_2}{2}$. The solution (4.3.8), in the terminology of Zel'dovich and Raizer (1967), is a self-similar solution of the first kind since it becomes fully known in terms of the data, the end conditions u_1 and u_2. It is also an intermediate asymptote to which a class of solutions arising out of a certain special set of initial conditions with the asymptotically same end conditions converge as $t \to \infty$. This can be shown with the help of the following example. Consider the initial conditions

$$u(x, 0) = f(x) = \begin{cases} u_1 & x > 0 \\ u_2 & x < 0 \end{cases} \tag{4.3.9}$$

where $u_2 > u_1$. In view of the Cole-Hopf transformation (see Section 6.2), the conditions for the corresponding transformed (heat) equation become

$$\phi(x, 0) = \begin{cases} e^{-u_1 x/2D} & x > 0 \\ e^{-u_2 x/2D} & x < 0. \end{cases} \tag{4.3.10}$$

Substituting (4.3.10) in the general solution of the heat equation and using the Cole-Hopf transformation again, we arrive, after some rearrangement (see Sachdev (1987), p. 25), at the solution

$$u = u_1 + \frac{u_2 - u_1}{1 + \left\{\exp\left[\frac{u_2 - u_1}{2D}\left(x - \frac{u_1 + u_2}{2}t\right)\right]\right\} \frac{\int_{-(x - u_1 t)}^{\infty} e^{-v^2/4Dt}\, dy}{\int_{x - u_2 t}^{\infty} e^{-v^2/4Dt}\, dy}}. \tag{4.3.11}$$

If we let $t \to \infty$ in the integrals in (4.3.11) such that $u_1 < x/t < u_2$, the lower limits become large and negative so that the ratio of the integrals tends to 1. Under this limiting process, the solution (4.3.11) tends to the travelling wave solution (4.3.8). Here is one set of initial conditions (4.3.9) subject to which the solution of the Burgers equation (4.3.1) tends in the limit $t \to \infty$ to the travelling wave solution. This result was more rigorously proved by Oleinik (1957).

The solution (4.3.8) is referred to as the "Taylor shock structure," since it was first derived by G.I. Taylor (1910). This solution plays an important role in the construction of matched asymptotic solutions for generalised Burgers equations by Crighton and his collaborators (see Crighton and Scott (1979), for example).

Instead of the nonlinear convective term, if we add a reaction term to the linear diffusion equation we get, instead of (4.3.1), the most well-known reaction-diffusion equation

$$u_t = ku(1-u) + Du_{xx} \qquad (4.3.12)$$

where k and D are positive parameters. Equation (4.3.12) was proposed by Fisher (1937) as a deterministic model for the spatial spread of a favoured gene in a population, and is one of the best studied equations for travelling waves. The most classical paper in this context is that of Kolomogoroff, Petrovsky, and Piscounoff (1937), who investigated a more general model with $f(u)$ replacing $ku(1-u)$ but retaining all the properties of the latter function.

Although (4.3.12) is referred to as the "Fisher's equation," there is much history to it (see Murray (1989)), particularly in the context of chemical waves.

If we introduce the variables

$$t^* = kt, \quad x^* = x \left(\frac{k}{D} \right)^{1/2} \qquad (4.3.13)$$

into (4.3.12) and omit asterisks, we obtain the equation

$$u_t = u(1-u) + u_{xx} \qquad (4.3.14)$$

without any parameters. In the spatially homogeneous case ($u_{xx} = 0$), (4.3.14) has two steady states $u = 0, u = 1$ which can be shown to be respectively stable and unstable. Since $u > 0$ from physical description, we seek a travelling wave solution of (4.3.14) for which $0 \le u \le 1$. We seek a travelling wave solution of (4.3.14) over the whole real line; we shall later discuss how this wave actually arises from a certain class of initial conditions. Putting

$$u(x,t) = U(z), \quad z = x \quad ct \qquad (4.3.15)$$

where c is a real constant, which, for convenience, we choose to be positive. The wave speed c is unknown at this stage and must be determined from the analysis. Substituting (4.3.15) into (4.3.14), we have

$$-cU' - U'' = U(1-U), \quad ' = \frac{d}{dz}. \qquad (4.3.16)$$

Equation (4.3.16) cannot be solved in a closed form, but, since it is autonomous, it can be discussed in the phase plane (It also admits considerable approximate analysis). Writing (4.3.16) as the system

$$U' = V$$
$$V' = -cV - U(1 - U), \tag{4.3.17}$$

we immediately observe that it has two critical points in the U-V plane: $P : (0,0)$ and $Q : (1,0)$. Linearising the system (4.3.12) about each point and finding the corresponding eigenvalues etc. (see Sachdev (1991)), we find that the eigenvalues of the linear systems are as follows. For $P(0,0)$, we have

$$\lambda_\pm = \frac{1}{2}\left[-c \pm (c^2 - 4)^{1/2}\right], \tag{4.3.18}$$

therefore, P is a stable node if $c^2 \geq 4$ and a stable spiral if $c^2 < 4$. The point $Q(1,0)$ has eigenvalues

$$\lambda_\pm = \frac{1}{2}\left[-c \pm (c^2 + 4)^{1/2}\right] \tag{4.3.19}$$

and, therefore, is always a saddle point. The phase portrait for $c > 2$ is shown in Figure 4.1. It can be seen that the unique separatrix joining the saddle point $Q(1,0)$ with the node $P(0,0)$ gives a unique solution. This solution in the (U, z) plane is shown in Figure 4.2. It is well known (see Sachdev (1991)) that the independent variable z must tend either to $+\infty$ or $-\infty$ as the trajectory enters or leaves a critical point. As the direction of the trajectory shows, $U \to 1$ as $z \to -\infty$ as the point Q is approached while $U \to 0$ as $z \to +\infty$ as the point P is attained. Moreover, $U' < 0$ and tends to 0 as

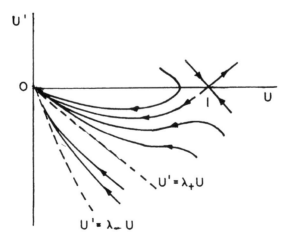

Figure 4.1. Phase plane trajectories for equation (4.3.16) for the travelling wave front solution with $c^2 > 4$.

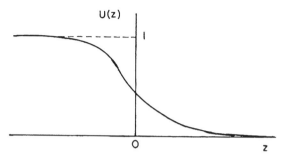

Figure 4.2. Travelling wave front for equation (4.3.16) where $c \geq 2$.

$|z| \to \infty$, thus we have a monotonic solution joining the end points $u = 1, 0$, as shown in Figure 4.2.

Since the physical model (describing chemical reaction or population growth) demands $U > 0$, the other case $c < 2$, which gives P as a stable spiral, leading both to positive and negative values of U, is unrealistic and is therefore ignored.

As for the travelling wave solution of the Burgers equation, we enquire what kind of initial conditions for $u(x, 0)$ will evolve to the travelling waves discussed here. This question was answered by Kolmogoroff et al. (1937): if $u(x, 0)$ has compact support

$$u(x,0) = u_0(x) \geq 0, \quad u_0(x) = \begin{cases} 1 & x \leq x_1 \\ 0 & x \geq x_2 \end{cases} \tag{4.3.20}$$

where $x_1 < x_2$ and $u_0(x)$ is continuous in $x_1 < x < x_2$, then the solution $u(x, t)$ of (4.3.14) evolves to a travelling wave front solution $U(z)$ with $z = x - 2t$, i.e., it evolves to the wave solution with minimum speed $c_{\min} = 2$. For initial data other than (4.3.20), the asymptotic (wave) solution depends critically on the behaviour of $u(x, 0)$ as $x \to \pm\infty$.

The critical dependence of the solution on the initial data at infinity can be seen easily by the following simple argument due to Mollison (1977). At the leading edge of the wave $(x \to \infty)$, the wave amplitude u is small, therefore the u^2 term in (4.3.14) may be neglected and we have the linear PDE

$$u_t = u + u_{xx} \tag{4.3.21}$$

Considering the initial condition

$$u(x,0) \sim Ae^{-ax} \quad \text{as} \quad x \to +\infty \tag{4.3.22}$$

where $a, A > 0$ are arbitrary, we may easily find a travelling wave solution of (4.3.21) in the form

$$u(x,t) = Ae^{-a(x-ct)}. \tag{4.3.23}$$

The form (4.3.23) satisfies (4.3.21) provided

$$c = a + \frac{1}{a}. \tag{4.3.24}$$

From (4.3.24) it follows that c as a function of a attains its minimum at $a = 1$, i.e., $c_{min} = 2$. For all other values of $a(> 0)$ the wave speed $c > 2$. Let us consider $\min[e^{-ax}, e^{-x}]$ for x large and positive. Two cases arise. If $a < 1$, then $e^{-ax} > e^{-x}$, and the velocity (4.3.24) of propagation with asymptotic initial conditions like (4.3.22) will depend on the leading edge of the wave. If, on the other hand, $a > 1$, then e^{-ax} is bounded above by e^{-x} (corresponding to $a = 1$); the wave front speed in this case is $c = 2$. To summarise,

$$c = a + \frac{1}{a}, \quad 0 < a \leq 1; \quad c = 2, \quad a \geq 1. \tag{4.3.25}$$

These results have been proved more rigorously by Mckean (1975) and Larson (1978) and have been verified numerically by Manoranjan and Mitchell (1983).

Equation (4.3.16) admits a rather neat perturbation analysis (Logan (1994)) based on the fact that $c \geq 2$. We seek solution of (4.3.16) over $-\infty < z < \infty$ with the boundary condition

$$U(-\infty) = 1, \quad U(+\infty) = 0. \tag{4.3.26}$$

Since (4.3.16) is autonomous, if $U(z)$ is a solution, so is $U(z + z_0)$. We translate the solution curve suitably such that

$$U(0) = \frac{1}{2}. \tag{4.3.27}$$

Now identify the small parameter

$$\epsilon = \frac{1}{c^2}. \tag{4.3.28}$$

In terms of ϵ, Equation (4.3.16) becomes

$$\sqrt{\epsilon}U'' + U' + \sqrt{\epsilon}U(1 - U) = 0. \tag{4.3.29}$$

To the lowest order (4.3.29) gives $U =$ constant, an appropriate form in the neighbourhood of $z = +\infty$ or $-\infty$ (see Equation (4.3.26)). To shrink the infinite domain we introduce the variable

$$s = \sqrt{\epsilon}z = \frac{z}{c} \tag{4.3.30}$$

so that the solution has the form

$$g(s) = U\left(\frac{s}{\sqrt{\epsilon}}\right). \tag{4.3.31}$$

Equation (4.3.29) now becomes

$$\epsilon g'' + g' + g(1 - g) = 0, \quad ' = \frac{d}{ds}. \tag{4.3.32}$$

Although (4.3.32) looks like a singular ODE, the last two terms, which dominate, have the appropriate form for the boundary conditions:

$$g(-\infty) = 1, \quad g(0) = \frac{1}{2}, \quad g(+\infty) = 0. \tag{4.3.33}$$

Writing the solution in the (regular) perturbation form

$$g(s) = g_0(s) + \epsilon g_1(s) + \epsilon^2 g_2(s) + \ldots \tag{4.3.34}$$

and substituting it in (4.3.32) and (4.3.33), we get

$$g_0' = -g_0(1 - g_0), g_0(0) = \frac{1}{2}, g_0(-\infty) = 1, g_0(+\infty) = 0 \tag{4.3.35}$$

$$g_1' = -g_1(1 - g_0) + g_0'', g_1(-\infty) = g_1(0) = g_1(+\infty) = 0, \text{ etc.}$$

The system (4.3.35) can be solved explicitly:

$$
\begin{aligned}
g_0(s) &= (1 + e^s)^{-1} \\
g_1(s) &= e^s(1 + e^s)^{-2} \ln \frac{4e^s}{(1 + e^s)^2}, \text{ etc.}
\end{aligned}
\tag{4.3.36}
$$

In terms of the original variable z, we can write

$$U(z) = \frac{1}{1 + e^{z/c}} + \frac{1}{c^2} e^{z/c}(1 + e^{z/c})^{-2} \ln \frac{4e^{z/c}}{(1 + e^{z/c})^2} + O\left(\frac{1}{c^4}\right). \tag{4.3.37}$$

The solution (4.3.37) is found to be remarkably accurate even for $c = 2$ for all z, for which it is supposed to be least accurate. The first term $(1+e^{z/c})^{-1}$ itself is found to be remarkably close to a numerically-computed solution, even for $c = 2$ (Logan (1994)).

A similar analysis for (4.3.16) was carried out in the phase plane by Canosa (1973). He showed that the travelling wave solutions of the Fisher's equation are stable to perturbations of compact support, that is, to perturbations which are zero outside a finite domain including the wave front; the perturbations are also required to be in the moving frame of reference. On the other hand, the simple solution (4.3.23) for large x also shows that the speed of propagation of the wavefront solutions depends sensitively on the explicit behaviour of the initial conditions as $|x| \to \infty$, implying that the travelling wave solutions are unstable to perturbations in the far field (see also Sachdev (1987)). Since most of the numerical simulations are carried out with a finite extent of the initial conditions, and since travelling waves of Fisher's equation are stable to disturbances on finite domains, the large time solutions of IVP come out to be stable wave front solutions with speed $c = 2$.

Fisher's equation has been generalised in diverse ways. We first consider a couple of simple generalisations which admit exact solutions (Murray (1989)). The equation

$$u_t = u(1 - u^q) + u_{xx}, q > 0 \tag{4.3.38}$$

is one such extension. Since $u = 0$ and $u = 1$ are steady-state solutions of (4.3.38), we look for its travelling wave solutions

$$u(x, t) = U(z), z = x - ct, \quad U(-\infty) = 1, U(\infty) = 0 \tag{4.3.39}$$

where the wave speed $c > 0$ must be found as part of the solution. Substituting (4.3.39) into (4.3.38), we get

$$U'' + cU' + U(1 - U^q) = 0. \tag{4.3.40}$$

Equation (4.3.40), like (4.3.16) for the Fisher's equation, may be studied in the phase plane. Inspired by the good accuracy of the first term in the solution (4.3.37) for the Fisher's equation, we seek a solution of (4.3.40) in the form

$$U(z) = \frac{1}{(1 + ae^{bz})^s} \tag{4.3.41}$$

where a, b, and s are positive constants to be found. The conditions (4.3.26) for (4.3.40) are automatically satisfied by (4.3.41). Substituting (4.3.41) into (4.3.40) we get

$$(1 + ae^{bz})^{-s-2} \left\{ [s(s + 1)b^2 - sb(b + c) + 1]a^2 e^{2bz} \right.$$
$$\left. + [2 - sb(b + c)]ae^{bz} + 1 - [1 + ae^{bz}]^{2-sq} \right\}. \tag{4.3.42}$$

The coefficients of e^0, e^{bz}, and e^{2bz} in the curly bracket must identically be zero. This gives the balances

$$2 - sq = 0, 1 \text{ or } 2, \tag{4.3.43}$$

that is,

$$s = \frac{2}{q}, \frac{1}{q} \text{ or } sq = 0. \tag{4.3.44}$$

Since s and q are positive, $sq = 0$ is not admissible. Considering $s = 1/q$, we have the following coefficients of e^{bz} and e^{2bz}:

$$e^{bz}: \quad 2 - sb(b + c) - 1 = 0 \tag{4.3.45}$$
$$e^{2bz}: \quad s(s + 1)b^2 - sb(b + c) + 1 = 0. \tag{4.3.46}$$

Equations (4.3.45) and (4.3.46) imply that

$$s(s+1)b^2 = 0 \qquad (4.3.47)$$

or $b = 0$ since $s > 0$. This case also is not possible since we must have $b > 0$. If $s = 2/q$, the coefficients are

$$
\begin{array}{ll}
e^{bz} & : \quad sb(b+c) = 2 \\
e^{2bz} & : \quad s(s+1)b^2 = 1
\end{array}
\qquad (4.3.48)
$$

which together yield

$$s = \frac{2}{q}, \quad b = [s(s+1)]^{-1/2}, \quad c = \frac{2}{sb} - b. \qquad (4.3.49)$$

In terms of q we have

$$s = \frac{2}{q}, \quad b = q[2(q+2)]^{-1/2}, \quad c = (q+4)[2(q+2)]^{-1/2}, \qquad (4.3.50)$$

the last in (4.3.50) defining the wave speed c uniquely. Let us see what the special solution (4.3.41) with (4.3.50) reduces to for the special case $q = 1$, corresponding to the Fisher's equation. In this case we get

$$s = 2, \quad b = \frac{1}{\sqrt{6}}, \quad c = \frac{5}{\sqrt{6}}. \qquad (4.3.51)$$

The arbitrary constant a in (4.3.41) can be chosen so that $z = 0$ corresponds to $1/2$; this leads to the value $a = \sqrt{2} - 1$ and the exact solution (4.3.41) for the Fisher's equation becomes

$$U(z) = \frac{1}{[1 + (\sqrt{2} - 1)e^{z/\sqrt{6}}]^2}. \qquad (4.3.52)$$

The expansion solution (4.3.37), in contrast, is so different, although the first term therein is a good approximation to the exact solution. The wave speed here is $\frac{5}{\sqrt{6}} \approx 2.04$ in contrast to 2 for (4.3.37). It is not clear what the role of the special (exact) solution (4.3.52) is in the class describing the asymptotic character of the solutions of IVPs for the Fisher's equation.

It is interesting to note (and it may be verified as for (4.3.38)) that

$$U(z) = \frac{1}{(1 + ae^{bz})^s}, \quad s = \frac{1}{q}, \quad b = \frac{q}{(q+1)^{1/2}}, \quad c = (q+1)^{-1/2} \quad (4.3.53)$$

is an exact solution of the generalised Fisher's equation

$$u_t = u^{q+1}(1 - u^q) + u_{xx}. \qquad (4.3.54)$$

Sherratt and Marchant (1996) have considered the case

$$u_t = u_{xx} + u^2(1-u) \tag{4.3.55}$$

in some detail. They find that the travelling wave solutions of (4.3.55) have algebraically decaying tails and wavefronts with varying shape and speed. This was demonstrated mainly by the numerical solution of the appropriate initial value problem.

We now discuss the model which combines both the above effects, namely nonlinear convection and simplest reaction, that is, the Burgers-Fisher equation

$$u_t + kuu_x = u(1-u) + u_{xx}. \tag{4.3.56}$$

The parmeteric value $k = 0$ reduces (4.3.56) to the Fisher equation. We seek the travelling wave solution of (4.3.56),

$$u(x,t) = U(z), \quad z = x - ct, \tag{4.3.57}$$

with c as the wave speed to be determined subject to the boundary conditions as in (4.3.39). Equation (4.3.56) in the variable z reduces to

$$U'' + (c - kU)U' + U(1-U) = 0 \tag{4.3.58}$$

and the boundary conditions become

$$\lim_{z \to \infty} U(z) = 0 \quad \lim_{z \to -\infty} U(z) = 1. \tag{4.3.59}$$

With $V = U'$, (4.3.58) in the (U,V) plane becomes

$$\frac{dV}{dU} = \frac{-(c - kU)V - U(1-U)}{V}. \tag{4.3.60}$$

In the present case we look for the relation $c = c(k)$ such that (4.3.56) has a monotonic solution with $0 \le U \le 1$ and $U'(z) \le 0$. Usual linear analysis shows that the condition $c \ge 2$ guarantees that $(0,0)$ is a node while $(1,0)$ is a saddle point. Murray (1977, 1989) has shown that a solution for (4.3.60) subject to (4.3.59) exists for all $c = c(k)$ where

$$c(k) = \begin{cases} 2 & 2 > k > -\infty \\ \frac{k}{2} + \frac{2}{k} & 2 \le k < \infty. \end{cases} \tag{4.3.61}$$

We conclude this section by transforming (4.3.56) into a form where the diffusion term becomes less important. With $k > 0$, we set

$$\epsilon = \frac{1}{k^2}, \quad y = \frac{x}{k} = \epsilon^{1/2}x \tag{4.3.62}$$

so that (4.3.56) becomes

$$u_t + uu_y = u(1-u) + \epsilon u_{yy}. \tag{4.3.63}$$

If $u(x, t)$ is a solution of (4.3.56), then $u(ky, t)$ is a solution of (4.3.63). The solution $U(x - ct)$ of (4.3.56) satisfying $U(-\infty) = 1$, $U(\infty) = 0$ becomes the solution $U(ky - ct)$ of (4.3.63) with the wave speed $\lambda = c/k = c\epsilon^{1/2}$. Therefore, using the wave speed estimates (4.3.61), we find that (4.3.63) has travelling wave solutions for all

$$\lambda \geq \lambda(\epsilon) = \frac{c(k)}{k} = c(\epsilon^{-1/2})\epsilon^{1/2} \tag{4.3.64}$$

and so

$$\lambda(\epsilon) = \begin{cases} 2\epsilon^{1/2} & \epsilon > \frac{1}{4} \\ \frac{1}{2} + 2\epsilon & \frac{1}{4} \geq \epsilon > 0. \end{cases} \tag{4.3.65}$$

In the limit $\epsilon \to 0$, (4.3.63) becomes

$$u_t + uu_y = u(1 - u). \tag{4.3.66}$$

Equation (4.3.66) can develop discontinuous (shock) solutions. For ϵ small, although (4.3.63) can develop (steep) shock-like solutions, it does not have discontinuous travelling wave solutions (see Murray (1989)).

Another generalisation of the Fisher's equation

$$u_t = \frac{\partial}{\partial x}(D(u)u_x) + f(u) \tag{4.3.67}$$

has attracted much attention; here, typically $D(u) = D_0 u^m$ and $f(u) = ku^p(1-u^q)$, where D_0, m, p, and q are positive constants (Murray (1989)). Here we summarize the results for a special case of (4.3.67), namely

$$u_t = (uu_x)_x + u(1 - u) \tag{4.3.68}$$

(where the coefficients have been suitably scaled out) due to Sherrat and Marchant (1996) and Rothe (1978). Rothe (1978) proved that if $u(x, 0)$ decreases with x and tends to 1 as $x \to -\infty$, and $u(x, 0) = O(e^{-\xi x})$ as $x \to \infty$, then the solution tends to a travelling wave as $t \to \infty$; the speed of the wave is related to the initial conditions as follows:

$$c = \begin{cases} 2 & \text{if } \xi > 1 \\ \xi + \frac{1}{\xi} & \text{if } \xi < 1. \end{cases} \tag{4.3.69}$$

Sherrat and Marchant (1996), with Rothe's results in view, solved the IVP

$$u(x, 0) = \begin{cases} 1 & x < 0 \\ e^{-\xi x} & x > 0 \end{cases} \tag{4.3.70}$$

numerically. The solution did evolve to a travelling wave for $\xi \geq 2$. The travelling wave speed approaches the value $c \approx \dfrac{1}{\sqrt{2}}$, the minimum speed of the travelling wave solutions, according to their analysis, which have a sharp front. Unlike Fisher's equation, (4.3.68) admits both smooth and sharp-fronted solutions. We recall that sharp-fronted solutions are those for which u_x is discontinuous at some finite point x.

4.4 Travelling Waves for Higher-Order Diffusive Systems

We consider an example of a fourth-order nonlinear diffusive system which not only admits travelling waves but other phenomena, including some chaotic regimes. We follow closely the work of Merkin and Sadiq (1996). The system

$$a_t = D_A a_{xx} + m_A(a_0 - a) - kab^2, \qquad (4.4.1)$$

$$b_t = D_B b_{xx} + kab^2 - m_B b, \qquad (4.4.2)$$

describes in plane geometry a flow reactor, where a and b are the concentrations of reactant A and autocatalyst B, and k is a constant. It is assumed that the reactant A is flowing over the reactor at a constant concentration a_0, with A initially present at the same uniform concentration everywhere in the reactor. The reaction is then assumed to be initiated by some local input of autocatalyst. It is further assumed that the interchange of reactant A between the flow and the reactor and the loss of autocatalyst B from the reactor can both be described by linear diffusive mechanisms with mass-transfer coefficients m_A and m_B, respectively. Thus, the system (4.4.1)-(4.4.2) is to be solved subject to the following initial and boundary conditions:

$$a = a_0 \ (t = 0, -\infty < x < \infty)$$

$$b = \begin{cases} b_0 g(x) & \text{if } |x| < l \\ 0 & \text{if } |x| > l, \end{cases} \text{ at } t = 0 \qquad (4.4.3)$$

$a \to a_0$ and $b \to 0$ as $|x| \to \infty$ $(t \geq 0)$ where a_0 and b_0 are constants and $g(x)$ is continuous and differentiable on $|x| < l$ with a maximum value of unity.

Taking a_0, the concentration of A in the flow, as the control variable, we nondimensionalise (4.4.1)-(4.4.3) according to

$$\bar{a} = \frac{a}{a_0}, \bar{b} = \frac{b}{a_0}, \bar{t} = m_A t, \bar{x} = \left(\frac{m_A}{D}\right)^{1/2} x \qquad (4.4.4)$$

and drop the bars in the sequel. We get

$$a_t = a_{xx} + 1 - a - \mu ab^2 \qquad (4.4.5)$$

$$b_t = b_{xx} + \mu ab^2 - \phi b \qquad (4.4.6)$$

where

$$\mu = \frac{ka_0^2}{m_A}, \quad \phi = \frac{m_B}{m_A} \qquad (4.4.7)$$

are the nondimensional parameters. The bifurcation parameter in the following is μ since only this parameter involves a_0, the controlling parameter.

The initial and boundary conditions become

$$a = 1 \; (-\infty < x < \infty, t = 0) \tag{4.4.8}$$

$$b = \begin{cases} \beta_0 g(x) & \text{if } |x| < \sigma \\ 0 & \text{if } |x| > \sigma \end{cases} \; (t = 0) \tag{4.4.9}$$

$$b \to 0 \text{ and } a \to 1 \text{ as } |x| \to \infty (t \geq 0) \tag{4.4.10}$$

where $\beta_0 = b_0/a_0$ and $\sigma = \left(\dfrac{m_A l^2}{D}\right)^{1/2}$ are other dimensionless parameters. The behaviour of the solution is assumed to be symmetric about $x = 0$; it suffices, therefore, to discuss the region $x > 0$, with the symmetry conditions

$$a_x = b_x = 0 \quad \text{on } x = 0, t > 0. \tag{4.4.11}$$

Although our main concern is with travelling waves, we also carry out some other (simple) analysis which throws light on the system (4.4.5)-(4.4.6), generally. This is quite instructive and may be applied to other systems.

First we discuss the dynamical behaviour of (4.4.5)-(4.4.6) which is spatially homogeneous; we have the ODEs

$$\dot{a} = 1 - a - \mu a b^2, \tag{4.4.12}$$

$$\dot{b} = \mu a b^2 - \phi b, \tag{4.4.13}$$

in $a, b \geq 0, \mu > 0, \phi \geq 0$.

The system (4.4.12)-(4.4.13) has the stationary state

$$a_s = 1, \quad b_s = 0 \tag{4.4.14}$$

for all values of μ and ϕ. For $\phi = 0$, a simple balancing argument shows that the behaviour of (4.4.12)-(4.4.13) for $t \to \infty$ is

$$a \sim \frac{1}{\mu t^2} \quad \text{and} \quad b \sim t. \tag{4.4.15}$$

For $\phi > 0$, apart from the (trivial) stationary solution (4.4.14), we also have nontrivial stationary states

$$b_s^+ = \frac{\mu + (\mu^2 - 4\mu\phi^2)^{1/2}}{2\mu\phi}, \quad a_s^+ = \frac{\mu - (\mu^2 - 4\mu\phi^2)^{1/2}}{2\mu} \tag{4.4.16}$$

$$b_s^- = \frac{\mu - (\mu^2 - 4\mu\phi^2)^{1/2}}{2\mu\phi}, \quad a_s^- = \frac{\mu + (\mu^2 - 4\mu\phi^2)^{1/2}}{2\mu} \tag{4.4.17}$$

provided $\mu \geq 4\phi^2$. Usual linear analysis shows that (4.4.16) is a stable node for all μ and ϕ, while (4.4.17) is an (unstable) saddle point for all $\mu > 4\phi^2$.

The (local) stability of stationary state (a_s^+, b_s^+) is determined by the eigenvalues given by

$$\lambda^2 + \left(\frac{\mu}{\phi}b_s^+ - \phi\right)\lambda + \mu b_s^+ - 2\phi = 0. \qquad (4.4.18)$$

It then follows from (4.4.16) and (4.4.18) that there is a saddle-node bifurcation at

$$\mu_s = 4\phi^2 \quad (\phi > 0) \qquad (4.4.19)$$

and a Hopf bifurcation at

$$\mu_H = \frac{\phi^4}{\phi - 1} \quad (\phi > 2) \qquad (4.4.20)$$

with a double-zero-eigenvalue point at

$$\mu = 16, \quad \phi = 2. \qquad (4.4.21)$$

Our concern here is not with bifurcation analysis of (4.4.12)-(4.4.13). Merkin and Sadiq (1996) discuss how two limit cycles are created and subsequently destroyed at the periodic orbit bifurcation.

We now consider the PDE system (4.4.5)-(4.4.6) for the case $\phi = 0$. The asymptotic form (4.4.15) and the numerical solutions suggest the introduction of the variables

$$a = A(y, t)/t^2, b = tB(y, t), y = x/t. \qquad (4.4.22)$$

Substitution of (4.4.22) into (4.4.5)-(4.4.6) and expansion in inverse powers of t leads to the asymptotic solution

$$A - \frac{1}{\mu(1 - c_0 y)^2} - \frac{2c_1 t^{-1}}{\mu(1 - c_0 y)^2} + \dots \qquad (4.4.23)$$

$$B = (1 - c_0 y) + c_1 t^{-1} + \dots \qquad (4.4.24)$$

with constants c_0 and c_1 to be determined. The solution (4.4.23)-(4.4.24) does not hold near $x = 0$; it becomes singular at $y = \frac{1}{c_0}$, i.e., $x = \frac{t}{c_0}$. To consider that domain, we write, using a balancing argument again, the solution as

$$a = f(\eta, t)/t^2, b = tg(\eta, t), \eta = x/t^{1/2}. \qquad (4.4.25)$$

The functions f and g may be expanded in the form

$$f(\eta, t) = f_0(\eta) + t^{-1/2}f_1(\eta) + \dots \qquad (4.4.26)$$
$$g(\eta, t) = g_0(\eta) + t^{-1/2}f_1(\eta) + \dots \qquad (4.4.27)$$

The resulting equations must be solved and matched with (4.4.23) and (4.4.24) for $\eta \to \infty$, that is,

$$f \quad \sim \quad \frac{1}{\mu}(1 + 2c_0\eta t^{-1/2} + \ldots), \tag{4.4.28}$$

$$g \quad \sim \quad 1 - c_0\eta t^{-1/2} + \ldots, \tag{4.4.29}$$

as $\eta \to \infty$. After all the details are worked out, the result is

$$g_0 \quad = \quad 1, \quad f_0 = \frac{1}{\mu}, \tag{4.4.30}$$

$$g_1 \quad = \quad \frac{c_0}{\pi^{1/2}} \left(\eta \int_\eta^\infty \exp\left(-\frac{1}{4}s^2\right) ds - 2\exp\left(-\frac{1}{4}\eta^2\right) - \eta\pi^{1/4} \right), \tag{4.4.31}$$

$$f_1 \quad = \quad -\frac{2}{\mu}g_1. \tag{4.4.32}$$

Equations (4.4.25)-(4.4.32) give

$$a(0,t) \quad \sim \quad \frac{1}{\mu t^2}\left(1 - \frac{4c_0}{\pi^{1/2}}t^{-1/2} + \ldots\right) \tag{4.4.33}$$

$$b(0,t) \quad \sim \quad t\left(1 - \frac{2c_0}{\pi^{1/2}}t^{-1/2} + \ldots\right) \tag{4.4.34}$$

as $t \to \infty$.

To determine the constant c_0, we must consider the solution in the reaction-diffusion front, $x \sim t/c_0$. For this purpose, we introduce the moving coordinate

$$\xi = x - t/c_0 \tag{4.4.35}$$

leaving a and b unscaled. The system (4.4.5)-(4.4.6) (with $\phi = 0$) reduces to

$$a'' \quad + \quad c_0^{-1}a' + 1 - a - \mu ab^2 = 0, \tag{4.4.36}$$

$$b'' \quad + \quad c_0^{-1}b' + \mu ab^2 = 0, \tag{4.4.37}$$

where $' \equiv \dfrac{d}{d\xi}$. The system (4.4.36)-(4.4.37) must be solved subject to the conditions

$$a \to 1 \quad \text{and} \quad b \to 0 \quad \text{as} \quad \xi \to \infty \tag{4.4.38}$$

and matched with (4.4.23), that is,

$$a \sim \frac{1}{\mu c_0^2\xi^2} + \ldots \quad \text{and} \quad b \sim -c_0\xi + \ldots \quad \text{as} \quad \xi \to -\infty. \tag{4.4.39}$$

This would determine the wave speed c_0^{-1}.

It is clear from (4.4.39) that this form will not be accurate for small μ. For this case μ one must solve (4.4.36)-(4.4.39) in a different manner. Writing

$$b = \mu^{-1/2}\bar{b}, \quad c_0 = \mu^{-1/4}\bar{c}_0, \quad \bar{\xi} = \mu^{1/4}\xi, \tag{4.4.40}$$

(4.4.36)-(4.4.37) reduce to

$$1 - a - a\bar{b}^2 + \mu^{1/2}(a'' - (\bar{c}_0)^{-1}a') = 0 \tag{4.4.41}$$

$$\bar{b}'' + \bar{c}_0^{-1}\bar{b}' + a\bar{b}^2 = 0. \tag{4.4.42}$$

To leading order, (4.4.41) gives $a = 1/(1 + \bar{b}^2)$, which, on substitution in (4.4.42), leads to

$$\bar{b}'' + c_0^{-1}\bar{b}' + \frac{\bar{b}^2}{1 + \bar{b}^2} = 0. \tag{4.4.43}$$

Equation (4.4.43) must be solved subject to

$$\bar{b} \to 0 \text{ as } \bar{\xi} \to \infty, \quad \bar{b} \sim -\bar{c}_0\bar{\xi} \text{ as } \bar{\xi} \to -\infty. \tag{4.4.44}$$

The numerical solution of (4.4.43)-(4.4.44) gives $\bar{c}_0 = 0.9239$. Thus,

$$c_0^{-1} \sim 1.0824\mu^{1/4} + \dots \text{ as } \mu \to 0. \tag{4.4.45}$$

This expression agrees well with the numerical solution and also shows the singular nature of the solution for $\mu \to 0$.

Merkin and Sadiq (1996) also give some qualitative results for the general case $\phi > 0$. Now we turn to our main concern, namely the travelling wave solutions of (4.4.5)-(4.4.6). Merkin and Sadiq (1996) call a subset of this class "permanent-form travelling waves:" these are nonnegative, nontrivial solutions to the ODEs

$$a'' + va' + 1 - a - \mu ab^2 = 0 \tag{4.4.46}$$

$$b'' + vb' + \mu ab^2 - \phi b = 0 \tag{4.4.47}$$

obtained from (4.4.5)-(4.4.6) by writing $y = x - vt$ as the only independent variable. The accent in (4.4.46)-(4.4.47) denotes derivative with respect to y. The boundary conditions at the front of the wave are

$$a \to 1, \quad b \to 0 \quad \text{as} \quad y \to \infty, \tag{4.4.48}$$

while at the rear there are two possibilities:

$$a \to a_s^+ \text{ and } b \to b_s^+ \text{ as } y \to -\infty \tag{4.4.49}$$

or

$$a \to a_s^- \text{ and } b \to 0 \text{ as } y \to -\infty. \tag{4.4.50}$$

We shall have to examine which of the BCs are applicable in different parametric regimes. The conditions (4.4.49) are applicable only when $\mu \geq 4\phi^2$

(see the statement following (4.4.17)). The boundary conditions (4.4.48)-(4.4.49) would give a front wave, joining the initial unreacted state to the fully reacted, nontrivial state (a_s^+, b_s^+). With the BC (4.4.50), the system returns to its unreacted state $(1,0)$ behind a propagating nonzero reaction-diffusion event. This kind of wave is called a "pulse" (or a "single hump").

We consider the special case with $\phi = 1$ in (4.4.47) in some detail, since it is relatively simple, and is reflective of results for general ϕ. In this case, if we add (4.4.46) and (4.4.47), we easily conclude that the only solution which satisfies (4.4.48) and either (4.4.49) or (4.4.50) has $a+b = 1$. Eliminating a from (4.4.47) with the help of this relation we get

$$b'' + vb' + \mu b^2 (1 - b) - b = 0, \tag{4.4.51}$$

which must satisfy (4.4.48). Multiplying (4.4.51) by b', integrating and applying the condition (4.4.48), we get

$$b'^2 - 2v \int_y^\infty b'^2 \, dy + \frac{b^2}{6}[\mu(4b - 3b^2) - 6] = 0. \tag{4.4.52}$$

For (4.4.52) to satisfy $b' \to 0$ as $y \to -\infty$, we must have $v = 0$. And then

$$b \to b_s^+ = \frac{\mu + (\mu^2 - 4\mu)^{1/2}}{2\mu} \quad \text{as } y \to -\infty \tag{4.4.53}$$

requiring that $\mu \geq 4$. This shows that only front waves are possible. In the limit $y \to -\infty$, (4.4.52) gives

$$v \int_{-\infty}^\infty b'^2 \, dy = \frac{(b_s^+)^2}{12}[\mu - 6 + (\mu^2 - 4\mu)^{1/2}] \geq 0. \tag{4.4.54}$$

For the RHS of (4.4.54) to be positive, we must have

$$\mu \geq \frac{9}{2}. \tag{4.4.55}$$

Besides $\phi = 1$, we must have $\mu \geq 9/2$ for a front wave to emerge as a long-time solution of the initial value problem (4.4.5)-(4.4.11).

To get an idea of the solution for $\mu > 9/2$, we write

$$\mu = \frac{9}{2} + \delta \ (0 < \delta << 1) \tag{4.4.56}$$

$$b(y; \delta) = b_0(y) + \delta b_1(y) + \delta^2 b_2(y) + \dots, v = v_1 \delta + \dots \tag{4.4.57}$$

Equation (4.4.51) then gives

$$b_0(y) = \frac{2e^{-y}}{1 + 3e^{-y}} \tag{4.4.58}$$

to leading order. Equation (4.4.58) shows that $b_0 \to 0$ as $y \to \infty$ and $b_0 \to \frac{2}{3}$ as $y \to -\infty$ (see (4.4.53)). At $O(\delta)$ we get

$$b_1'' + \left[\frac{9}{2}(2b_0 - 3b_0^2) - 1\right] b_1 = -v_1 b_0' - b_0^2(1 - b_0) \qquad (4.4.59)$$

subject to

$$b_1 \to 0 \text{ as } y \to \infty, b_1 \to \frac{4}{27} \text{ as } y \to -\infty, \qquad (4.4.60)$$

the latter by putting (4.4.56) into (4.4.53), etc. We observe that b_0' is a complementary solution of (4.4.59) satisfying homogeneous boundary conditions; a solution can be obtained only by meeting a compatibility condition. This gives $v_1 = 2/3$.

The above analysis is based on regular perturbation technique. Although (4.4.51) also admits solution for negative value of v, these do not appear as long-time solutions to initial value problems, as demonstrated by the numerical solution.

Numerical solution of (4.4.51) shows that a solution exists with $v > 0$ for all $\mu > 9/2$.

For general values of ϕ, both pulse waves and front waves exist for different sets of (ϕ, μ). About the pulse wave, the general result is that they will be formed when $\mu = O(\phi)$ for large ϕ. Pulse waves do not seem to exist for smaller values of $\phi(\leq 3)$. The discussion of ODEs governing permanent waves is supplemented by Merkin and Sadiq (1996) by the numerical solutions of IVP for original PDE to determine which kind of travelling waves finally emerge as large-time asymptotics. This treatment also reveals the possibility of a stable travelling wave propagating through the system, leaving behind a temporally unstable stationary state: under these conditions, spatiotemporal chaotic behaviour is seen to develop after the passage of the wave.

4.5   Simple Wave Flows in Multidimensional Systems of Homogeneous Partial Differential Equations

The following discussion is due to Schindler (1970), specifically directed to multidimensional gas flows, although the argument goes through for other nonlinear homogeneous and autonomous systems of PDEs. It generalises the 1-dimensional case discussed in Section 4.2.

The system governing n-dimensional gas flows in a nonviscous, nonconducting compressible medium is

$$\rho_t + \sum_{j=1}^{n} (\rho u_j)_{x_j} = 0 \tag{4.5.1}$$

$$\rho \left[(u_i)_t + \sum_{j=1}^{n} u_j (u_i)_{x_j} \right] + p_{x_i} = 0, \quad i = 1, 2, \cdots, n \tag{4.5.2}$$

$$S_t + \sum_{j=1}^{n} u_j S_{x_j} = 0 \tag{4.5.3}$$

$$S = S(p\rho^{-\gamma}). \tag{4.5.4}$$

Here, p and ρ are pressure and density, respectively, at the spatial point $(x_1, \cdots, x_n)$ and time t. $u_1, \cdots, u_n$ are components of the particle velocity. S is the specific entropy, a function of the combination $p\rho^{-\gamma}$ of pressure and density; γ is the ratio of specific heats.

Generalising the argument of Section 4.1, we let the $(n + 2)$ dependent variables $p, \rho, \ u_i, \cdots, u_n$ be functions of one generating function $q = q(x_1, \cdots, x_n, t)$:

$$p = P(q) \tag{4.5.5}$$
$$\rho = \Omega(q) \tag{4.5.6}$$
$$u_i = U_i(q). \tag{4.5.7}$$

Putting (4.5.5)-(4.5.7) into (4.5.1)-(4.5.3), we get a linear homogeneous algebraic system in the $(n + 1)$ derivatives $q_t, q_{x_i}, i = 1, 2, \cdots, n$:

$$\Omega' q_t + \sum_{j=1}^{n} (\Omega U_j)' q_{x_j} = 0 \tag{4.5.8}$$

$$\Omega U_i' \left[q_t + \sum_{j=1}^{n} U_j q_{x_j} \right] + P' q_{x_i} = 0, \quad i = 1, 2, \cdots, n \tag{4.5.9}$$

$$\left[P' - \frac{\gamma P}{\Omega} \Omega' \right] \left[q_t + \sum_{j=1}^{n} U_j q_{x_j} \right] = 0. \tag{4.5.10}$$

The prime in the above denotes differentiation with respect to q. The second factor in (4.5.10) gives a constant flow and is therefore of no interest. The first factor, on integration, gives an isentropic flow:

$$P\Omega^{-\gamma} = p\rho^{-\gamma} = p_0 \rho_0^{-\gamma} = \text{constant}. \tag{4.5.11}$$

(In Section 4.1, for historical reasons, we had assumed the one-dimensional flow to be isentropic. Here, isentropy follows from the analysis.) We rewrite (4.5.8)-(4.5.9) to make the form of algebraic system in q_t and $q_{x_i}, i = 1, 2, \cdots, n$ more explicit:

$$\Omega' q_t + \sum_{j=1}^{n} \left(\Omega U_j' + \Omega' U_j \right) q_{x_j} = 0 \qquad (4.5.12)$$

$$\Omega U_i' q_t + \sum_{j=1}^{n} \left(\Omega U_i' U_j + \delta_{ij} P' \right) q_{x_j} = 0 \qquad (4.5.13)$$

where δ_{ij} is Kronecker's delta. The homogeneous algebraic system (4.5.12)-(4.5.13) for the derivatives of q has a nontrivial solution only if the determinant of coefficient matrix vanishes identically:

$$D_{n+1} = \begin{vmatrix} \Omega' & \Omega U_1' + \Omega' U_1 & \Omega U_2' + \Omega' U_2 & \cdots & (\Omega U_n' + \Omega' U_n) \\ \Omega U_1' & (\Omega U_1' U_1 + P') & \Omega U_1' U_2 & \cdots & \Omega U_1' U_n \\ \Omega U_2' & \Omega U_2' U_1 & \cdots & \cdots & \cdots \\ \cdots & \cdots & \cdots & \cdots & \cdots \\ \Omega U_n' & \Omega U_n' U_1 & \cdots & \cdots & \Omega(U_n' U_n + P') \end{vmatrix} = 0$$

$$(4.5.14)$$

There is considerable algebra in simplifying D_{n+1}. One may multiply the first column by U_k and subtract the same from the $(k+1)$st column, $k = 1, \cdots n$. Then, D_{n+1} is obtained in a symmetrical form:

$$D_{n+1} = \begin{vmatrix} \Omega' & \Omega U_1' & \Omega U_2' & \cdots & \Omega U_n' \\ \Omega U_1' & P' & & & \\ \Omega U_2' & & \ddots & 0 & \\ \cdots & & 0 & \ddots & \\ \Omega U_n' & & & & P' \end{vmatrix} = 0 \qquad (4.5.15)$$

Multiplying the first row by P' and subtracting from it $\Omega U_n'$ times the last row, we have

$$D_{n+1} = \frac{1}{P'} \cdot \begin{vmatrix} \Omega' P' - \Omega^2 U_n'^2 & \Omega P' U_1' & \cdots & \Omega P' U_{n-1}' \\ \Omega U_1' & & P' & 0 \\ \vdots & & & \ddots \\ \Omega U_n' & & 0 & P' \end{vmatrix}$$

$$= \begin{vmatrix} \Omega' P' - \Omega^2 U_n'^2 & \Omega P' U_1' & \cdots & \Omega P' U_{n-1}' \\ \Omega U_1' & P' & \ddots & 0 \\ \vdots & & \ddots & \\ \Omega U_{n-1}' & 0 & & P' \end{vmatrix} = 0 \quad (4.5.16)$$

Applying this procedure $n - 1$ times, we can arrive at the result

$$D_{n+1} = (P')^{n-1} \left[\Omega' P' - \Omega^2 \sum_{j=1}^{n} (U_j')^2 \right] = 0. \tag{4.5.17}$$

If the condition (4.5.17) is satisfied, the system (4.5.12)-(4.5.13) can be solved for derivatives of q provided the matrix of coefficients has rank n. For example, the determinant obtained from D_{n+1} in (4.5.14) by crossing out the first row and the first column, namely

$$\Delta_n = \begin{vmatrix} \Omega U_1' U_1 + P' & \Omega U_1' U_2 & \Omega U_1' U_3 & \cdots & \Omega U_1' U_n \\ \Omega U_2' U_1 & (\Omega U_2' U_2 + P') & \Omega U_2' U_3 & \cdots & \Omega U_2' U_n \\ \vdots & \vdots & & \ddots & \\ \Omega U_n' U_1 & & \cdots & & (\Omega U_n' U_n + P') \end{vmatrix}, \tag{4.5.18}$$

after considerable manipulation, becomes

$$\Delta_n = P' \Delta_{n-1}. \tag{4.5.19}$$

Repeating this reduction procedure $n - 2$ times, we get

$$\Delta_n = (P')^{n-1} \left[\Omega \sum_{j=1}^{n} U_j U_j' + P' \right]. \tag{4.5.20}$$

The case $P' = 0$ is clearly inadmissible since it leads to constant pressure and density. The other factor

$$\begin{aligned} G &= \Omega \sum_{j=1}^{n} U_j U_j' + P' \\ &= \Omega \frac{d}{dq} \left[\frac{1}{2} \sum_{j=1}^{n} U_j^2 + \frac{\gamma}{\gamma - 1} \frac{P}{\Omega} \right] \end{aligned} \tag{4.5.21}$$

in view of (4.5.11). For $G = 0$, we must have

$$\frac{1}{2} \sum_{j=1}^{n} U_j^2 + \frac{\gamma}{\gamma - 1} \frac{P}{\Omega} = \text{constant}. \tag{4.5.22}$$

Equation (4.5.22) expresses Bernoulli's law for steady flow of a polytropic gas, which, in general, will not hold for an unsteady flow. Therefore, $G \neq 0$, and, hence, $\Delta_n \neq 0$ generally for the flows that we consider. After further manipulation (Schindler (1970)), the system (4.5.9) can be solved for $\dfrac{\partial q}{\partial x_i}$ in terms of $\dfrac{\partial q}{\partial t}$. For example,

$$\frac{\partial q}{\partial x_1} = -\frac{\Omega U_1'}{\Omega \sum_{j=1}^{n} U_j U_j' + P'} \cdot \frac{\partial q}{\partial t}. \tag{4.5.23}$$

The same formula holds for $\dfrac{\partial q}{\partial x_i}, i = 2, \cdots, n$ if the subscript 1 in (4.5.23) is replaced by i. The system of PDEs (4.5.8)-(4.5.10) reduces to

$$\left(\Omega \sum_{j=1}^{n} U_j U_j' + P' \right) \frac{\partial q}{\partial x_i} + \Omega U_i' \frac{\partial q}{\partial t} = 0, \quad i = 1, \cdots, n. \tag{4.5.24}$$

Equation (4.5.24) must hold along with the isentropy condition (4.5.11) and the necessary condition in (4.5.17):

$$P = p_0 \left(\frac{\Omega}{\rho_0} \right)^\gamma \tag{4.5.25}$$

$$\Omega' P' - \Omega^2 \sum_{j=1}^{n} (U_j')^2 = 0. \tag{4.5.26}$$

Introducing the new variable (a nondimensional sound speed)

$$\bar{q} = \left[\frac{\Omega(q)}{\rho_0} \right]^{(\gamma-1)/2} = \left[\frac{P(q)}{p_0} \right]^{(\gamma-1)/2\gamma} = \frac{1}{c_0} \left[\frac{\gamma P(q)}{\Omega(q)} \right]^{1/2}, \tag{4.5.27}$$

where

$$c_0^2 = \frac{\gamma p_0}{\rho_0}, \tag{4.5.28}$$

$$U_i(q) = \bar{U}_i(\bar{q}) \tag{4.5.29}$$

and equivalent expressions for the derivatives, we find that

$$\frac{d\bar{q}}{dq} = \left(\frac{\gamma-1}{2} \right) \frac{1}{c_0^2} \frac{1}{\bar{q}} \frac{P'(q)}{\Omega(q)} = \left(\frac{\gamma-1}{2c_0} \right) \left[\frac{\Omega'(q) P'(q)}{\Omega^2(q)} \right]^{1/2} \tag{4.5.30}$$

and, hence, (4.5.24)-(4.5.26) can be written as

$$\Omega \left\{ \left[\sum_{j=1}^{n} \bar{U}_j \frac{d\bar{U}_j}{d\bar{q}} + \left(\frac{2}{\gamma-1} \right) c_0^2 \bar{q} \right] \frac{\partial \bar{q}}{\partial x_i} + \frac{d\bar{U}_i}{d\bar{q}} \frac{\partial \bar{q}}{\partial t} \right\} = 0, i = 1, \cdots, n$$

$$P = p_0 \bar{q}^{(2\gamma)/(\gamma-1)}$$

$$\Omega = \rho_0 \bar{q}^{2/(\gamma-1)} \tag{4.5.31}$$

$$\Omega^2 \left(\frac{d\bar{q}}{dq} \right)^2 \left\{ \left(\frac{2c_0}{\gamma-1} \right)^2 - \sum_{j=1}^{n} \left(\frac{d\bar{U}_j}{d\bar{q}} \right)^2 \right\} = 0.$$

Suppressing the bar, Equations (4.5.31) may be rewritten as

$$\left[\sum_{j=1}^{n} U_j U_j' + \left(\frac{2}{\gamma-1}\right) c_0^2 q\right] \frac{\partial q}{\partial x_i} + U_i' \frac{\partial q}{\partial t} = 0, \qquad i = 1, \cdots, n$$

$$P = p_0 q^{(2\gamma)/(\gamma-1)}$$

$$\Omega = \rho_0 q^{2/(\gamma-1)} \qquad\qquad (4.5.32)$$

$$\sum_{j=1}^{n} (U_j')^2 = \left(\frac{2c_0}{\gamma-1}\right)^2$$

where

$$q = \left(\frac{c}{c_0}\right) = \frac{1}{c_0}\left[\frac{\gamma P}{\Omega}\right]^{1/2} = \frac{1}{c_0}\left(\frac{\gamma p}{\rho}\right)^{1/2}.$$

The first of (4.5.32) is a quasilinear system of equations for q which do not involve x_i or t in their coefficients. Hence, the complete integral is linear in these variables. Thus, we may write the solution for q as

$$\left[\sum_{j=1}^{n} U_j(q) U_j'(q) + \frac{2}{\gamma-1} c_0^2 q\right] t - \sum_{j=1}^{n} U_j'(q) x_j = A(q) \qquad (4.5.33)$$

where $A(q)$ is an arbitrary function of q. We can finally write the simple wave solutions of the original n-dimensional system of gasdynamic equations as

$$p = P(q) = p_0 q^{(2\gamma)/(\gamma-1)} \qquad\qquad (4.5.34)$$
$$\rho = \Omega(q) = \rho_0 q^{2(\gamma-1)} \qquad\qquad (4.5.35)$$
$$u_i = U_i(q), \ i = 1, \cdots, n \qquad\qquad (4.5.36)$$
$$\sum_{j=1}^{n} [U_j'(q)]^2 = \left(\frac{2c_0}{\gamma-1}\right)^2 \qquad\qquad (4.5.37)$$

$$\left[\sum_{j=1}^{n} U_j(q) U_j'(q) + \frac{2}{\gamma-1} c_0^2 q\right] t - \sum_{j=1}^{n} U_j'(q) x_j = A(q). \qquad (4.5.38)$$

The number of arbitrary functions of q in (4.5.34)-(4.5.38) is n, namely $U_j(q), j = 1, \cdots n$, subject to the constraint (4.5.37), and $A(q)$. Specific flow problems may be solved with information in hand.

Schindler (1970) considered some general properties of the solution (4.5.34)-(4.5.38). For example, the surface of constant q are $(n-1)$ dimensional planes (4.5.38) moving in the n-dimensional space. This puts a restriction on the types of flows one may consider. There are several gasdynamic applications appended to this work, but they are not particularly relevant to our discussion here.

4.6 Travelling Waves for Nonhomogeneous Hyperbolic or Dispersive Systems

The systems of nonlinear PDEs describing physical problems, particularly in geophysics, have (unlike the homogeneous systems developed in Section 4.5) some inhomogeneous term(s) on the right-hand sides arising from external forces such as gravity and frictional effects.

One such system was first considered by Seshadri and Sachdev (1977) where travelling waves (or quasi-simple waves as these were denominated to contrast them with simple waves) were considered in an undisturbed isothermal atmosphere. The system of PDEs governing this phenomenon, like that in Section 4.4, is autonomous, that is, it does not explicitly involve independent variables in its coefficients. This work was developed more systematically by Sachdev and Gupta (1990), where an approach similar to that of Schindler (1970) not only helped to reduce the systems of PDEs to those of ODEs, but also yielded first integrals when they existed. We consider two examples to illustrate this method: the travelling acoustic gravity waves in an isothermal atmosphere, and hydromagnetic travelling waves in the following section.

Exact travelling waves in an isothermal atmosphere

For such an atmosphere, the basic equation governing an inviscid compressible flow under gravity are

$$\rho_t + u\rho_x + w\rho_z + \rho(u_x + w_z) = 0 \tag{4.6.1}$$

$$u_t + uu_x + wu_z + \frac{1}{\rho}p_x = 0 \tag{4.6.2}$$

$$w_t + uw_x + ww_z + \frac{1}{\rho}p_z + g = 0 \tag{4.6.3}$$

$$p_t + up_x + wp_z + \gamma p(u_x + w_z) = 0. \tag{4.6.4}$$

Here, u and w are fluid velocity components in the horizontal (x) and vertical (z) directions, respectively; p is the pressure, ρ is the density, and γ is the ratio of specific heats. g is the acceleration due to gravity.

We first render the system (4.6.1)-(4.6.4) nondimensional by introducing the capital variables

$$
\begin{aligned}
X &= \frac{x}{H}, \quad Z = \frac{z}{H}, \quad T = (g/H)^{1/2}t, \\
U &= \frac{u}{(gH)^{1/2}}, \quad W = \frac{w}{(gh)^{1/2}}, \\
P &= \frac{p}{p_0 e^{-z/H}}, \quad R = \frac{\rho}{\rho_0 e^{-z/H}}, \\
\frac{\gamma p_0}{\rho_0} &= c_0^2 = \gamma gH.
\end{aligned}
\tag{4.6.5}
$$

The nondimensionalising of P and R also scales out exponentially decreasing pressure and density in the undisturbed atmosphere where $p = p_0 e^{-z/H}, \rho = \rho_0 e^{-z/H}$.

The new system of PDEs is

$$U_T + UU_X + WU_Z + \frac{1}{R}P_X = 0 \qquad (4.6.6)$$

$$W_T + UW_X + WW_Z + \frac{1}{R}P_Z + \left(1 - \frac{P}{R}\right) = 0 \qquad (4.6.7)$$

$$P_T + UP_X + WP_Z + \gamma PU_X + \gamma PW_Z - WP = 0 \qquad (4.6.8)$$

$$R_T + UR_X + WR_Z + RU_X + RW_Z - WR = 0. \qquad (4.6.9)$$

We seek solutions of the system (4.6.6)-(4.6.9) which depend only on a function $\phi = \phi(X, Z, T)$, which itself must be determined. In terms of ϕ, the system (4.6.6)-(4.6.9) becomes

$$U_\phi \phi_T + UU_\phi \phi_X + WU_\phi \phi_Z + \frac{1}{R}P_\phi \phi_X = 0 \qquad (4.6.10)$$

$$W_\phi \phi_T + UW_\phi \phi_X + WW_\phi \phi_Z + \frac{1}{R}P_\phi \phi_Z = \frac{P}{R} - 1 \qquad (4.6.11)$$

$$P_\phi \phi_T + UP_\phi \phi_X + WP_\phi \phi_Z + \gamma PU_\phi \phi_X + \gamma PW_\phi \phi_Z = WP \qquad (4.6.12)$$

$$R_\phi \phi_T + UR_\phi \phi_X + WR_\phi \phi_Z + RU_\phi \phi_X + RW_\phi \phi_Z = WR \qquad (4.6.13)$$

where suffixes denote partial derivatives. To determine the form of $\phi(X, Z, T)$ we treat (4.6.10)-(4.6.13) as a system of four inhomogeneous algebraic equations for the derivatives $\phi_T, \phi_X,$ and ϕ_Z. Considering any three of these, say, (4.6.10), (4.6.12), and (4.6.13), the determinant of the associated matrix, namely

$$\text{Det} = W_\phi P_\phi (\gamma PR_\phi - RP_\phi)/R, \qquad (4.6.14)$$

is not zero except when either the flow is isentropic or W and P (and similarly R) are constants. Disallowing these cases, we solve the system (4.6.10)-(4.6.13) and find that the derivatives are functions of ϕ only:

$$\phi_T = \phi_T(\phi), \quad \phi_X = \phi_X(\phi), \quad \phi_Z = \phi_Z(\phi). \qquad (4.6.15)$$

Since $\phi_{TX} = \phi_{XT}$ etc., we get

$$\frac{d\phi_T}{d\phi}\phi_X = \frac{d\phi_X}{d\phi}\phi_T$$
$$\frac{d\phi_T}{d\phi}\phi_Z = \frac{d\phi_Z}{d\phi}\phi_T. \qquad (4.6.16)$$

Treating each of (4.6.16) as a PDE for $\phi(X, Z, T)$, we conclude that the general solution for ϕ is

$$\phi = \lambda_0 T + \lambda_1 X + \lambda_2 Z \qquad (4.6.17)$$

where λ_0, λ_1, and λ_2 are arbitrary constants. Substituting for ϕ from (4.6.17) into (4.6.10)-(4.6.13), we get the following system of algebraic equations for the derivatives,

$$A_{ij} \frac{dU_j}{d\phi} = B_i \quad (i, j = 1, 2, 3, 4), \tag{4.6.18}$$

where the matrices in (4.6.18) are

$$U = \begin{bmatrix} U \\ W \\ R \\ P \end{bmatrix}, \quad A = \begin{bmatrix} \delta & 0 & 0 & \lambda_1/R \\ 0 & \delta & 0 & \lambda_2/R \\ \lambda_1 R & \lambda_2 R & \delta & 0 \\ \gamma\lambda_1 P & \gamma\lambda_2 P & 0 & \delta \end{bmatrix} \tag{4.6.19}$$

$$B = \begin{bmatrix} 0 \\ \frac{P}{R} - 1 \\ WR \\ WP \end{bmatrix},$$

and

$$\delta = \lambda_0 + \lambda_1 U + \lambda_2 W. \tag{4.6.20}$$

The basic idea of the present approach is that we may consider (4.6.18) as a linear inhomogeneous algebraic system in $dU_j/d\phi$, $j = 1, 2, 3, 4$. If the determinant of coefficient matrix, $|A_{ij}|$, is nonzero, we may uniquely solve for the coefficients $\dfrac{dU_j}{d\phi}$. If, however, $|A_{ij}| = 0$, the system (4.6.18) has a rank less than 4 and we must use the Kronecker-Capelli theorem, namely that the system (4.6.18) is consistent if the rank of the matrix A is equal to that of the augmented matrix $[A, B]$. The system does not have a solution if the rank of A is less than that of $[A, B]$.

The following cases arise

(a) First of all, it may be checked that the conditions for the coefficient matrix A_{ij} to be of maximum rank are

$$\delta \neq \begin{cases} 0 \\ \epsilon \left(\gamma n \frac{P}{R}\right)^{1/2} \end{cases}, \quad \epsilon = \pm 1, \quad n = (\lambda_1^2 + \lambda_2^2)^{1/2} \tag{4.6.21}$$

(see (4.6.20) for the definition of δ).

For A_{ij} to be of rank 3, one of the following sets of conditions must be satisfied:

(b) $$\delta = \epsilon \left(\gamma n \frac{P}{R}\right)^{1/2} \quad \text{and} \quad W\delta = \gamma\lambda_2 \left(\frac{P}{R} - 1\right), \tag{4.6.22}$$

(c) $$\delta = 0 \quad \text{and} \quad \frac{dP}{d\phi} \neq 0 \tag{4.6.23}$$

(d) $$\delta = 0 \quad \text{and} \quad \frac{dP}{d\phi} = 0, \quad R = P \qquad (4.6.24)$$

We discuss each of these cases separately.

(a) $$\delta \neq 0 \quad \text{or} \quad \epsilon \left(\gamma n \frac{P}{R} \right)^{1/2}; \epsilon = \pm 1; n = \left(\lambda_1^2 + \lambda_2^2 \right)^{1/2}.$$

In this case, solving the system (4.6.18) for $(U_j)_\phi, j = 1, 2, 3, 4$, we get

$$\frac{dU}{d\phi} = -\lambda_1 \frac{P/R}{\delta(\delta^2 - \gamma n P/R)} \left[\gamma \lambda_2 \left(1 - \frac{P}{R} \right) + W\delta \right] \qquad (4.6.25)$$

$$\frac{dW}{d\phi} = \frac{1}{\delta(\delta^2 - \gamma n P/R)} \left[\left(1 - \frac{P}{R} \right) \left(\gamma \lambda_1^2 \frac{P}{R} - \delta^2 \right) - \lambda_2 W \delta \frac{P}{R} \right] \qquad (4.6.26)$$

$$\frac{d \log R}{d\phi} = \frac{1}{\delta(\delta^2 - \gamma n P/R)} \left[W\delta^2 + \lambda_2 \delta \left(1 - \frac{P}{R} \right) - (\gamma - 1) n W \frac{P}{R} \right] \qquad (4.6.27)$$

$$\frac{d \log P}{d\phi} = \frac{1}{(\delta^2 - \gamma n P/R)} \left[W\delta + \gamma \lambda_2 \left(1 - \frac{P}{R} \right) \right]. \qquad (4.6.28)$$

Introducing $K = \dfrac{P}{R}$, essentially the sound speed square, and suitably combining (4.6.27) and (4.6.28), we get

$$\frac{d \log K}{d\phi} = \frac{\gamma - 1}{\delta(\delta^2 - \gamma n K)} [n K W + \lambda_2 \delta(1 - K)]. \qquad (4.6.29)$$

Manipulating (4.6.25), (4.6.26), and (4.6.29), we get the intermediate integral

$$\delta K^{1/(\gamma-1)} = \text{const} = -\sigma, \quad \text{say}, \qquad (4.6.30)$$

where σ is positive. Now using (4.6.30) in (4.6.26) and dividing the resulting equation by (4.6.29), we get a first-order equation in the $W - K$ plane:

$$\frac{dW}{dK} = [\lambda_2 \sigma W K^{1 - 1/(\gamma - 1)} + (1 - K)(\lambda_1^2 \gamma K - \sigma^2 K^{-2/(\gamma-1)}] \\ \times \{(\gamma - 1) K [n W K + \lambda_2 \sigma(K - 1) K^{-1/(\gamma-1)}]\}^{-1}. \qquad (4.6.31)$$

Equation (4.6.31) has the following singular points:

(i) $$W = 0, \quad K = 1 \qquad (4.6.32)$$

(ii)
$$K \equiv K_b = \left(\frac{\sigma^2}{\gamma n}\right)^{(\gamma-1)/(\gamma+1)},$$

$$W = W_b \equiv (1 - K_b)(\sigma K_b^{-1/(\gamma-1)})\left(\frac{nK_b}{\lambda_2}\right)^{-1}. \tag{4.6.33}$$

The points (i)-(ii) coincide when $K_b = 1$.

A rather complete $W - K$ phase plane study of (4.6.31) was carried out by Seshadri and Sachdev (1977), wherein the parameter $\sigma^2/\gamma n$ plays a crucial role. Not unexpectedly, it was shown that (since basically the system in the linearized form is dispersive) no shock solutions are possible. For $\frac{\sigma^2 \lambda_1}{\gamma} < 1$ and $\lambda_2 \to 0$, there exist periodic waves propagating horizontally with an arbitrary speed λ_1.

(b)
$$\delta = \epsilon(\gamma n P/R)^{1/2}, W\delta = \gamma\lambda_2(P/R - 1)$$

Combining the conditions (4.6.22) with (4.6.6)-(4.6.8), we have

$$\frac{dK}{d\phi} = \frac{(\gamma - 1)K}{\gamma}\left[\frac{d\log P}{d\phi} - \frac{\lambda_2}{\gamma n}(\gamma - 1)(K - 1)\right]. \tag{4.6.34}$$

On differentiating the second of (4.6.22), we get

$$\frac{dW}{d\phi} = \frac{\gamma\lambda_2}{2}\frac{(K + 1)}{K\delta}\frac{dK}{d\phi}. \tag{4.6.35}$$

Equation (4.6.11), on use of (4.6.35), gives

$$\frac{d\log P}{d\phi} = \frac{1}{\lambda_2 K}\left[K - 1 - \frac{\gamma\lambda_2(K + 1)}{2K}\frac{dK}{d\phi}\right] \tag{4.6.36}$$

or, equivalently via (4.6.34),

$$\frac{dK}{d\phi} = \frac{IK(K - 1)}{(1 + KJ)} \tag{4.6.37}$$

where

$$I = \frac{2\lambda_1^2}{\gamma n\lambda_2}, \quad J = \frac{\gamma + 1}{\gamma - 1}. \tag{4.6.38}$$

Using (4.6.37) in (4.6.10), (4.6.35), (4.6.36), and (4.6.12), we get the system

$$\frac{dU}{d\phi} = -\frac{\lambda_1\psi}{\epsilon}\frac{K - 1}{(\gamma n K)^{1/2}}\frac{1 + \theta K}{1 + JK} \tag{4.6.39}$$

$$\frac{dW}{d\phi} = \frac{\lambda_1^2}{\epsilon}\frac{K^2 - 1}{(\gamma n^2 K)^{1/2}}\frac{1}{1 + JK} \tag{4.6.40}$$

$$\frac{d}{d\phi}\log R = \frac{K - 1}{1 + JK}\left[(1 + \theta K)\frac{\psi}{K} - I\right] \tag{4.6.41}$$

$$\frac{d\log P}{d\phi} = \frac{(K - 1)\psi}{K}\frac{1 + \theta K}{1 + JK} \tag{4.6.42}$$

where

$$\psi = \frac{\lambda_2}{n} \quad \text{and} \quad \theta = \frac{2\lambda_1^2}{\lambda_2^2} + J. \tag{4.6.43}$$

The special solution of the system (4.6.39)-(4.6.43), namely, $K = 1$ and U, W, R, and P constants, represents an isothermal atmosphere moving with a constant speed.

For $K \neq 1$, we can rewrite the system (4.6.39)-(4.6.43) as

$$\frac{dU}{dK} = -\frac{1}{2\epsilon} \left(\frac{\gamma}{nK^3} \right)^{1/2} \frac{\lambda_2^2}{\lambda_1} (1 + \theta K) \tag{4.6.44}$$

$$\frac{dW}{dK} = \frac{\lambda_2}{2\epsilon} \left(\frac{\gamma}{nK^2} \right)^{1/2} (1 + K) \tag{4.6.45}$$

$$\frac{d}{dK} \log R = \frac{1}{IK} \left[(1 + \theta K) \frac{\psi}{K} - I \right] \tag{4.6.46}$$

$$\frac{d}{dK} \log P = \frac{\psi}{IK^2} (1 + \theta K). \tag{4.6.47}$$

The system (4.6.44)-(4.6.47) can be explicitly integrated in terms of K:

$$U = \frac{1}{\epsilon} \left(\frac{\gamma}{nK} \right)^{1/2} \frac{\lambda_2^2}{\lambda_1} [\theta K + (1 - \theta) K^{1/2} - 1] \tag{4.6.48}$$

$$W = \frac{\lambda_2}{\epsilon} \left(\frac{\gamma}{nK} \right)^{1/2} (K - 1) \tag{4.6.49}$$

$$R = \exp \left[\frac{\psi}{I} \left(1 - \frac{1}{K} \right) \right] K^{(\theta\psi/I)-1} \tag{4.6.50}$$

$$P = \exp \left[\frac{\psi}{I} \left(1 - \frac{1}{K} \right) \right] K^{\theta\psi/I}. \tag{4.6.51}$$

The solution (4.6.48)-(4.6.50) satisfies the initial conditions $U = W = 0$, $P = R = 1$ (see Equation (4.6.5)) pertaining to a quiescant isothermal atmosphere. To find the distribution of K (and hence of other variables) with respect to ϕ, we put $K = 1 + \xi$ in (4.6.37) and obtain

$$\frac{d\xi}{d\phi} = \frac{\gamma - 1}{\lambda_2 n} \left(\frac{\lambda_1}{\gamma} \right)^2 \frac{\xi(1 + \xi)}{1 + (\gamma + 1)/2\gamma} \tag{4.6.52}$$

which integrates to yield

$$(K + 1) K^{(1-\gamma)/(2\gamma)} = C_1 \exp(C\phi) \tag{4.6.53}$$

where C_1 is the constant of integration and

$$C = \frac{\gamma - 1}{\lambda_2 n} \left(\frac{\lambda_1}{\gamma} \right)^2. \tag{4.6.54}$$

It is easy to check from (4.6.37) and its solution (4.6.53) that K has a monotonic behaviour and joins the equilibrium states $K = 1$ and $K = 0$ corresponding to $\phi = -\infty$ and $\phi = +\infty$, respectively, provided $C > 0$. The transition between these states is described by (4.6.52).

(c) $$\delta = 0 \quad \text{and} \quad \frac{dP}{d\phi} \neq 0.$$

In this case, the solution of the system is easily found to be

$$U = U(\phi), \quad W = -\frac{\lambda_0}{\lambda_2}, \quad R = P - \lambda_2 \frac{dP}{d\phi}, \quad P = P(\phi), \qquad (4.6.55)$$

provided that $\gamma = 1$. Here, $U(\phi)$ and $P(\phi)$ are arbitrary functions of

$$\phi = \lambda_0 T + \lambda_2 Z. \qquad (4.6.56)$$

In terms of the original variables, we can write (4.6.55) as

$$u = (gH)^{1/2} U(\phi), \quad w = -\frac{\lambda_0}{\lambda_2} (gH)^{1/2}, \quad p = p_0 e^{-z/H} P(\phi),$$

$$\rho = -\frac{\lambda_2}{gH} \frac{dP}{d\phi}, \quad \gamma = 1 \qquad (4.6.57)$$

where

$$\phi = \lambda_0 \left(\frac{g}{H} \right)^{1/2} t + \frac{\lambda_2}{H} Z. \qquad (4.6.58)$$

If, for example, we let $P(\phi) = \sin \phi$, then $p = p_0 e^{-z/H} \sin \phi$ and $\rho = \rho_0 e^{-z/H} \left[\cos \phi - \frac{1}{\lambda_2} \sin \phi \right]$. This solution describes the propagation of acoustic gravity waves in an isothermal atmosphere with constant vertical velocity equal to $(-\lambda_0/\lambda_2)(gH)^{1/2}$ and a variable horizontal velocity equal to $(gH)^{1/2} U(\phi)$. The pressure and density distribution are $p_0 e^{-z/H} \sin \phi$ and $\rho_0 e^{-z/H} [\cos \phi - \frac{1}{\lambda_2} \sin \phi]$, respectively.

(d) $$\hat{\delta} = 0, \quad \frac{dP}{d\psi} = 0, \quad R = P$$

This case yields the isothermal equilibrium state

$$U = -\frac{\lambda_0}{\lambda_1}, \quad W = 0, P = R = 1, \quad \gamma = 1 \qquad (4.6.59)$$

which, in original variables, is

$$u = -\frac{\lambda_0}{\lambda_1} (gH)^{1/2}, \quad w = 0, \quad p = p_0 e^{-z/H}$$
$$\rho = \rho_0 e^{-z/H}, \quad \gamma = 1. \qquad (4.6.60)$$

This describes an isothermal atmosphere moving with a constant horizontal speed $-\lambda_0/\lambda_1$.

The paper of Sachdev and Gupta (1990) describes several other geophysical models and their travelling wave solutions; these include inertial waves, internal waves, and nondivergent Rossby waves in stratified fluids.

4.7 Exact Hydromagnetic Travelling Waves

The system of nonlinear PDEs to be presented is another neat example wherein the application of the analysis similar to that in Section 4.6 leads to a rather tractable second-order ODE. The original system of nonlinear PDEs is a coupled eighth-order one; thus, the exact reduction to a single second-order ODE is rather unusual. We consider the motion of a two-dimensional compressible, nonviscous, and perfectly electrically-conducting fluid which is stratified in the vertical direction (Venkatachalappa, Rudraiah, and Sachdev (1992)). The governing equations of motion are

$$\rho \frac{D\vec{q}}{Dt} = -\nabla P + \rho \vec{g} + \mu (\vec{H} \cdot \nabla) \vec{H}, \tag{4.7.1}$$

$$\frac{\partial \rho}{\partial t} + (\vec{q} \cdot \nabla) \rho = 0 \tag{4.7.2}$$

$$\nabla \cdot \vec{q} = 0 \tag{4.7.3}$$

$$\frac{D\vec{H}}{Dt} = (\vec{H} \cdot \nabla) \vec{q} \tag{4.7.4}$$

$$\nabla \cdot \vec{H} = 0 \tag{4.7.5}$$

where $\frac{D}{Dt} = \frac{\partial}{\partial t} + (\vec{q} \cdot \nabla), \vec{q} = (u, v)$ is velocity vector with components in the horizontal and vertical directions, $\vec{H} = (H_x, H_z)$ is the magnetic field, ρ is the fluid density, $P = p + \frac{\mu H^2}{2}$ is the pressure head, p is the hydrodynamic pressure, $\vec{g}$ is the acceleration due to gravity, and μ the magnetic permeability. This model has been much studied in literature (see Acheson (1972), for example).

The undisturbed fluid is at rest with mass density $\rho_0(z)$ and horizontal magnetic field $H_0(z)$:

$$\rho_0(z) = \rho_c \exp(-z/\bar{H}) \tag{4.7.6}$$

$$H_0(z) = H_c \exp(-z/2\bar{H}) \tag{4.7.7}$$

where $\bar{H}$ is the constant scale height, and ρ_c and H_c are the reference density and magnetic field at $z = 0$. The structure (4.7.6)-(4.7.7) has been very useful in the investigation of magnetoatmospheric waves (see Nye and Thomas (1976a, 1976b)). From (4.7.2) and (4.7.6) we find that the corresponding pressure distribution is given by

$$P_0(z) = P_c \exp(-z/\bar{H}) \tag{4.7.8}$$

where $P_c = g\rho_c \bar{H}$.

We may reduce the system (4.7.1)-(4.7.5) by the use of the scales $\bar{H}$, $(\bar{H}/g)^{1/2}$, $(g\bar{H})^{1/2}$, $P_c \exp(-z/\bar{H})$, $\rho_c \exp(-z/\bar{H})$, and $H_c \exp(-z/2\bar{H})$ for length, time, velocity, pressure, density, and magnetic field, respectively, and write it more explicitly as

$$
\frac{\partial u}{\partial t} + u\frac{\partial u}{\partial x} + w\frac{\partial u}{\partial z} + \frac{1}{\rho}\frac{\partial P}{\partial x} - A^2(H_x/\rho)\frac{\partial H_x}{\partial x}
$$
$$
- A^2(H_z/\rho)\frac{\partial H_x}{\partial x} + A^2 H_z H_x/(2\rho) = 0 \tag{4.7.9}
$$

$$
\frac{\partial w}{\partial t} + u\frac{\partial w}{\partial x} + w\frac{\partial w}{\partial z} + \frac{1}{\rho}\frac{\partial P}{\partial z} - \frac{P}{\rho} + 1
$$
$$
- A^2(H_x/\rho)\frac{\partial H_z}{\partial x} - (A^2 H_z/\rho)\frac{\partial H_z}{\partial z} + A^2 H_z^2/(2\rho) = 0, \tag{4.7.10}
$$

$$
\frac{\partial \rho}{\partial t} + u\frac{\partial \rho}{\partial x} + w\frac{\partial \rho}{\partial z} - \rho w = 0 \tag{4.7.11}
$$

$$
\frac{\partial u}{\partial x} + \frac{\partial w}{\partial z} = 0 \tag{4.7.12}
$$

$$
\frac{\partial H_x}{\partial t} + u\frac{\partial H_x}{\partial x} + w\frac{\partial H_x}{\partial z} - H_x\frac{\partial u}{\partial x} - H_z\frac{\partial u}{\partial z} - H_x\frac{w}{2} = 0 \tag{4.7.13}
$$

$$
\frac{\partial H_z}{\partial t} + u\frac{\partial H_z}{\partial x} + w\frac{\partial H_z}{\partial z} - H_x\frac{\partial w}{\partial x} - H_z\frac{\partial w}{\partial z} - H_z\frac{w}{2} = 0 \tag{4.7.14}
$$

$$
\frac{\partial H_x}{\partial x} + \frac{\partial H_z}{\partial z} - \frac{H_z}{2} = 0. \tag{4.7.15}
$$

Here, $A = \sqrt{\mu H_c^2/(\rho_c g \bar{H})}$ is the nondimensional Alfvén velocity. The system (4.7.9)-(4.7.15) is of the type discussed in Section 4.6. We seek travelling wave solution of the form

$$
u = u(\phi), w = w(\phi), \rho = \rho(\phi), P = P(\phi), H_x = H_x(\phi), H_z = H_z(\phi) \tag{4.7.16}
$$

where $\phi = \phi(x, z, t) = kx + mz - t$, as may be confirmed in the manner of Section (4.6). k and m are wave numbers in the horizontal and vertical directions, respectively. These waves may be referred to as "Alfvén gravity waves." We examine whether the system (4.7.9)-(4.7.15) admits this type of wave subject to the (normalised) initial conditions

$$
u = w = H_z = 0, P = \rho = H_x = 1. \tag{4.7.17}
$$

Putting (4.7.16) into (4.7.9)-(4.7.15) and remembering that $\phi = kx + mz - t$, we get the following system of nonlinear ODEs for u, w, ρ, P, H_z, and H_x :

$$
(-1 + ku + mw)u_\phi + (k/\rho)P_\phi - (A^2/\rho)
$$
$$
\times (kH_x + mH_z)(H_x)_\phi + \frac{A^2}{2}H_x H_z/\rho = 0 \tag{4.7.18}
$$
$$
(-1 + ku + mw)w_\phi + (m/\rho)P_\phi - P/\rho - (A^2/\rho)
$$
$$
\times (kH_x + mH_z)(H_z)_\phi + \frac{A^2}{2}H_z^2/\rho + 1 = 0 \tag{4.7.19}
$$

$$(-1 + ku + mw)\rho_\phi - \rho w = 0 \qquad (4.7.20)$$

$$ku_\phi + mw_\phi = 0 \qquad (4.7.21)$$

$$(-1 + ku + mw)(H_x)_\phi - (kH_x + mH_z)u_\phi - (H_x/2)w = 0 \qquad (4.7.22)$$

$$(-1 + ku + mw)(H_z)_\phi - (kH_x + mH_z)w_\phi - (H_z/2)w = 0 \qquad (4.7.23)$$

$$k(H_x)_\phi + m(H_z)_\phi - H_z/2 = 0. \qquad (4.7.24)$$

Integrating (4.7.21), we have

$$ku + mw = 0 \qquad (4.7.25)$$

if we use the IC (4.7.17). The constant on RHS of (4.7.25) may be chosen to be different from zero if we wish to consider (constant) winds in the ambient state. On use of (4.7.25), Equations (4.7.18)-(4.7.20) and (4.7.22)-(4.7.24) simplify:

$$u_\phi - kP_\phi/\rho - A^2(kH_x + mH_z)(H_x)_\phi/2 - (A^2 H_x H_z)/(2\rho) = 0 \qquad (4.7.26)$$

$$w_\phi - kP_\phi/\rho + P/\rho + (A^2/2)(kH_x + mH_z)(H_z)_\phi$$
$$-(A^2 H_z^2)/(2\rho) - 1 = 0 \qquad (4.7.27)$$

$$\rho_\phi + \rho w = 0 \qquad (4.7.28)$$

$$(H_x)_\phi + (kH_x + mH_z)u_\phi + H_x w/2 = 0 \qquad (4.7.29)$$

$$(H_z)_\phi + (kH_x + mH_z)w_\phi + H_z w/2 = 0 \qquad (4.7.30)$$

$$k(H_x)_\phi + m(H_z)_\phi - H_z/2 = 0. \qquad (4.7.31)$$

Multiplying (4.7.29) by k and (4.7.30) by m, adding them and using (4.7.21) and (4.7.31), we get an integral

$$H_z + (kH_x + mH_z)w = 0. \qquad (4.7.32)$$

Again multiplying (4.7.26) by k, (4.7.27) by m, adding them, and making use of (4.7.21) and (4.7.31), we have

$$P_\phi/P + \left(\frac{m}{\bar{n}}\right)\left(1 - \frac{P}{\rho}\right)\left(\frac{P}{\rho}\right) = 0 \qquad (4.7.33)$$

where

$$\bar{n} = k^2 + m^2 \qquad (4.7.34)$$

is the effective wave number. Writing (4.7.28) as

$$\frac{\rho_\phi}{\rho} + w = 0 \qquad (4.7.35)$$

and combining it with (4.7.33) suitably, we get

$$K_\phi = Kw - m(1 - K)/\bar{n} \qquad (4.7.36)$$

where

$$K = P/\rho. \tag{4.7.37}$$

On using (4.7.30) and (4.7.33), we can write (4.7.27) as

$$w_\phi = (1 - K)k^2/\bar{n}(1 - A^2 H_z^2/\rho w^2). \tag{4.7.38}$$

Equation (4.7.30) can now be simplified as

$$(H_z)_\phi = \frac{k^2(H_z/w)(1 - K)}{\bar{n}(1 - A^2 H_z^2/\rho w^2)} - \frac{H_z w}{2}. \tag{4.7.39}$$

Combining (4.7.35), (4.7.38), and (4.7.39), we may write

$$\frac{d}{d\phi}\left(\frac{H_z^2}{\rho w^2}\right) = 0 \tag{4.7.40}$$

yielding the second first integral

$$H_z^2/\rho w^2 = \text{ constant } = Q, \text{ say.} \tag{4.7.41}$$

Thus, we are able to reduce the discussion of the system of ODEs (4.7.18)-(4.7.24) with the help of first integrals (4.7.25) and (4.7.41) to just two equations, namely

$$w_\phi = k^2(1 - K)/[\bar{n}(1 - E)], \tag{4.7.42}$$
$$K_\phi = m[\bar{n}Kw/m - (1 - K)]/\bar{n} \tag{4.7.43}$$

where $E = A^2 Q$ (see (4.7.41) and below (4.7.15)). E can be shown to be the ratio of kinetic to magnetic energy in the vertical direction and is constant in the present case.

We shall first study the system (4.7.42)-(4.7.43) in the (w, K) plane:

$$\frac{dw}{dK} = mk^2(1 - K)/m[(1 - E)\{\bar{n}Kw - (1 - K)\}]. \tag{4.7.44}$$

The only singular point of (4.7.44) corresponds to the isothermal atmosphere at rest:

$$w = 0, K = 1. \tag{4.7.45}$$

To determine the nature of this point, we linearise (4.7.42)-(4.7.43) about $(0, 1)$:

$$\frac{dw}{d\phi} = -k^2 K/[\bar{n}(1 - E)] \tag{4.7.46}$$

$$\frac{dK}{d\phi} = [\bar{n}w + mK]/\bar{n}. \tag{4.7.47}$$

The eigenvalues of the linear sysem (4.7.46)-(4.7.47) are found to be

$$\lambda_{1,2} = \frac{m}{k^2 + m^2} \pm \left[\frac{m^2}{(k^2 + m^2)^2} - \frac{4k^2}{(k^2 + m^2)(1 - E)} \right]^{1/2}, \qquad (4.7.48)$$

λ_1, λ_2 are real if $E > 1$ or $< 1 - 4k^2\bar{n}/m^2$; they are complex if $1 - 4k^2\bar{n}/m^2 < E < 1$. They are pure imaginary in the limit $m \to 0$ or $k \to \infty$, $E < 1$. For $m \to 0$, the waves propagate horizontally (recall that $\phi = kx + mz - t$) in a periodic fashion. As we show later, this is true even when nonlinear terms are retained. The system (4.7.46)-(4.7.47) can in this case be integrated to give

$$\frac{w^2}{c} + \frac{K^2}{[c(1 - E)]} = 1 \qquad (4.7.49)$$

where c is the constant of integration. The curves (4.7.49) are ellipses if $c > 0$, and $E < 1$, and hyperbolas if $c > 0$ and $E > 1$.

If we let $m \to 0$ in the nonlinear system (4.7.42)-(4.7.43) and write the latter in the (w, K) plane, we have

$$\frac{dw}{dK} = (1 - K)/[(1 - E)Kw]. \qquad (4.7.50)$$

Equation (4.7.50) immediately integrates to yield

$$w^2 + \frac{2}{(1 - E)}(K - \log K) = D, \quad \text{say.} \qquad (4.7.51)$$

It may be observed that the curves represented by (4.7.51) are closed in (w, K) plane provided $K > 0$ and $E < 1$ (see Figure 4.3). Remembering that $\bar{n} = k^2 + m^2$ and letting $k \to \infty$ in (4.7.42)-(4.7.44), etc., we again arrive at Equation (4.7.50) in the phase plane, showing the existence of periodic solutions in this limit as well. In the limit $E \to 0$, that is, when there is no magnetic field, Equations (4.7.42)-(4.7.43) show that there is no qualitative change in the solution, and periodic solutions exist whenever they do for the case with magnetic field.

It is interesting to note that the system (4.7.42)-(4.7.43) reduces to the single ODE

$$w_{\phi\phi} - \frac{m}{\bar{n}}w_\phi + \frac{k^2 w}{\bar{n}(1 - E)} - ww_\phi = 0 \qquad (4.7.52)$$

in w. If we introduce the scaling

$$w = -[\bar{n}(1 - E)/k^2]^{1/2}y \qquad (4.7.53)$$
$$\phi = [\bar{n}(1 - E)/k^2]^{-1/2}\tau, \qquad (4.7.54)$$

(4.7.52) becomes

$$y'' + 2\alpha y' + y + yy' = 0, \qquad (4.7.55)$$

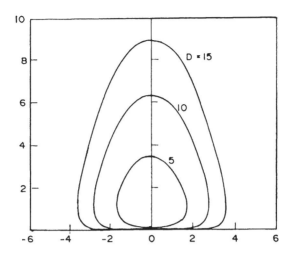

Figure 4.3. Phase portrait for equation (4.7.50) with $E = 0.1$ and $k \to \infty$.

where $\alpha = m/[2\bar{n}(\bar{n}(1 - E)/k^2)^{1/2}]$ and the prime in (4.7.55) denotes differentiation with respect to τ. Equation (4.7.55) appears quite frequently in many other contexts: internal waves, Rossby waves, and topographic Rossby waves where the parameter α represents stratification, rotation, or topography (see Odulo et al (1977)). For $\alpha = 0$, Equation (4.7.55) can be solved in terms of the amplitude parameter a as,

$$y = a \sin \psi + \frac{1}{6}a^2 \left(1 + \frac{1}{144}a^2\right) \sin 2\psi + \frac{1}{32}a^3 \sin 3\psi$$

$$+ \frac{13}{2160}a^4 \sin 4\psi + O(a^4) \qquad (4.7.56)$$

$$\psi = \left[1 - \frac{1}{24}a^2 + \dots\right] \tau + \psi_0, \psi_0 = \text{constant} \qquad (4.7.57)$$

by adopting the well-known Poincaré approach (Kevorkian and Cole (1996)). Numerical results show that the solution (4.7.56)-(4.7.57) gives good description for a as large as 2. Venkatachalappa, Rudraiah, and Sachdev (1992) have computed typical periodic and nonperiodic solutions for a variety of parametric values (particularly when $\alpha \neq 0$ in (4.7.55)) and compared them with the analytic ones. The agreement is excellent.

Following Sachdev and Gupta (1990), Venkatachalappa et al. (1992) have also shown that no nontrivial travelling wave solutions exist for the system (4.7.26)-(4.7.31) when the rank of the coefficient matrix is lower than the maximum; one arrives at equilibrium states only.

4.8 Exact Simple Waves on Shear Flows in a Compressible Barotropic Medium

We discuss an example for which the travelling wave speed is not constant but changes with the evolving flows. We consider equations of motions describing shear flows in a compressible barotropic, homentropic (mixed) atmosphere in the hydraulic or shallow-water approximation,

$$u_t + uu_x + wu_z + \frac{1}{\rho}p_x = 0 \tag{4.8.1}$$

$$\frac{1}{\rho}p_z + \bar{g} = 0 \tag{4.8.2}$$

$$\rho_t + u\rho_x + w\rho_z + \rho(u_x + w_z) = 0, \tag{4.8.3}$$

where x is the horizontal distance, z is the vertical distance measured from the ground, $\bar{g}$ is the acceleration due to gravity, and

$$\rho^{-1} = \frac{di}{dp} \tag{4.8.4}$$

where

$$i = i(p) \tag{4.8.5}$$

is the enthalpy of the atmospheric gas.

We consider the case when the flow domain is bounded below by a rigid boundary $z = 0$ and above by a free surface $z = h(x, t)$ which evolves with the flow. The pressure at the free surface is assumed to be zero. Thus, the boundary conditions are

$$w = 0 \quad \text{on} \quad z = 0 \tag{4.8.6}$$

$$p = 0 \quad \text{and} \quad w = h_t + uh_x \quad \text{on} \quad z = h(x, t). \tag{4.8.7}$$

For isentropic flow of a polytropic gas we have

$$i = \frac{\gamma}{\gamma - 1}\frac{p_{0\infty}}{\rho_{0\infty}}\left(\frac{p}{p_{0\infty}}\right)^{(\gamma-1)/\gamma} = \frac{\gamma}{\gamma - 1}RT \tag{4.8.8}$$

where $p_{0\infty}$ and $\rho_{0\infty}$ denote the equilibrium pressure and density at the ground $(z = 0)$ ahead of the wave, and γ is the ratio of specific heats. Equations (4.8.4) and (4.8.8) give

$$\rho_0 = \rho_{0\infty}\left(\frac{p_0}{p_{0\infty}}\right)^{1/\gamma}, \quad T_0 = T_{0\infty}\left(\frac{p_0}{p_{0\infty}}\right)^{(\gamma-1)/\gamma} \tag{4.8.9}$$

where $p_0(x,t), \rho_0(x,t)$, and $T_0(x,t)$ are the ground pressure, density, and temperature, respectively. Away from the ground, we can find the distribution by considering (4.8.2) and (4.8.8):

$$p = p_0 \left(1 - \frac{z}{A}\right)^{\gamma/(\gamma-1)}, \quad \rho = \rho_0 \left(1 - \frac{z}{A}\right)^{1/(\gamma-1)}, T = T_0 \left(1 - \frac{z}{A}\right) \tag{4.8.10}$$

where

$$A(x,t) = \frac{\gamma}{\gamma-1} \frac{p_{0\infty}}{\bar{g}\rho_{0\infty}} \left(\frac{p_0}{\rho_{0\infty}}\right)^{(\gamma-1)/\gamma} = \left(\frac{p_0}{p_{0\infty}}\right)^{(\gamma-1)/\gamma} A_\infty.$$

After further transformations for which we refer the reader to Varley et al. (1977) and Sachdev and Philip (1988), the basic system (4.8.1) and (4.8.3) and the BCs (4.8.6)-(4.8.7) reduce to

$$u_t + u u_x + v u_y + \frac{(p_{0\infty})^{1/\gamma}}{\rho_{0\infty}} h^{-1/\gamma} \left(1 - \frac{y}{h}\right)^{-1/\gamma} h_x = 0 \tag{4.8.11}$$

$$u_x + v_y = 0 \tag{4.8.12}$$

$$v = 0 \quad \text{on} \quad y = 0 \tag{4.8.13}$$

$$v = h_t + u h_x \quad \text{on} \quad y = h(x,t). \tag{4.8.14}$$

Introducing the nondimensional variables

$$\bar{x} = \frac{x}{l_0}, \quad \bar{t} = \frac{t}{t_0}, \quad \bar{u} = \frac{u}{u_0}, \quad \bar{v} = \frac{v}{v_0}$$

$$\bar{y} = \frac{y}{p_{0\infty}} \quad \text{and} \quad \bar{h} = \frac{h}{p_{0\infty}} \tag{4.8.15}$$

where

$$l_0 = \frac{p_{0\infty}}{\bar{g}\rho_{0\infty}}, \quad t_0 = \frac{1}{\bar{g}}\left(\frac{p_{0\infty}}{\rho_{0\infty}}\right)^{1/2},$$

$$u_0 = \left(\frac{p_{0\infty}}{\rho_{0\infty}}\right)^{1/2} \quad \text{and} \quad v_0 = \bar{g}(p_{0\infty}\rho_{0\infty})^{1/2}, \tag{4.8.16}$$

Equations (4.8.11)-(4.8.12) reduce to

$$u_t + u u_x + v u_y + h^{-1/\gamma}\left(1 - \frac{y}{h}\right)^{-1/\gamma} h_x = 0 \tag{4.8.17}$$

$$u_x + v_y = 0 \tag{4.8.18}$$

where we have dropped the bars. The system (4.8.17)-(4.8.18) reduces to that describing shear waves in an incompressible medium in the hydraulic approximation in the limit $\gamma \to \infty$ (see Freeman (1972) and Sachdev and Philip (1986)). The boundary conditions (4.8.13)-(4.8.14) remain unchanged.

We now seek simple wave solution of (4.8.17) and (4.8.18) such that the wave speed (in the horizontal direction) depends on the height of the free surface:

$$u = u(x - ct, y) \tag{4.8.19}$$
$$h = h(x - ct) \tag{4.8.20}$$

where

$$c = c(h(x, t)). \tag{4.8.21}$$

Equations (4.8.19) and (4.8.20) imply that

$$u_t + cu_x = 0 \tag{4.8.22}$$
$$h_t + ch_x = 0. \tag{4.8.23}$$

Using (4.8.22), (4.8.17) becomes

$$(u - c)u_x + vu_y + h^{-1/\gamma}\left(1 - \frac{y}{h}\right)^{-1/\gamma} h_x = 0. \tag{4.8.24}$$

Introducing the variable $\xi = x - ct$ into (4.8.24) and (4.8.18), we get

$$(u - c)u_\xi + \tilde{w}u_y + h^{-1/\gamma}\left(1 - \frac{y}{h}\right)^{-1/\gamma} \frac{dh}{d\xi} = 0 \tag{4.8.25}$$
$$u_\xi + \tilde{w}_y = 0 \tag{4.8.26}$$

where

$$\tilde{w} = v(\xi_x)^{-1}. \tag{4.8.27}$$

The boundary conditions (4.8.13)-(4.8.14) become

$$\tilde{w} = 0 \quad \text{on} \quad y = 0 \tag{4.8.28}$$
$$\tilde{w} = (u - c)\frac{dh}{d\xi} \quad \text{on} \quad y = h. \tag{4.8.29}$$

Since $h = h(\xi)$, we may use y and h as the new independent variables so that (4.8.25)-(4.8.26) can be written as

$$(u - c)u_h + w_1 u_y + h^{-1/\gamma}\left(1 - \frac{y}{h}\right)^{-1/\gamma} = 0 \tag{4.8.30}$$
$$u_h + w_{1y} = 0 \tag{4.8.31}$$

where

$$w_1 = \tilde{w}\left(\frac{dh}{d\xi}\right)^{-1} = v\left(\frac{\partial h}{\partial x}\right)^{-1}. \tag{4.8.32}$$

The BCs (4.8.28)-(4.8.29) become

$$w_1 = 0 \quad \text{on} \quad y = 0 \tag{4.8.33}$$

$$w_1 = u - c \quad \text{on} \quad y = h. \tag{4.8.34}$$

Eliminating u_h from (4.8.30) and (4.8.31), we get

$$\frac{\partial}{\partial y}\left(\frac{w_1}{u - c}\right) = \frac{h^{-1/\gamma}(1 - y/h)^{-1/\gamma}}{(u - c)^2}. \tag{4.8.35}$$

Integration of (4.8.35) with respect to y and use of (4.8.33) give

$$w_1 = (u - c)h^{-1/\gamma}\int_0^y \frac{(1 - y/h)^{-1/\gamma}}{(u - c)^2}dy. \tag{4.8.36}$$

Using the other BC (4.8.34) in (4.8.36) we have

$$h^{-1/\gamma}\int_0^h \frac{(1 - y/h)^{-1/\gamma}}{(u - c)^2}dy = 1. \tag{4.8.37}$$

Eliminating w_1 from (4.8.30) with the help of (4.8.36), we have

$$u_h + u_y h^{-1/\gamma}\int_0^y \frac{(1 - y/h)^{-1/\gamma}}{(u - c)^2}dy + h^{-1/\gamma}\frac{(1 - y/h)^{-1/\gamma}}{u - c} = 0. \tag{4.8.38}$$

Introducing the function

$$I = \int_0^y \frac{(1 - y/h)^{-1/\gamma}}{(u - c)^2}dy \tag{4.8.39}$$

(see (4.8.37)) into (4.8.38), we get an equation for I:

$$I_{yh} + h^{-1/\gamma}II_{yy} = 2I_y\left[h^{-1/\gamma}I_y - c'(h)(I_y)^{1/2}\right]$$
$$+ \frac{1}{\gamma h^2}\left(1 - \frac{y}{h}\right)^{-1}I_y\left(1 + Ih^{1-1/\gamma}\right) \tag{4.8.40}$$

where

$$c' = \frac{dc}{dh}. \tag{4.8.41}$$

Introducing the scaled variable $\eta = \dfrac{y}{h}$ into (4.8.40), we have

$$hI_{\eta h} + (h^{-1/\gamma}I - \eta)I_{\eta\eta} = I_\eta + 2I_\eta\left[h^{-1/\gamma}I_\eta - c'h^{1/2}(I_\eta)^{1/2}\right]$$
$$+ \frac{1}{\gamma h}(1 - \eta)^{-1}(1 + h^{1-1/\gamma}I)I_\eta. \tag{4.8.42}$$

Now, assuming that

$$I = h^{1/\gamma}J(\eta) \tag{4.8.43}$$

where J is a function of η alone and putting it into (4.8.42), we get on ODE for $J(\eta)$:

$$(J - \eta)J'' = \frac{\gamma - 1}{\gamma}J' + 2J'\left[J' - c'(h)h^{(\gamma+1)/2\gamma}(J')^{1/2}\right]$$
$$+ \frac{h^{(1/\gamma)-1}}{\gamma}(1 - \eta)^{-1}[1 + hJ]J'. \qquad (4.8.44)$$

Equation (4.8.44) reduces to the case considered by Freeman (1972) in the incompressible limit $\gamma \to \infty$. It is instructive to seek similarity solution of (4.8.30)-(4.8.31) directly. It is easy to check that we may write

$$\begin{aligned} u &= y^{(\gamma-1)/2\gamma}P(\eta), \quad w_1 = y^{(\gamma-1)/2\gamma}Q(\eta) \\ c &= c(h) = c_0 h^{(\gamma-1)/2\gamma} \end{aligned} \qquad (4.8.45)$$

where $\eta = y/h$ and c_0 is a dimensionless constant. Substituting (4.8.45) into (4.8.30)-(4.8.31) we get the ODEs

$$Q = \frac{\eta^2 PP' - c_0\eta^{(3\gamma+1)/2\gamma}P' - \eta^{1/\gamma}(1 - \eta)^{-1/\gamma}}{[(\gamma - 1)/2\gamma]P + \eta P'} \qquad (4.8.46)$$

$$\eta^2 P' - \frac{\gamma - 1}{2\gamma}Q - \eta Q' = 0 \qquad (4.8.47)$$

where prime denotes differentiation with respect to η. Writing

$$U = \frac{u}{h^{(\gamma-1)/2\gamma}} = \eta^{(\gamma-1)/2\gamma}P(\eta) \qquad (4.8.48)$$

in (4.8.46) and (4.8.47), we get

$$Q = \frac{\eta(U - c_0)U' - \frac{\gamma-1}{2\gamma}\left(U^2 - c_0 U + \frac{2\gamma}{\gamma-1}(1 - \eta)^{-1/\gamma}\right)}{\eta^{(\gamma-1)/2\gamma}U'} \qquad (4.8.49)$$

$$\eta^2 U' - \frac{\gamma - 1}{2\gamma}\eta U - \frac{\gamma - 1}{2\gamma}\eta^{(\gamma-1)/2\gamma}Q - \eta^{(3\gamma-1)/2\gamma}Q' = 0. \qquad (4.8.50)$$

Eliminating Q from (4.8.49) and (4.8.50), we get a single second-order ODE for U:

$$\left(U^2 - c_0 U + \frac{2\gamma}{\gamma - 1}(1 - \eta)^{-1/\gamma}\right)\frac{d^2U}{d\eta^2} + \frac{\gamma + 1}{\gamma - 1}(U - c_0)\left(\frac{dU}{d\eta}\right)^2$$
$$- \frac{2}{\gamma - 1}(1 - \eta)^{-(1/\gamma)-1}\frac{dU}{d\eta} = 0. \qquad (4.8.51)$$

The BC (4.8.34) in terms of the similarity variables P and Q in (4.8.45) becomes

$$P(1) - c_0 = Q(1). \qquad (4.8.52)$$

The integral condition (4.8.37) on writing $w_1 = u - c$ on $y = h$ and using (4.8.45), etc., becomes

$$I_1 = \int_0^1 \frac{(1 - \eta)^{-1/\gamma}}{(P - c_0)^2} d\eta = 1. \tag{4.8.53}$$

The other condition $w_1 = 0$ on $y = 0$ is automatically satisfied (see (4.8.36)).

Unlike for the incompressible case (see Freeman (1972) and Sachdev and Philip (1986)), it does not seem possible to solve (4.8.51) in a closed form. We write it as the system

$$\frac{dU}{d\eta} = V \tag{4.8.54}$$

$$\frac{dV}{d\eta} = \frac{\frac{2}{\gamma-1}(1 - \eta)^{-(1/\gamma)-1} - \frac{\gamma+1}{\gamma-1}(U - c_0)V^2}{U^2 - c_0 U + \frac{2\gamma}{\gamma-1}(1 - \eta)^{-1/\gamma}}. \tag{4.8.55}$$

To get the behaviour for $\eta \sim 1$ and $\gamma \sim 1$, Equation (4.8.51) is approximated under these assumptions and, hence, integrated; we get

$$U \approx D - \frac{\gamma}{\gamma - 1} C(1 - \eta)^{(\gamma-1)/\gamma}$$

where D and C are integration constants. We fix D and find C such that I_1 is equal to 1. Thus, we get a single parameter family of solutions. The other constant c_0 (see (4.8.45)) is the second parameter. A typical member of this two-parameter family of solutions is shown in Figure 4.4. It depicts the normalised horizontal velocity $\bar{U} = \dfrac{U - U_0}{U_1 - U_0}$, where U_0 and U_1 are the

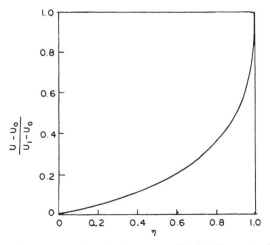

Figure 4.4. A typical wave profile for Equation (4.8.51).

scaled horizontal velocities at the bottom and free surface, respectively. The parameters chosen for Figure 4.4 are $D = 2$, $C_0 = 0.067089$, and $\gamma = 1.4$. The solution has the qualitative behaviour of an incomplete Beta function.

The main point of this example is to show that the wave speed is a function of the disturbed water height $h(x, t)$ (see (4.8.45)); the function $h(x, t)$ itself remains arbitrary in the solution of the travelling wave.

Chapter 5

Exact Linearisation of Nonlinear PDEs

5.1 Introduction

Nonlinear PDEs generally do not allow exact linearisation; one must deal with them directly either by finding similarity solutions or their more generalised forms (see Chapter 8). However, there is a small class of nonlinear PDEs which do admit linearisation either to PDEs with variable coefficients or to those with constant coefficients; sometimes it becomes feasible to change the (transformed) PDEs with variable coefficients to those with constant coefficients. Then, the whole arsenal developed to solve linear PDEs with constant coefficients over the last 150 years or so becomes available for use for the original nonlinear PDEs (see Section 5.2). On the other hand, the linear PDEs with variable coefficients are not much easier than the nonlinear ones as far as their explicit solutions are concerned; their major advantage is that they enjoy the principle of linear superposition.

There is another complicating factor: even when we may linearise a nonlinear PDE, the corresponding initial/boundary conditions generally transform in such a cumbersome way that the solution of the (exactly) linearised problem is rendered difficult. For example, the hodograph transformation for 1-dimensional, time-dependent, isentropic gasdynamic equations or two-dimensional, isentropic steady-flow equations, even though exactly linearisable, are not much useful for solving physical problems; there may be additional complications in the neighbourhood of points or curves where the hodograph transformation breaks down (see Sections 5.5 and 5.7). There are a few honourable exceptions such as the plane Burgers equation or shallow-water equations describing flow up a uniformly sloping beach, which admit neat (and readily usable) transformation of initial/boundary conditions (see Sections 5.3 and 5.7). It is the purpose of this chapter to bring out some of these features with the help of some interesting model equations. We shall carry out the analysis of these equations only to a limited extent, the main purpose here being elucidation of the exact linearisation process.

5.2   Comments on the Solution of Linear PDEs

Since much of the present chapter is devoted to transforming nonlinear PDEs to linear PDEs exactly, some comments on the solution of the latter will be in order. If the linear PDEs have constant coefficients, and the relevant ICs/BCs are not too complicated, the integral transform techniques usually suffice to solve them. This is a consequence of the principle of linear superposition. However, if the linear PDEs have variable coefficients, integral transform techniques can still be employed but finding the inverse transform is not always feasible. One may still elicit some useful information (usually of an asymptotic nature) about the behaviour of the dependent variable for a particular range of one of the independent variables.

A good introduction to analytic methods for linear PDEs may be found in Chapter 7 of Williams (1980). For a special class of PDEs, for example,

$$a_1 u_{xx} + a_2 u_x + a_3 u + b_1 u_{yy} + b_2 u_y + b_3 u = 0 \qquad (5.2.1)$$

where a_i's are functions of x only and b_i's are functions of y only, clear steps are given how one may solve initial-boundary value problems. It is easy to see that (5.2.1) permits separable solutions $u = X(x)Y(y)$ where

$$a_1 X_{xx} + a_2 X_x + a_3 X = \lambda X \qquad (5.2.2)$$
$$b_1 Y_{yy} + b_2 Y_y + b_3 Y = -\lambda Y, \qquad (5.2.3)$$

λ being the separation constant. The next step is to find from the boundary conditions an infinitely denumerable set of values of $\lambda : (\lambda_1, \lambda_2, \cdots)$ such that the solution can be represented, using the principle of linear superposition, as

$$u = \sum_{n=1}^{\infty} X_n(x) Y_n(y). \qquad (5.2.4)$$

In (5.2.4), one of the sets of functions $X_n(X)$ or $Y_n(y)(n = 1, 2, \cdots)$ can be specified uniquely and the representation (5.2.4) is such that some of the boundary conditions on u are satisfied automatically. The third step is to use the information $u(x, y_0)$ on one of the lines, $y = y_0$, say,

$$u(x, y_0) = \sum_{n=1}^{\infty} X_n(x) Y_n(y_0), \qquad (5.2.5)$$

to be able to find $Y_n(y_0)$. If the set $\{X_n\}$ is a complete orthogonal set, then $Y_n(y_0)$ can be found uniquely. The final step is to fully determine $Y_n(y)$ and hence the complete series solution. Williams (1980) clearly gives conditions on the nature of PDEs and the related boundary and initial conditions such that these steps can be gone through fully.

Another possibility is to find transformations such that linear PDEs with variable coefficients can be changed to those with constant coefficients, and hence have access to the large literature that is available for the latter. Varley and Seymour (1988) considered the class

$$\alpha(x)\frac{\partial^2 g}{\partial x^2} + \beta(x)\frac{\partial g}{\partial x} + \gamma(x)g = a\frac{\partial g}{\partial t} + b\frac{\partial^2 g}{\partial t^2} \tag{5.2.6}$$

where α, β, and γ are functions of x while a and b are arbitrary constants. Equation (5.2.6) is quite general and has many important linear equations as special cases (see Varley and Seymour (1988); Bluman and Kumei (1987)). It was first shown that (5.2.6) can be reduced to the simpler form

$$C(X)\frac{\partial}{\partial X}\left(\frac{1}{C(X)}\frac{\partial f}{\partial X}\right) = a\frac{\partial f}{\partial t} + b\frac{\partial^2 f}{\partial t^2} \tag{5.2.7}$$

through the transformation

$$X = \int^x |\alpha(\tau)|^{-1/2}\,d\tau \quad \text{and} \quad f(X,t) = \frac{g(x,t)}{g_0(x)} \tag{5.2.8}$$

provided $g = g_0(X)$ is a nonzero equilibrium solution of (5.2.6) ($a = b = 0$ therein). The function $C(X)$ in (5.2.7) is fully determined by the coefficients $\alpha(x), \beta(x)$, and $\gamma(x)$ appearing in (5.2.6). The main result of Varley and Seymour (1988) is that for certain forms of $C(X)$, any solution of (5.2.7) may be expressed in terms of solution of the PDE

$$\frac{\partial^2 F(X,t)}{\partial X^2} = a\frac{\partial F(X,t)}{\partial t} + b\frac{\partial^2 F(X,t)}{\partial t^2} \tag{5.2.9}$$

with constant coefficients in the form

$$f(X,t) = \sum_{n=0}^{N} f_n(X)\frac{\partial^n F(X,t)}{\partial X^n}, \quad f_0(X) \equiv \text{ a constant} \tag{5.2.10}$$

where the functions $f_n(X)(n = 1, 2, \cdots, N)$ and $C(X)$ satisfy a certain system of coupled nonlinear ordinary differential equations. It was also shown that by choosing N in (5.2.10) sufficiently large, it is possible to approximate any given $C(X)$ in (5.2.7) by a function which, together with $f_n(X)$, satisfies this system. The relation (5.2.10) is a Bäcklund transformation (BT) connecting the solutions of Equations (5.2.7) and (5.2.9).

The work of Varley and Seymour (1988) was generalised by Sachdev and Mayil Vaganan (1992) in several significant ways. The latter dispensed with the requirement that an equilibrium solution $g_0(X)$ of (5.2.6) must be known. The transformation

$$X = \int^x |\alpha(\tau)|^{-1/2} d\tau, \quad f(X, t) = g(x, t) \tag{5.2.11}$$

changes (5.2.6) to the form

$$\frac{\partial^2 f(X, t)}{\partial X^2} + \alpha(X) \frac{\partial f(X, t)}{\partial X} + \beta(X) f(X, t) = a \frac{\partial f(X, t)}{\partial t} + b \frac{\partial^2 f(X, t)}{\partial t^2}$$
$$\tag{5.2.12}$$

which includes (5.2.7) as a special case with $\alpha(X) = -C'(X)/C(X)$ and $\beta(X) \equiv 0$. A Bäcklund transformation was found in a systematic way which connects solutions of the PDE (5.2.12) to those of (5.2.9) with constant coefficients. This BT includes that found earlier by Varley and Seymour (1988) as a special case. The theory developed was profusely illustrated with examples describing some realistic linear problems.

It is clear from the brief discussion above that the reduction of PDEs with variable coefficients to those with constant coefficients is a difficult task and may not always be accomplished.

5.3   Burgers Equation in One and Higher Dimensions

The Burgers equation in one dimension

$$u_t + u u_x = \frac{\delta}{2} u_{xx} \tag{5.3.1}$$

is the most canonical model equation, describing a balance between (simplest) nonlinear convection and constant viscous diffusion. Here, u is the excess wavelet speed in the context of gasdynamics and δ is (small) coefficient of viscous dissipation. t and x are time and space coordinates, respectively. The derivation of (5.3.1) in gasdynamic context may be found in Sachdev (1987), where its properties and solutions are also detailed. This equation and its various generalisations appear in many applications and again may be found in Sachdev (1987) (see also Crighton (1979)).

Equation (5.3.1) is exactly linearized by the well-known Cole-Hopf transformation (Cole (1951); Hopf (1950)); the motivation for the latter comes from the first-order nonlinear ordinary differential equation

$$y'(x) = a(x) y^2(x) + b(x) y(x) + c(x). \tag{5.3.2}$$

Equation (5.3.2) is linear if $a(x) = 0$, and is of Bernoulli type and therefore easily solvable if $c(x) = 0$. There are no general ways of solving (5.3.2) as it is. Assuming, thus, that $a(x) \neq 0$, $c(x) \neq 0$, we make the substitution

$$y(x) = -\frac{w'(x)}{a(x)w(x)} \tag{5.3.3}$$

in (5.3.2) to get a second-order linear equation

$$w''(x) - \left[\frac{a'(x)}{a(x)} + b(x)\right] w'(x) + a(x)c(x)w(x) = 0. \tag{5.3.4}$$

Equation (5.3.2) has been exactly linearised to the form (5.3.4) but its order has been raised by one. If (5.3.4) can be solved explicitly, so can (5.3.2), and vice versa. A simple (nontrivial) solution of (5.3.4) helps one to find its general solution by a multiplicative transformation. The same holds for (5.3.2) except that the transformation now is additive (see Sachdev (1991)).

Now we turn to Burgers equation (5.3.1). In the present case, there are no variable coefficients. So we proceed with (5.3.3) as if $a = 1$ and employ it in two steps. Write

$$u = \psi_x \tag{5.3.5}$$

so that the order of (5.3.1) increases by one. It is fortunate that we can then exactly integrate it once. Ignoring the function of integration, we obtain the equation

$$\psi_t + \frac{1}{2}\psi_x^2 = \frac{\delta}{2}\psi_{xx} \tag{5.3.6}$$

for ψ. Now transform (5.3.6) by the second step

$$\psi = -\delta(\ln\phi) \tag{5.3.7}$$

to arrive at the well-known heat equation for the function ϕ :

$$\phi_t = \frac{\delta}{2}\phi_{xx}. \tag{5.3.8}$$

Combining (5.3.5) and (5.3.7), we observe that the transformation

$$u = -\delta(\ln\phi)_x \tag{5.3.9}$$

changes the (nonlinear) Burgers equation (5.3.1) to the linear heat equation (5.3.8). In Chapter 6, we shall discuss in detail initial value problems for (5.3.1) via (5.3.9) and (5.3.8).

It is interesting to observe what happens if the Cole-Hopf transformation is extended in a simple-minded way to the other famous model equation, namely the Korteweg-deVries equation

$$\eta_t + \sigma\eta\eta_x + \eta_{xxx} = 0 \tag{5.3.10}$$

where σ is a constant. Equation (5.3.10) describes a balance between the simplest forms of nonlinear convection and dispersion (Whitham (1974)).

The first step in the transformation is the same as before. We write $\eta = p_x$ and integrate once with respect to x, ignoring again the function of integration. We get

$$p_t + \frac{1}{2}\sigma p_x^2 + p_{xxx} = 0. \tag{5.3.11}$$

The second step is to write

$$\sigma p = 12(\log F)_x; \tag{5.3.12}$$

the factor 12 can be determined by some guesswork. The equation for F can be found to be

$$F(F_t + F_{xxx})_x - F_x(F_t + F_{xxx}) + 3(F_{xx}^2 - F_x F_{xxx}) = 0. \tag{5.3.13}$$

Equation (5.3.10) gets further nonlinearised, but two features become prominent. Equation (5.3.13) is homogeneous of degree two in F and its derivatives, and the linear operator $\left(\dfrac{\partial}{\partial t} + \dfrac{\partial^3}{\partial x^3} \right)$ comes in bold relief. Whitham (1974) has provided some reasoning for the writing of (5.3.12) by reference to the solitary wave solution and its comparison with the steady shock solution for the Burgers equation: "The working rule for finding exact solutions in this area is to consider transformations which make the special solitary wave solutions appear as simple exponentials." Thus, if we write

$$\begin{aligned} F &= 1 + \exp\{-\alpha(x - s) + \alpha^3 t\} \\ &= 1 + \exp\{-(\theta - \theta_0)\} \end{aligned} \tag{5.3.14}$$

where $\theta = \alpha x - \alpha^3 t$ and $s = \theta_0/\alpha$, and θ_0 and α are parameters, we can easily check that the solitary wave solution is

$$\sigma\eta = 3\alpha^2 \,\text{sech}^2 \frac{\alpha x - \alpha^3 t - \theta_0}{2}. \tag{5.3.15}$$

Whitham (1974) also derived more general exact solutions of (5.3.13) describing N interacting solitary waves.

Now we show, following Nerney, Schmahl, and Musielak (1996), that the Cole-Hopf transformation goes through quite generally for the vector Burgers equation

$$\frac{\partial \vec{u}}{\partial t} + \vec{u} \cdot \nabla \vec{u} = \nu \nabla^2 \vec{u} \tag{5.3.16}$$

where ν is again the coefficient of viscosity.

First we observe that

$$\nabla^2 \vec{u} = \nabla(\nabla \cdot \vec{u}) - \nabla \times \nabla \times \vec{u} \tag{5.3.17}$$

and

$$\vec{u} \cdot \nabla \vec{u} = \nabla \frac{u^2}{2} - \vec{u} \times \nabla \times \vec{u}. \tag{5.3.18}$$

In view of (5.3.17) and (5.3.18), (5.3.16) simplifies greatly if we assume that

$$\vec{u} \times \nabla \times \vec{u} = \nu \nabla \times \nabla \times \vec{u}. \tag{5.3.19}$$

If we also assume that the flow is irrotational, and use Cartesian coordinates, then we have

$$\nabla \times \vec{u} = 0 \tag{5.3.20}$$

so that

$$\vec{u} = \nabla \phi. \tag{5.3.21}$$

Equation (5.3.21) extends (5.3.5) to three-dimensional flows.

Using (5.3.17)-(5.3.21), we can write (5.3.16) as

$$\nabla \left[\frac{\partial \phi}{\partial t} + \frac{(\nabla \phi)^2}{2} - \nu \nabla^2 \phi \right] = 0. \tag{5.3.22}$$

Since the gradient of the function within the bracket is zero, it must be a function of t alone:

$$\frac{\partial \phi}{\partial t} + \frac{(\nabla \phi)^2}{2} - \nu \nabla^2 \phi = E(t). \tag{5.3.23}$$

The function $E(t)$ may be suitably absorbed by writing $\phi_1 = \phi - \int E \, dt$, etc. Now, observing the vector identity for a scalar function α,

$$\nabla \alpha = \frac{\partial \alpha}{\partial \theta} \nabla \theta, \tag{5.3.24}$$

writing $\alpha = \phi$ in (5.3.24), and taking its divergence, we have

$$\nabla^2 \phi = \frac{\partial^2 \phi}{\partial \theta^2} (\nabla \theta)^2 + \frac{\partial \phi}{\partial \theta} \nabla^2 \theta \tag{5.3.25}$$

where we have used (5.3.24) again with $\alpha = \dfrac{\partial \phi}{\partial \theta}$. Let θ satisfy the heat equation in several variables :

$$\frac{\partial \theta}{\partial t} = \nu \nabla^2 \theta. \tag{5.3.26}$$

Writing $\phi = \phi(\theta(x, t))$ and using (5.3.23), (5.3.25), and (5.3.26), we finally arrive at the generalized Cole-Hopf transformation

$$\vec{u} = -\frac{2\nu}{\theta} \nabla \theta. \tag{5.3.27}$$

We recall that $\vec{u} = \nabla \phi$.

We shall return to the discussion of solution of (5.3.16) in cylindrical coordinates in Section 6.3.

5.4 Nonlinear Degenerate Diffusion Equation $u_t = [f(u)u_x^{-1}]_x$

The equation in the title

$$u_t = [f(u)u_x^{-1}]_x \tag{5.4.1}$$

is a special case of a much-studied, more general nonlinear diffusion equation

$$
\begin{aligned}
u_t &= \frac{\partial}{\partial x}(f(u)g(u_x)) \\
&= f(u)g'(u_x)u_{xx} + f'(u)g(u_x)u_x
\end{aligned} \tag{5.4.2}
$$

where f and g are functions of u and u_x, respectively. The case of special interest in (5.4.2) is one for which f is strictly positive and g is a strictly increasing function with $\lim_{s\to\infty} g(s) = g_\infty < \infty$. In this case (5.4.2) is a strongly degenerate parabolic equation and has been used as a model for heat and mass transfer in a turbulent fluid.

Goard, Broadbridge, and Arrigo (1996) have studied (5.4.2) for its symmetries and have tabulated all functions $f(u)$ and $g(u_x)$ which admit reduction of (5.4.1) to an ODE with explicit solutions.

Our interest here is limited to a special case of (5.4.2), namely Equation (5.4.1), which enjoys infinite dimensional classical symmetry and, hence, exact linearisation (Goard, Broadbridge, and Arrigo (1996)).

Indeed, if we introduce the hodograph transformation $x = x(u,t)$ so that

$$u_x = \frac{1}{x_u},$$

$$u_t = -\left(\frac{x_t}{x_u}\right), \quad u_{xx} = -\frac{1}{x_u^3}x_{uu}, \text{ etc., (5.4.1) immediately linearises to}$$

$$f(u)x_{uu} + f'(u)x_u + x_t = 0 \tag{5.4.3}$$

or

$$x_{uu} + \frac{f'(u)}{f(u)}x_u + \frac{x_t}{f(u)} = 0. \tag{5.4.4}$$

Actually, (5.4.4) can be changed to an equation with constant coefficients so that standard transform techniques may be used to solve it.

If we introduce a point transformation

$$
\begin{aligned}
x_1 &= x_1(u,t) \\
x_2 &= x_2(u,t) \\
z &= H(u,t)x,
\end{aligned} \tag{5.4.5}
$$

then (5.4.4) can be transformed to (Bluman and Kumei (1989))

$$\frac{\partial^2 z}{\partial x_1^2} + \frac{\partial z}{\partial x_2} + w(x_1, x_2)z = 0 \tag{5.4.6}$$

where w is some function of x_1 and x_2. To obtain the transformation explicitly we substitute (5.4.5) into (5.4.6) and compare the resulting equation with (5.4.4) to get

$$x_1 = \int \frac{1}{\sqrt{\alpha f(u)}} du, \quad \alpha f > 0$$
$$x_2 = t \tag{5.4.7}$$
$$z = \beta(f(u))^{1/4} x$$

where α and β are constants. The function w is simply

$$w = \alpha \left[\frac{(f'(u))^2}{16f(u)} - \frac{f''(u)}{4} \right] \tag{5.4.8}$$

which, in view of first of (5.4.7), is a function of x_1 alone. In the special case when w is a quadratic in x_1,

$$w = a_0 + a_1 x_1 + a_2 x_1^2, \tag{5.4.9}$$

Equation (5.4.6) can be transformed to a linear PDE with constant coefficients (Bluman and Kumei (1989)). Considering (5.4.7) and (5.4.8), this would require that

$$\alpha \left\{ \frac{(f'(u))^2}{16f(u)} - \frac{f''(u)}{4} \right\} = a_0 + a_1 \int \frac{1}{\sqrt{\alpha f(u)}} du + a_2 \left(\int \frac{1}{\sqrt{\alpha f(u)}} du \right)^2.$$
$$\tag{5.4.10}$$

If the function $f(u)$ satisfies (5.4.10), we can reduce the nonlinear PDE (5.4.1) to a linear PDE with constant coefficients. Goard et al. (1996) give several examples to illustrate this interesting case of exactly linearisable equations. We consider two of them: the first leads to the standard heat equation while the second requires the solution of a Sturm-Liouville eigenvalue problem for a second-order linear ODE.

(i) If we choose $f(u) = u^{4/3}$ in (5.4.1), then (5.4.10) is satisfied if $a_0 = a_1 = a_2 = 0$ so that $w(x_1) = 0$. In this case (5.4.1) in the hodograph plane becomes

$$x_t + \frac{4}{3} u^{1/3} x_u + u^{4/3} x_{uu} = 0. \tag{5.4.11}$$

Through the change of variables (see (5.4.7))

$$x_1 = 3u^{1/3}$$
$$x_2 = t \tag{5.4.12}$$
$$z = u^{1/3} x,$$

Equation (5.4.11) transforms to the standard heat equation (with opposite sign):

$$\frac{\partial^2 z}{\partial x_1^2} + \frac{\partial z}{\partial x_2} = 0. \tag{5.4.13}$$

(ii) In the present example, we attempt to solve linear PDE (5.4.4) with variable coefficients directly by reducing it to a Sturm-Liouville problem. If we put $x(u, t) = \phi(u)T(t)$ in (5.4.4) we get

$$(f\phi')' - \lambda\phi = 0 \qquad (5.4.14)$$

and

$$T(t) = ce^{-\lambda t} \qquad (5.4.15)$$

where λ is the separation constant.

For (5.4.14) to be treated as a regular Sturm-Liouville problem for $u_1 < u < u_2$, we must have $f < 0$ on this interval, and the solution should satisfy the homogeneous boundary conditions

$$\beta_1 \phi(u_1) + \beta_2 \frac{d\phi}{du}(u_1) = 0$$

$$\beta_3 \phi(u_2) + \beta_4 \frac{d\phi}{du}(u_2) = 0, \qquad (5.4.16)$$

where $\beta_1, \beta_2, \beta_3$, and β_4 are constants (Haberman (1987)).

As a special case we let $f(u) = -u^2$ so that (5.4.3) becomes

$$u^2 x_{uu} + 2u x_u - x_t = 0. \qquad (5.4.17)$$

We choose the initial and boundary conditions to be

$$x(u, 0) = \frac{u^3}{3} - \frac{5u^2}{2} + 4u \qquad (5.4.18)$$

$$\frac{dx}{du}(1, t) = 0, \quad \frac{dx}{du}(4, t) = 0 \qquad (5.4.19)$$

in the domain $(1, 4) \times [0, \infty)$ of $x(u, t)$. For the nonlinear problem (5.4.1) with $f(u) = -u^2$, the BCs (5.4.19) correspond to zero flux conditions on free boundaries $x_1(t)$ and $x_2(t)$ where u assumes values 1 and 4, respectively.

Using the standard techniques of Sturm-Liouville theory (Haberman (1987)), we arrive at the solution

$$x(u, t) = \sum_{k=1}^{\infty} c_k e^{-\left(\frac{1}{4} + \frac{k^2 \pi^2}{4(\ln 2)^2}\right)t} u^{-\frac{1}{2}} \left\{ \frac{k\pi}{\ln 2} \cos\left(\frac{k\pi}{2\ln 2} \ln u\right) \right.$$

$$\left. + \sin\left(\frac{k\pi}{2\ln 2} \ln u\right) \right\} + A \qquad (5.4.20)$$

where

$$c_k = \frac{\int_1^4 \left(\frac{u^3}{3} - \frac{5u^2}{2} + 4u\right) \left[u^{-\frac{1}{2}} \left\{\frac{k\pi}{\ln 2} \cos\left(\frac{k\pi}{2\ln 2} \ln u\right) + \sin\left(\frac{k\pi}{2\ln 2} \ln u\right)\right\}\right] du}{\int_1^4 \frac{1}{u} \left[\frac{k\pi}{\ln 2} \cos\left(\frac{k\pi}{2\ln 2} \ln u\right) + \sin\left(\frac{k\pi}{2\ln 2} \ln u\right)\right]^2 du}$$

$$\qquad (5.4.21)$$

and $A = -0.416$.

The numerical evaluation of the solution shows that $x \to -0.416$ as t increases; u develops an infinite gradient at this point.

Goard et al. (1996) give several other examples where the original PDE (5.4.1) can either be linearised to one with constant coefficients or is solvable by reduction to a Sturm-Liouville problem.

5.5   One-Dimensional Motion of an Ideal Compressible Isentropic Gas in the Hodograph Plane

Hodograph methods are historically associated with gasdynamics. These methods have not been of great practical use, as we shall have occasion to explain, but have some mathematical interest. Courant and Friedrichs (1948) considered a system of two nonlinear PDEs

$$A_1 u_x + B_1 u_y + C_1 v_x + D_1 v_y + E_1 = 0 \qquad (5.5.1a)$$

$$A_2 u_x + B_2 u_y + C_2 v_x + D_2 v_y + E_2 = 0 \qquad (5.5.1b)$$

where $A_1, A_2, \ldots, E_2$ are known functions of x, y, u, and v alone. When $E_1 = E_2 = 0$ so that system (5.5.1) is homogeneous, and $A_1, \ldots, D_2$ are functions of u and v alone, the system is said to be reducible and the so-called hodograph transformation (see (5.5.9)-(5.5.11) below) reduces (5.5.1) to a linear form which is more amenable to analysis.

Here, we consider a special case of (5.5.1) which describes time-dependent isentropic motion of an (ideal) compressible gas in one dimension (Here t and x replace x and y, respectively). The Euler equations describing such a flow are

$$\frac{\partial \rho}{\partial t} + v \frac{\partial \rho}{\partial x} + \rho \frac{\partial v}{\partial x} = 0 \qquad (5.5.2)$$

$$\rho \left(\frac{\partial v}{\partial t} + v \frac{\partial v}{\partial x} \right) + \frac{\partial p}{\partial x} = 0 \qquad (5.5.3)$$

where ρ, v, and p are density, particle velocity, and pressure of the gas at any point (x, t), respectively. We use the thermodynamic relation

$$dw = T ds + V dp \qquad (5.5.4)$$

where w is the enthalpy per unit mass, V is the specific volume, T is the temperature, and s is entropy. Since we assume that the flow is isentropic and $s = $ constant, we have

$$dw = \frac{dp}{\rho}. \qquad (5.5.5)$$

From the equation of state, ρ is a function of two other state variables, s and w, say, and since s is constant, we have $\rho = \rho(w)$. The sound speed in this case is given by $c_s = \sqrt{p'(\rho)}$. In view of (5.5.5), we have

$$\frac{d\rho}{dw} = \frac{d\rho}{dp}\frac{dp}{dw} = \frac{\rho}{c_s^2}. \tag{5.5.6}$$

Equations (5.5.2) and (5.5.3) can now be written in terms of w and v:

$$\frac{\partial w}{\partial t} + v\frac{\partial w}{\partial x} + c_s^2\frac{\partial v}{\partial x} = 0 \tag{5.5.7}$$

$$\frac{\partial v}{\partial t} + v\frac{\partial v}{\partial x} + \frac{\partial w}{\partial x} = 0 \tag{5.5.8}$$

where c_s is a function of w via $\rho = \rho(w)$. The system (5.5.7)-(5.5.8) is a little simpler than (5.5.2)-(5.5.3). As we observed in Section 5.4, the hodograph method, in principle, interchanges the role of dependent and independent variables so that the given equations become linear with co-efficients functions of the (new) independent variables. The easiest way to accomplish this task is to write the derivatives in (5.5.7) and (5.5.8) in terms of Jacobians:

$$\frac{\partial(w, x)}{\partial(t, x)} - v\frac{\partial(w, t)}{\partial(t, x)} + c_s^2\frac{\partial(t, v)}{\partial(t, x)} = 0 \tag{5.5.9}$$

$$-\frac{\partial(x, v)}{\partial(t, x)} + v\frac{\partial(t, v)}{\partial(t, x)} - \frac{\partial(w, t)}{\partial(t, x)} = 0. \tag{5.5.10}$$

Now, if we multiply (5.5.9) and (5.5.10) by the Jacobian of the transformation

$$\frac{\partial(t, x)}{\partial(w, v)} = \frac{\partial t}{\partial w}\frac{\partial x}{\partial v} - \frac{\partial t}{\partial v}\frac{\partial x}{\partial w} \neq 0 \tag{5.5.11}$$

and remember that the Jacobians "behave" like fractions, we immediately arrive at the linear form of the system (5.5.2)-(5.5.3), namely

$$\frac{\partial x}{\partial w} + \frac{\partial t}{\partial v} - v\frac{\partial t}{\partial w} = 0 \tag{5.5.12}$$

$$\frac{\partial x}{\partial v} - v\frac{\partial t}{\partial v} + h(w)\frac{\partial t}{\partial w} = 0 \tag{5.5.13}$$

where $h(w) = c_s^2(w)$. Rewriting (5.5.12) in the conservation form

$$\frac{\partial t}{\partial v} - \frac{\partial}{\partial w}(x - vt) = 0, \tag{5.5.14}$$

we are led to the "potential" function ϕ:

$$t = \frac{\partial\phi}{\partial w}, \qquad x = v\frac{\partial\phi}{\partial w} - \frac{\partial\phi}{\partial v}. \tag{5.5.15}$$

Eliminating t and x from (5.5.13) with the help of (5.5.15), we arrive at the single equation for ϕ:

$$h(w)\frac{\partial^2 \phi}{\partial w^2} + \frac{\partial \phi}{\partial w} - \frac{\partial^2 \phi}{\partial v^2} = 0, \quad h(w) = c_s^2(w). \tag{5.5.16}$$

Equation (5.5.16) is well-known in gasdynamics, where for a polytropic gas $h(w) = (\gamma - 1)w$. Once (5.5.16) is solved for ϕ as a function of w and v, t and x can be found from (5.5.15).

While the task of linearising (5.5.7)-(5.5.8) to the form (5.5.16) has been accomplished quite easily, solving an IVP for the former via (5.5.16) is not straightforward. Suppose the initial conditions for (5.5.7)-(5.5.8) are prescribed in the form

$$w(x,0) = w_0(x), v(x,0) = v_0(x), x \in R. \tag{5.5.17}$$

Then (in principle), one may eliminate x from (5.5.17) to get the locus of the given curve in the (w,v) plane along which $t = 0$ and x are given. We have thus converted the IVP (5.5.7), (5.5.8), and (5.5.17) to a Cauchy problem for (5.5.16). In practice, such a task is generally difficult to accomplish and it may turn out easier to solve (5.5.7), (5.5.8), and (5.5.17) numerically using the method of characteristics or otherwise. For some problems, the Jacobian (5.5.11) may vanish along some curve, making the mapping from (x,t) plane to (u,v) plane rather complicated.

Carbonaro (1997) has studied in detail invariance properties of the system (5.5.12)-(5.5.13) as well as the single equation (5.5.16) using the group theoretic approach. The sound speed square c_s^2 is kept free and is exploited to maximize symmetries and, hence, reduction of PDEs to ODEs via a similarity transformation. Carbonaro (1997) has also discussed the potential symmetries of (5.5.12)-(5.5.13) via (5.5.16).

Logan (1994) adopted the same approach to linearise the related system

$$h_t + uh_x + hu_x = 0$$
$$u_t + uu_x + h_x = 0 \tag{5.5.18}$$

describing shallow-water waves. Here, h and u are depth and particle velocity at any point (t,x). The transformed equation for $t = t(u,h)$ comes out to be

$$t_{uu} = h^{-1}(h^2 t_h)_h. \tag{5.5.19}$$

In terms of the characteristic variables

$$\xi = 2h^{1/2} - u, \quad \eta = 2h^{1/2} + u, \tag{5.5.20}$$

(5.5.19) becomes

$$-4t_{\xi\eta} = 0 \tag{5.5.21}$$

with the general solution

$$
\begin{aligned}
t &= f(\xi) + g(\eta) \\
  &= f(2h^{1/2} - u) + g(2h^{1/2} + u)
\end{aligned}
\tag{5.5.22}
$$

where f and g are arbitrary functions of their arguments. The difficulties associated with solving an IVP for (5.5.18) via the solution (5.5.22), referred to earlier, apply here as well.

It is of some interest to find whether hodograph-type transformations are useful when the flow is nonisentropic. Such an attempt was made by Ardavan-Rhad (1970) to study the effect of a centered simple wave as it catches up with a plane shock. The one-dimensional, nonisentropic equations were written in terms of particle speed u, sound speed c, and entropy S:

$$
u_t + uu_x + \frac{2}{\gamma - 1}cc_x - \frac{c^2}{\gamma(\gamma - 1)c_v}S_x = 0
\tag{5.5.23}
$$

$$
c_t + uc_x + \frac{\gamma - 1}{2}cu_x = 0
\tag{5.5.24}
$$

$$
S_t + uS_x = 0
\tag{5.5.25}
$$

where γ is the ratio of specific heats. On introducing the variables

$$
\begin{aligned}
\xi &= \ln\left[c^{2/(\gamma-1)}\exp\left(\frac{-1}{\gamma(\gamma-1)}\frac{S - S_0}{c_v}\right)\right] \\
\eta &= \ln\, c^{2/(\gamma-1)},
\end{aligned}
\tag{5.5.26}
$$

(5.5.23)-(5.5.25) change to

$$
u_t + uu_x + e^{(\gamma-1)\eta}\xi_x = 0
\tag{5.5.27}
$$

$$
\xi_t + u\xi_x + u_x = 0
\tag{5.5.28}
$$

$$
\eta_t + u\eta_x + u_x = 0.
\tag{5.5.29}
$$

As for the isentropic system, we may write (5.5.27)-(5.5.29) in terms of Jacobians:

$$
\frac{\partial(u, x)}{\partial(t, x)} + u\frac{\partial(t, u)}{\partial(t, x)} + e^{(\gamma-1)\eta}\frac{\partial(t, \xi)}{\partial(t, x)} = 0
\tag{5.5.30}
$$

$$
\frac{\partial(\xi, x)}{\partial(t, x)} + u\frac{\partial(t, \xi)}{\partial(t, x)} + \frac{\partial(t, u)}{\partial(t, x)} = 0
\tag{5.5.31}
$$

$$
\frac{\partial(\eta, x)}{\partial(t, x)} + u\frac{\partial(t, \eta)}{\partial(t, x)} + \frac{\partial(t, u)}{\partial(t, x)} = 0.
\tag{5.5.32}
$$

It is clear from (5.5.26) that, for a nonisentropic flow, the variables ξ and η are independent. We therefore use the hodograph transformation which introduces (ξ, η) as the independent variables in place of (t, x). Since we assume that $\dfrac{\partial(t, x)}{\partial(\xi, \eta)} \neq 0$, we multiply each of (5.5.30)-(5.5.32) by this Jacobian and obtain

$$u_\xi x_\eta - u_\eta x_\xi + u(u_\eta t_\xi - u_\xi t_\eta) - e^{(\gamma-1)\eta} t_\eta = 0 \qquad (5.5.33)$$

$$x_\eta - u t_\eta + u_\eta t_\xi - u_\xi t_\eta = 0 \qquad (5.5.34)$$

$$x_\xi - u t_\xi - u_\eta t_\xi + u_\xi t_\eta = 0. \qquad (5.5.35)$$

Introducing the expression

$$u_\eta t_\xi - u_\xi t_\eta = \psi(\xi, \eta), \ \text{say}, \qquad (5.5.36)$$

as a new dependent variable and inserting x_η and x_ξ from (5.5.34) and (5.5.35) into (5.5.33), we arrive at the system

$$e^{(\gamma-1)\eta} t_\eta + (u_\xi + u_\eta)\psi = 0 \qquad (5.5.37)$$

$$x_\eta = u t_\eta - \psi \qquad (5.5.38)$$

$$x_\xi = u t_\xi + \psi. \qquad (5.5.39)$$

Assuming that $x_{\eta\xi} = x_{\xi\eta}$, we get from (5.5.38) and (5.5.39) a linear first-order PDE for ψ

$$\psi_\xi + \psi_\eta + \psi = 0 \qquad (5.5.40)$$

with the general solution

$$\psi = e^{-\eta} g(\xi - \eta) \qquad (5.5.41)$$

where g is an arbitrary function of its argument. If we know the function g, we may find x from (5.5.38) and (5.5.39) in terms of u and t. However, here we derive an equation for u with ξ and η as the independent variables. We derive from (5.5.36) and (5.5.41) the equation

$$u_\eta t_\xi - u_\xi t_\eta - e^{-\eta} g(\xi - \eta) = 0 \qquad (5.5.42)$$

and from (5.5.37) and (5.5.41)

$$e^{(\gamma-1)\eta} t_\eta + (u_\xi + u_\eta)e^{-\eta} g(\xi - \eta) = 0. \qquad (5.5.43)$$

Solving (5.5.42) and (5.5.43) for t_ξ and t_η, and using $t_{\xi\eta} = t_{\eta\xi}$, we arrive at a rather complicated equation for u alone with ξ and η as the independent variables:

$$\frac{\partial}{\partial \eta}\left[e^{-\gamma\eta} g\left(\frac{u_\xi(u_\xi + u_\eta) - e^{(\gamma-1)\eta}}{u_\eta} \right) \right] = \frac{\partial}{\partial \xi}\left[e^{-\gamma\eta} g(u_\xi + u_\eta) \right]. \qquad (5.5.44)$$

Equation (5.5.44), which involves all the dependent variables of the original system (5.5.23)-(5.5.25), namely $u, \xi,$ and η (see (5.5.26)), is difficult to solve generally. However, Ardavan-Rhad attempted a solution,

$$u = \frac{2}{\gamma - 1} e^{\frac{(\gamma-1)}{2} \eta} h(\xi - \eta) + \text{constant}, \qquad (5.5.45)$$

which introduces an arbitrary function $h(\xi - \eta)$ in the usual definition of a Riemann invariant for isentropic flows. $\xi - \eta$, it may be recalled, is essentially the entropy function (see (5.5.26)). Substitution of (5.5.45) into (5.5.44) leads to the following special solution (written in terms of the original variables) which describes a centered simple wave:

$$\begin{aligned}
x &= [u + ch(\sigma)]t \\
t &= c^{-(\gamma+1)/(\gamma-1)} f(\sigma) \\
f &= (h^2 - 1)^{-(\gamma+1)/2(\gamma-1)} \exp\left[\frac{\gamma+1}{2} \int \frac{d\sigma}{h^2 - 1}\right] \\
u &= \frac{2}{\gamma - 1} ch(\sigma) + \text{constant}
\end{aligned} \qquad (5.5.46)$$

where the entropy variable now is

$$\sigma = -\frac{1}{\gamma(\gamma - 1)} \frac{S - S_0}{c_v}. \qquad (5.5.47)$$

The solution (5.5.46) is essentially nonisentropic; it does not include $\sigma = $ constant, the isentropic flow, as a particular case since the entire derivation assumes that the variables ξ and η in (5.5.26) are independent.

The solution (5.5.46) was used by Ardavan-Rhad (1970) to describe the effect of a centered simple wave on an advancing plane shock. Although the solution was approximate, it gave quite satisfactory results. This work was generalised later by Sharma, Ram, and Sachdev (1987). Equation (5.5.44) needs to be investigated more generally.

We may refer here also to the work of Steketee (1976), who used the Lagrangian form of one-dimensional gas dynamic equations and, in the manner of Ardavan-Rhad, derived a host of new forms of gasdynamic equations. He also gave some special solutions of his new equations.

5.6 The Born-Infeld Equation

The Born-Infeld equation

$$(1 - \phi_t^2)\phi_{xx} + 2\phi_x \phi_t \phi_{xt} - (1 + \phi_x^2)\phi_{tt} = 0 \qquad (5.6.1)$$

has some interest of its own: for this equation the hodograph transformation, in conjuction with characteristic coordinates, leads to a rather neat

solution. Observe that (5.6.1) involves only derivatives of ϕ in the coefficients, and is also autonomous. It has travelling wave solutions of the form
$\phi = \phi(X)$ where $X = x - t$ or $x + t$. The ODE for $\phi(X)$ can be shown
to have single hump solutions, which look like solitary waves. It is easy to
check that (5.6.1) is of hyperbolic type if

$$\phi_x^2 \phi_t^2 + (1 + \phi_x^2)(1 - \phi_t^2) > 0,$$

that is,

$$1 + \phi_x^2 - \phi_t^2 > 0. \tag{5.6.2}$$

We shall assume that (5.6.2) is satisfied. In the sequel we closely follow
Whitham (1974). In the light of the above remarks, we introduce the
variables

$$\xi = x - t, \quad \eta = x + t,$$
$$u = \phi_\xi, \quad v = \phi_\eta \tag{5.6.3}$$

in (5.6.1). The new system of equations assumes a rather simple form:

$$u_\eta - v_\xi = 0 \tag{5.6.4}$$
$$v^2 u_\xi - (1 + 2uv)u_\eta + u^2 v_\eta = 0. \tag{5.6.5}$$

Equation (5.6.4) follows from (5.6.3$_3$) and (5.6.3$_4$), while (5.6.5) is the
new form of (5.6.1). Now we apply the hodograph transformation to (5.6.4)
-(5.6.5) so that the role of dependent and independent variables is inter-
changed. We get

$$\xi_v - \eta_u = 0 \tag{5.6.6}$$
$$v^2 \eta_v + (1 + 2uv)\xi_v + u^2 \xi_u = 0. \tag{5.6.7}$$

Eliminating η from (5.6.6) and (5.6.7) by differentiating, etc., we get a
single linear second-order PDE for ξ:

$$u^2 \xi_{uu} + (1 + 2uv)\xi_{uv} + v^2 \xi_{vv} + 2u\xi_u + 2v\xi_v = 0 \tag{5.6.8}$$

which, in view of our assumption regarding the original system, is also
hyperbolic. The characteristic directions for (5.6.8) are given by

$$u^2 (dv)^2 - (1 + 2uv)dudv + v^2 (du)^2 = 0$$

or

$$\frac{dv}{du} = \frac{1 + 2uv \pm [1 + 4uv]^{1/2}}{2u^2}. \tag{5.6.9}$$

On integration of (5.6.9) the characteristic curves are found to be
$r = $ constant, $s = $ constant, where

$$r = \frac{\sqrt{1 + 4uv} - 1}{2v}, \quad s = \frac{\sqrt{1 + 4uv} - 1}{2u}. \tag{5.6.10}$$

Introducing the characteristic variables r and s in (5.6.6) and (5.6.7), we get

$$r^2 \xi_r + \eta_r = 0 \qquad (5.6.11)$$

$$\xi_s + s^2 \eta_s = 0. \qquad (5.6.12)$$

Curiously, not only the hodograph method linearises the original system, the latter reduces to the rather simple form (5.6.11)-(5.6.12) via characteristic variables. The system (5.6.11)-(5.6.12) in ξ alone simply becomes

$$\xi_{rs} = 0. \qquad (5.6.13)$$

The general solution of (5.6.11)-(5.6.12) via that of (5.6.13) may be written as

$$x - t = \xi = F(r) - \int s^2 G'(s) ds \qquad (5.6.14)$$

$$x + t = \eta = G(s) - \int r^2 F'(r) dr \qquad (5.6.15)$$

where $F(r)$ and $G(s)$ are arbitrary functions of their arguments. We may now find

$$\phi_r = u\xi_r + v\eta_r = \frac{r}{1-rs}\xi_r + \frac{s}{1-rs}\eta_r = rF'(r) \qquad (5.6.16)$$

and, in the same manner,

$$\phi_s = sG'(s). \qquad (5.6.17)$$

We can therefore write

$$\phi = \int rF'(r) dr + \int sG'(s) ds. \qquad (5.6.18)$$

To interpret the solution as an interaction of two waves incident from $x = -\infty$ and $x = +\infty$, Whitham (1974) introduced the variables

$$F(r) = \rho, \qquad G(s) = \sigma$$
$$r = \Phi_1'(\rho), \qquad s = \Phi_2'(\sigma)$$

so that (5.6.18) becomes

$$\phi = \Phi_1(\rho) + \Phi_2(\sigma), \qquad (5.6.19)$$

and

$$x - t = \rho - \int_{-\infty}^{\sigma} \Phi_2'^2(\sigma) d\sigma \qquad (5.6.20)$$

$$x + t = \sigma + \int_{\rho}^{\infty} \Phi_1'^2(\rho) d\rho \qquad (5.6.21)$$

(see (5.6.14)-(5.6.15)). If, further, $\Phi_1(\rho)$ and $\Phi_2(\sigma)$ are localised so that they are nonzero only in $-1 < \rho < 0, 0 < \sigma < 1$, respectively, then

$$\phi = \Phi_1(x - t) + \Phi_2(x + t) \quad \text{for} \quad t < 0. \tag{5.6.22}$$

The solution (5.6.22) is a superposition of two waves, Φ_1 coming from $x = -\infty$ and Φ_2 from $x = +\infty$. In the limit $t \to +\infty$, the solution tends to

$$\phi = \Phi_1 \left\{ x - t + \int_{-\infty}^{\infty} \Phi_2'^2(\sigma)d\sigma \right\} + \Phi_2 \left\{ x + t - \int_{-\infty}^{\infty} \Phi_1'^2(\rho)d\rho \right\}. \tag{5.6.23}$$

The terms $\int_{-\infty}^{\infty} \Phi_i'^2(\tau)d\tau$ $(i = 1, 2)$ represent displacements in the direction opposite the direction of propagation of the respective waves.

5.7 Water Waves up a Uniformly Sloping Beach

The system of PDEs describing shallow-water waves up a uniformly sloping beach are

$$\frac{\partial v}{\partial t} + v\frac{\partial v}{\partial x} + \frac{\partial \eta}{\partial x} = 0 \tag{5.7.1}$$

$$\frac{\partial \eta}{\partial t} + \frac{\partial}{\partial x}[(\eta - x)v] = 0. \tag{5.7.2}$$

Here, all variables have been suitably nondimensionalised. v is the particle velocity, η is the height of the wave above the undisturbed shoreline, and x and t are space and time variables, respectively. The slope of the beach is rendered equal to 1 after appropriate scaling. The hyperbolic system of equations (5.7.1)-(5.7.2) has been derived and discussed in some detail by Stoker (1948). Here we follow the work of Carrier and Greenspan (1958) and some other investigators with respect to the transformation and exact solution of (5.7.1) and (5.7.2). As in the example of the Born-Infeld equation discussed in Section 5.6, the characteristics and hodograph transformation both play a crucial role in the linearisation and simplification of the system (5.7.1)-(5.7.2). An advantage of the present problem is that the free boundary, the instantaneous shoreline, on which the depth is zero becomes fixed in the new coordinate system and, therefore, the BC on it can be easily satisfied.

It is not difficult to check that the hyperbolic system (5.7.1)-(5.7.2) can be written in terms of the characteristic variables α and β, as

$$x_\beta - (v + c)t_\beta = 0 \tag{5.7.3}$$

$$x_\alpha - (v - c)t_\alpha = 0 \tag{5.7.4}$$

$$v_\beta + 2c_\beta + t_\beta = 0 \tag{5.7.5}$$

$$v_\alpha - 2c_\alpha + t_\alpha = 0 \tag{5.7.6}$$

where $c^2 = \eta - x$. Equations (5.7.3)-(5.7.4) define characteristic directions while (5.7.5) and (5.7.6) are the relations holding along them, respectively. Integration of (5.7.5) and (5.7.6) gives

$$v + 2c + t = \alpha \tag{5.7.7}$$
$$v - 2c + t = -\beta \tag{5.7.8}$$

where the "functions" of integration have been chosen to introduce some simplification. From (5.7.7) and (5.7.8) we get

$$v + t = (\alpha - \beta)/2 = \lambda/2 \tag{5.7.9}$$

and

$$c = (\alpha + \beta)/4 = \sigma/4 \tag{5.7.10}$$

where new independent variables λ and σ have been introduced. Equations (5.7.3)-(5.7.4) now become

$$x_\sigma - vt_\sigma + ct_\lambda = 0 \tag{5.7.11}$$
$$x_\lambda + ct_\sigma - vt_\lambda = 0. \tag{5.7.12}$$

Eliminating x from (5.7.11)-(5.7.12), we get a linear second-order PDE for t:

$$\sigma(t_{\lambda\lambda} - t_{\sigma\sigma}) - 3t_\sigma = 0. \tag{5.7.13}$$

Since $v + t = \lambda/2$, v also satisfies (5.7.13). Indeed, if we introduce the "potential" function ϕ via

$$v = \sigma^{-1}\phi_\sigma(\sigma, \lambda), \tag{5.7.14}$$

then ϕ satisfies the equation

$$(\sigma\phi_\sigma)_\sigma - \sigma\phi_{\lambda\lambda} = 0 \tag{5.7.15}$$

while other functions, by the use of (5.7.9)-(5.7.12), are expressed as

$$x = \phi_\lambda/4 - \sigma^2/16 - v^2/2 \tag{5.7.16}$$
$$\eta = c^2 + x = \phi_\lambda/4 - v^2/2 \tag{5.7.17}$$
$$t = \lambda/2 - v. \tag{5.7.18}$$

Not only do we have a linear PDE (5.7.13) or (5.7.15) replacing the original nonlinear system (5.7.1)-(5.7.2), the boundary condition on the (moving) instantaneous shoreline $c = 0$ has also been replaced by the fixed line $\sigma = 0$ in the (σ, λ) plane. Once Equation (5.7.15) for $\phi(\sigma, \lambda)$ has been solved, x, η, and t are found explicitly from (5.7.16)-(5.7.18) as functions of (σ, λ). To exclude the possibility of wave breaking, we assume that the Jacobian $\dfrac{\partial(x, t)}{\partial(\sigma, \lambda)} \neq 0$ for all $\sigma > 0$.

Before we pose an IVP for (5.7.13) or (5.7.15), we observe that a simple product solution of (5.7.15) is

$$\phi = A J_0(w\sigma) \cos(w\lambda - \psi) \tag{5.7.19}$$

where J_0 is a Bessel function of order zero. We choose the constants in (5.7.19) to be $\psi = 0$ and $w = 1$ without loss of generality. It may be easily checked that $J = \dfrac{\partial(x,t)}{\partial(\sigma,\lambda)}$ for (5.7.19) is nonzero for $\sigma > 0$ if $A \leq 1$. Therefore, no breaking of the wave takes place under this condition. The solution $A J_0(\sigma) \cos \lambda$ shows that the phenomenon is periodic in time and the wave shape far out at sea is like $J_0(4\sqrt{|x|})$. Near the shore, the wave is considerably distorted. The shoreline, where the depth is zero, is given by $\sigma = 0$ (see (5.7.10)). Accordingly, the maximum penetration of the wave is $x(\lambda, 0) = (\phi_\lambda/4) - \dfrac{u^2}{2}$ is $A/4$ (see (5.7.16)).

Now we pose a simple problem: the release of a mound of water at time $t = 0$. Here the initial shape of the mound $\eta(x,0)$ is prescribed; it is also assumed to be at rest so that $v(x,0) = 0$ everywhere. From the relation $v + t = \lambda/2$, we find that $v = 0$, $t = 0$ imply that $\lambda = 0$ initially. We must solve (5.7.13) for v (replacing t) with the conditions that $v = 0$ and v_λ prescribed on $\lambda = 0$, corresponding to $t = 0$. We also require that v must remain finite on the free boundary $\sigma = 0$ corresponding to $c = 0$. To find v_λ, we note the relation

$$[\eta(x,0) - x]^{1/2} = c(x,0) = \frac{1}{4}\sigma. \tag{5.7.20}$$

Since the initial height $\eta(x,0)$ is prescribed, we may solve (5.7.20) for x and use the relation (5.7.11) to write

$$x_\sigma = -c t_\lambda$$

on $\lambda = 0$ where $v = 0$. Now, from (5.7.18) and (5.7.11), we have, for $\lambda = 0$,

$$v_\lambda = \frac{1}{2} - t_\lambda = \frac{1}{2} + 4\sigma^{-1}x_\sigma = f(\sigma), \quad \text{say}, \tag{5.7.21}$$

where x_σ (and hence $f(\sigma)$) is obtained from (5.7.20). Thus, all the data required for the equation

$$\sigma(v_{\lambda\lambda} - v_{\sigma\sigma}) - 3v_\sigma = 0 \tag{5.7.22}$$

become available. We solve the linear PDE (5.7.22) by the Laplace transform techniques. Writing

$$\bar{v} = \int_0^\infty e^{-s\lambda} v \, d\lambda,$$

Equation (5.7.22) and the conditions $v = 0, v_\lambda = f(\sigma)$ on $\lambda = 0$ give

$$\sigma \bar{v}_{\sigma\sigma} + 3\bar{v}_\sigma - s^2 \sigma \bar{v} = -\sigma f(\sigma). \tag{5.7.23}$$

With $z = s\sigma$ and $p = z\bar{v}$, (5.7.23) changes to

$$(zp')' - \frac{p}{z} - zp = -\frac{z^2}{s^2} f(z/s). \tag{5.7.24}$$

Now we use the Hankel transform

$$\bar{p} = \int_0^\infty z J_1(\xi z) p(z) dz$$

in (5.7.24) and, after some manipulation, obtain

$$\bar{p} = \frac{1}{1 + \xi^2} \int_0^\infty \frac{\beta^2}{s^2} J_1(\beta s) f(\beta/s) d\beta. \tag{5.7.25}$$

Setting $\xi = s\tau$ and using the inverse Hankel transformation, we arrive at the solution

$$v = \int_0^\infty \sigma^{-1} J_1(\tau\sigma) \sin \tau\lambda \, d\tau \int_0^\infty \sigma_0^2 J_1(\tau\sigma_0) f(\sigma_0) d\sigma_0. \tag{5.7.26}$$

Recalling the relation (5.7.14), we obtain from (5.7.26) the potential function

$$\phi = -\int_0^\infty \tau^{-1} J_0(\tau\sigma) \sin \tau\lambda d\tau \int_0^\infty \sigma_0^2 J_1(\tau\sigma_0) f(\sigma_0) d\sigma_0. \tag{5.7.27}$$

Carrier and Greenspan (1958) choose particular forms of initial profile $\eta(x, 0)$ and, hence, $f(\sigma)$ to find some explicit solutions (5.7.26) and illustrate the evolution of this profile, starting from rest, under the shallow-water equations (5.7.1)-(5.7.2). The results are shown graphically.

Spielvogel (1975), following closely the analysis of Carrier and Greenspan (1958), considered the inverse problem: with the run-up data as the initial state, he tried to identify the wave that gave rise to such a run-up. In this case, an initial velocity equal to zero is interpreted to mean that all the kinetic energy of the wave has been transformed into potential energy of the run-up. The solution is used to discover the possibilities of high shoreline amplification and run-up. The implications in the context of generation of tsunamis were also discussed.

The system (5.7.1)-(5.7.2), when transformed to (5.7.15), has some interesting structures, as was shown by Sachdev and Narasimha Chari (1982). Indeed, (5.7.15) has multinomial solutions. When the number of terms in the multinomials is allowed to tend to infinity, the solutions for specific initial value problems for (5.7.15) may be found. Here, we discuss the multinomial solutions and recover solution of an IVP found earlier by Spielvogel (1975) using transform techniques.

First we observe that (5.7.15) has the symmetry $\phi(-\sigma, \lambda) = \phi(\sigma, \lambda)$. Therefore, we seek multinomial solutions of degree $2n$ in σ and degree n in λ:

$$\phi_n(\sigma, \lambda) = \text{constant} + \sum_{i=0}^{n} \sum_{j=0}^{n} a_{i,j} \sigma^{2i} \lambda^j. \tag{5.7.28}$$

Substituting (5.7.28) into (5.7.15) and using the zero initial condition $v_n(\sigma, o) = 0$ (see (5.7.14)) lead to

$$\phi_n(\sigma, \lambda) = \text{constant} + \sum_{i=0}^{n} \sigma^{2i} \left(\sum_{j=0}^{n-i} a_{i,j} \lambda^{2j+1} \right) \tag{5.7.29}$$

where

$$a_{i+1,j} = \frac{(2j+3)(2j+2)}{(2i+2)^2} a_{i,j+1}. \tag{5.7.30}$$

The solution (5.7.29)-(5.7.30) corresponds to the initial profile

$$\eta_n(\sigma, 0) = \frac{(\phi_n)_\lambda(\sigma, 0)}{4} = \sum_{i=0}^{n} \frac{a_{i,0} \sigma^{2i}}{4} \tag{5.7.31}$$

with

$$x = \eta_n(\sigma, 0) - \frac{\sigma^2}{16} \tag{5.7.32}$$

(see (5.7.16)-(5.7.17)). Successive use of (5.7.30) yields

$$\begin{aligned}
a_{i,j} &= \frac{(2i+2)^2(2i+4)^2 \ldots (2i+2j)^2}{(2j+1)(2j) \ldots 3.2.1} a_{i+j,0} \\
&= \frac{2^{2j}[(i+j)!]^2}{(i!)^2(2j+1)!} a_{i+j,0}
\end{aligned} \tag{5.7.33}$$

giving $a_{i,j}$ in terms of the coefficients $a_{i+j,0}$ appearing in the initial profile.

The physical solution can be found with the help of (5.7.14), (5.7.16), and (5.7.17). We give below the first few multinomial solutions and the corresponding profiles. Here, a_i are arbitrary constants.

$n = 1$.

$$\begin{aligned}
\phi &= a_0 + a_1\lambda + a_3\lambda^3 + \frac{3a_3}{2}\sigma^2\lambda \\
v &= 3a_3\lambda \\
\eta &= \frac{1}{4}\left(a_1 + 3a_3\lambda^2 + \frac{3a_3}{2}\sigma^2 \right) - \frac{(3a_3\lambda)^2}{2} \\
x &= \eta(\sigma, \lambda) - \frac{\sigma^2}{16}
\end{aligned}$$

The initial profile is parametrically given by

$$\eta(\sigma,0) = \frac{1}{4}\left(a_1 + \frac{3a_3}{2}\sigma^2\right)$$

$$x(\sigma,0) = r(\sigma,0) - \frac{\sigma^2}{16} \tag{5.7.34a}$$

$n = 2$.

$$\phi = a_0 + a_1\lambda + a_3\lambda^3 + a_5\lambda^5 + \sigma^2\left(\frac{3a_3}{2}\lambda + 5a_5\lambda^3\right) + \frac{15}{8}a_5\lambda\sigma^4$$

$$v = 3a_3\lambda + 10a_5\lambda^3 + \frac{15}{2}a_5\sigma^2\lambda$$

$$\eta = \frac{1}{4}\left(a_1 + 3a_3\lambda^2 + 5a_5\lambda^4 + \frac{3a_3}{2}\sigma^2 + 15a_5\sigma^2\lambda^2 + \frac{15}{8}a_5\sigma^4\right) - \frac{1}{2}v^2$$

$$x = \eta - \frac{\sigma^2}{16}.$$

The initial profile is

$$\eta(\sigma,0) = \frac{a_1}{4} + \frac{3a_3}{8}\sigma^2 + \frac{15}{32}a_5\sigma^4$$

$$x(\sigma,0) = \eta(\sigma,0) - \frac{\sigma^2}{16} \tag{5.7.34b}$$

$n = 3$.

$$\phi = a_0 + a_1\lambda + a_3\lambda^3 + a_5\lambda^5 + a_7\lambda^7$$
$$+\sigma^2\left(\frac{3a_3}{2}\lambda + 5a_5\lambda^3 + \frac{21}{2}a_7\lambda^5\right)$$
$$+\sigma^4\left(\frac{15a_5}{8}\lambda + \frac{105}{8}a_7\lambda^3\right) + \frac{35}{16}a_7\sigma^6\lambda$$

$$v = 3a_3\lambda + 10a_5\lambda^3 + 21a_7\lambda^5 + \sigma^2\left(\frac{15}{2}a_5\lambda + \frac{105}{2}a_7\lambda^3\right) + \frac{105}{8}a_7\sigma^4\lambda$$

$$\eta = \frac{1}{4}\left[a_1 + 3a_3\lambda^2 + 5a_5\lambda^4 + 7a_7\lambda^6 + \sigma^2\left(\frac{3a_3}{2} + 15a_5\lambda^2 + \frac{105}{2}a_7\lambda^4\right)\right.$$
$$\left.+\sigma^4\left(\frac{15a_5}{8} + \frac{315}{8}a_7\lambda^2\right) + \frac{35}{16}a_7\sigma^6\right] - \frac{1}{2}v^2$$

$$x = \eta(\sigma,\lambda) - \frac{\sigma^2}{16}.$$

The initial profile in this case is

$$\eta(\sigma,0) = \frac{a_1}{4} + \frac{3a_3}{8}\sigma^2 + \frac{15}{32}a_5\sigma^4 + \frac{35}{64}a_7\sigma^6$$

$$x = \eta(\sigma,0) - \frac{\sigma^2}{16} \tag{5.7.34c}$$

Sachdev and Narasimha Chari (1982) have checked that the multinomial solutions themselves do not give particularly useful physical solutions, but can be profitably used to validate the numerical scheme for a solution of an arbitrary IVP.

We conclude the discussion of multinomial solutions by recovering a solution of Spielvogel (1975), who considered the implicit initial profile

$$\eta_0 = \eta(x, 0) = A \, \exp\left[16p(x - \eta_0)\right] \tag{5.7.35}$$

or, in view of (5.7.16)-(5.7.17),

$$\eta_0 = \eta(\sigma, 0) = A \exp(-p\sigma^2) = A \sum_{k=0}^{\infty} \frac{(-p\sigma^2)^k}{k!}. \tag{5.7.36}$$

This initial condition generates the solution

$$\bar{\eta}(\sigma, \lambda) \equiv \eta(\sigma, \lambda) + \frac{\sigma^2}{16} = \frac{\phi_\lambda}{4}$$
$$= A \sum_{k=0}^{\infty} \sum_{s=0}^{\infty} \frac{(-4p\lambda^2)^s(-p\sigma^2)^k(s+k)!}{2s!(k!)^2}. \tag{5.7.37}$$

The choice

$$a_{k,0} = \frac{4A(-p)^k}{k!} \tag{5.7.38}$$

in (5.7.33) immediately gives

$$a_{k,s} = \frac{4A(k+s)!}{(2s+1)!(k!)^2}(-p)^k(-4p)^s. \tag{5.7.39}$$

The solution (5.7.37) easily follows from (5.7.28) as n is allowed to tend to infinity.

Sachdev and Narasimha Chari (1982) also rederived some of the solutions of Carrier and Greenspan (1958), obtained by them via integral transform techniques.

The work of Carrier and Greenspan (1958) was extended by Tuck and Hwang (1972) to take into account the ground motion where the bottom is uniformly sloping and the resulting wave propagates away into the deeper water. This situation is somewhat closer to common seismic tsunami-generating mechanisms. Tuck and Hwang (1972) also generalised the transformations due to Carrier and Greenspan (1958) in a small way for the case of no ground motion and uniformly sloping beach such that the solutions obtained for linear problem for small motions can be directly used for the large amplitude case with merely a change of notation.

We conclude this section by referring to a general result regarding linearising of a two-dimensional inhomogeneous system,

$$a_1 u_x + b_1 u_t + c_1 v_x + d_1 v_t + e_1 = 0 \tag{5.7.40}$$
$$a_2 u_x + b_2 u_t + c_2 v_x + d_2 v_t + e_2 = 0, \tag{5.7.41}$$

due to Seymour and Varley (1984); here a_1, a_2, b_1, b_2, etc. are functions
of (u, v, x, t). It is claimed that, for certain types of nonlinear systems
(5.7.40)-(5.7.41), if $x = X(u, v), t = T(u, v)$ is any particular solution, then
through the transformation

$$\bar{x} = x - X(u, v), \quad \bar{t} = t - T(u, v) \qquad (5.7.42)$$

the system (5.7.40)-(5.7.41) may be changed to a linear form. They illustrate this result with some examples. We consider the shallow-water
equations on a sloping beach, discussed earlier in this section, in the form

$$\eta_t + [u(\eta + \alpha x)]_x = 0 \qquad (5.7.43)$$
$$u_t + uu_x + g\eta_x = 0 \qquad (5.7.44)$$

where α and g are the (uniform) slope of the beach and acceleration due
to gravity, respectively. The system (5.7.43)-(5.7.44) is first rewritten in
terms of the variables

$$\bar{\eta} = \eta + u^2/2g \qquad (5.7.45)$$

and u:

$$u_t + g\bar{\eta}_x = 0 \qquad (5.7.46)$$
$$\bar{\eta}_t + u\bar{\eta}_x + \left(-\frac{u}{g}\right)u_t + \left(\alpha x + \bar{\eta} - \frac{3u^2}{2g}\right)u_x + \alpha u = 0. \qquad (5.7.47)$$

It is easy to see that Equations (5.7.43)-(5.7.44) possess the simple solution

$$\eta = -\alpha x, \quad u = g\alpha t. \qquad (5.7.48)$$

In terms of $\bar{\eta}$ and u, the solution (5.7.48) becomes

$$\bar{\eta} = -\alpha x + \frac{1}{2}g\alpha^2 t^2 \qquad (5.7.49)$$

$$u = g\alpha t \qquad (5.7.50)$$

which inverts to give

$$x = X(u, \bar{\eta}) = -\bar{\eta}/\alpha + u^2/2g\alpha \qquad (5.7.51)$$
$$t = T(u) = u/g\alpha. \qquad (5.7.52)$$

Now, defining the new variables according to (5.7.42), we have

$$\bar{x} = x - X(u, \bar{\eta}) = x + \bar{\eta}/\alpha - u^2/2g\alpha \qquad (5.7.53)$$
$$\bar{t} = t - T(u) = t - u/g\alpha. \qquad (5.7.54)$$

The system (5.7.46)-(5.7.47) in terms of the variables $\bar{x}$ and $\bar{t}$ becomes
linear:

$$u_{\bar{t}} + g\bar{\eta}_{\bar{x}} = 0 \qquad (5.7.55)$$

and

$$\overline{\eta}_{\overline{t}} + \alpha(\overline{x}u)_{\overline{x}} = 0. \tag{5.7.56}$$

In fact, this is the linearisation first obtained by Tuck and Hwang (1972). Seymour and Varley (1984) also discussed the linearisation of the (nonlinear) telegraph equation

$$\phi_{xx} - d(\phi_t)\phi_{tt} - e(\phi_t) = 0. \tag{5.7.57}$$

It is clear that the result enunciated by Seymour and Varley (1984) is not applicable to all equations of the form (5.7.40)-(5.7.41).

5.8 Simple Waves on Shear Flows

We discuss the present example for two reasons: first to show how even complicated nonlinear flows with free surfaces may have exact solutions, and second to demonstrate that while the original PDE system may not be linearisable, its changed ODE form in terms of similarity variable may admit exact linearisation.

The system of PDEs governing free surface flows in the hydraulic approximation is

$$u_t + uu_x + vu_y + gh_x = 0 \tag{5.8.1}$$
$$u_x + v_y = 0 \tag{5.8.2}$$

where (u, v) are velocity components in the x and y directions, the latter being measured along and perpendicular to the uniform bottom, respectively. The hydraulic approximation permits replacement of the pressure term in the original momentum equation in terms of "uniform" gravitation pressure $g\rho h$, where ρ is the density of water, g is acceleration due to gravity, and $h(x, t)$ is the depth of water in the channel. The boundary conditions on the flow are

$$v = h_t + uh_x \tag{5.8.3}$$
$$u_t + uu_x + gh_x = 0 \tag{5.8.4}$$

on the free surface $y = h(x, t)$, and

$$v = 0 \quad \text{on } y = 0 \tag{5.8.5}$$

at the bottom. The special (simple wave) solutions are sought such that u and h are constant on the lines

$$\frac{dx}{dt} = c(h(x, t)). \tag{5.8.6}$$

Thus,

$$u = u(x - ct, y) \tag{5.8.7}$$
$$h = h(x - ct). \tag{5.8.8}$$

The functions $c(h)$ and $h(x, t)$ are themselves unknown and must be found as part of the solution.

By introducing the variables

$$w_1 = v \left(\frac{\partial h}{\partial x} \right)^{-1}, \quad \xi = x - ct, \quad h = h(\xi) \tag{5.8.9}$$

in (5.8.1)-(5.8.2) and the boundary conditions (5.8.3)-(5.8.5), we arrive at the system

$$(u - c) \frac{\partial u}{\partial h} + w_1 \frac{\partial u}{\partial y} + g = 0 \tag{5.8.10}$$

$$\frac{\partial u}{\partial h} + \frac{\partial w_1}{\partial y} = 0 \tag{5.8.11}$$

$$w_1 = (u - c), (u - c) \frac{\partial u}{\partial h} + g = 0 \quad \text{on } y = h \tag{5.8.12}$$

and

$$w_1 = 0 \quad \text{on } y = 0. \tag{5.8.13}$$

Before we discuss the solution of the problem (5.8.10)-(5.8.13) by exact linearisation, we briefly describe how it was first solved by Freeman (1972) directly.

Eliminating $\frac{\partial u}{\partial h}$ from (5.8.10) and (5.8.11), we have

$$\frac{\partial}{\partial y} \left(\frac{w_1}{u - c} \right) = \frac{g}{(u - c)^2} \tag{5.8.14}$$

which, on integration, gives

$$w_1 = g(u - c) \int_0^y \frac{dy}{(u - c)^2} \tag{5.8.15}$$

where we have used the BC (5.8.13). The second BC (5.8.12$_1$) requires that

$$\int_0^h \frac{g \, dy}{(u - c)^2} = 1. \tag{5.8.16}$$

Taking a cue from (5.8.16), one introduces the function

$$I = g \int_0^y \frac{dy}{(u - c)^2} \tag{5.8.17}$$

so that the former may be easily satisfied. Expressing u in terms of I from (5.8.17) and using (5.8.15), Equation (5.8.10) transforms to

$$\frac{\partial^2 I}{\partial h \partial y} + I \frac{\partial^2 I}{\partial y^2} = 2 \frac{\partial I}{\partial y} \left[\frac{\partial I}{\partial y} - \frac{c'(h)}{g^{1/2}} \left(\frac{\partial I}{\partial y} \right)^{1/2} \right]. \tag{5.8.18}$$

From (5.8.16) and (5.8.17) it immediately follows that

$$I(h, h) = 1, \quad I(h, 0) = 0. \tag{5.8.19}$$

The condition $(5.8.12_2)$ is not imposed. The hyperbolic equation (5.8.18) subject to (5.8.19), with appropriate BC, $I = I(y, h_0)$ on $h = h_0$, seems difficult to solve generally. This set of supernumerary conditions must be imposed to find the function $c = c(h)$ in the process as well. Here we consider a special circumstance. First, we normalise y by introducing the variable $Y = \dfrac{y}{h}$ in (5.8.18). We obtain

$$h \frac{\partial^2 I}{\partial Y \partial h} + (I - Y) \frac{\partial^2 I}{\partial Y^2} = \frac{\partial I}{\partial Y} + 2 \frac{\partial I}{\partial Y} \left[\frac{\partial I}{\partial Y} - \frac{c'(h) h^{1/2}}{g^{1/2}} \left(\frac{\partial I}{\partial Y} \right)^{1/2} \right]. \tag{5.8.20}$$

If we assume that $\dfrac{c'(h) h^{1/2}}{g^{1/2}} = \alpha$, a constant, so that $c = 2\alpha(gh)^{1/2}$, and let I be a function of Y alone, we get a nonlinear ODE for I with Y as the independent variable:

$$(I - Y) \frac{d^2 I}{dY^2} = \frac{dI}{dY} + 2 \frac{dI}{dY} \left[\frac{dI}{dY} - \alpha \left(\frac{dI}{dY} \right)^{1/2} \right]. \tag{5.8.21}$$

The boundary conditions (5.8.19) now become

$$I(1) = 1, \quad I(0) = 0. \tag{5.8.22}$$

Equation (5.8.21) seems daunting but can be solved after several transformations. With $J = I - Y$, it can be shown to reduce to the first-order ODE

$$J J' \frac{dJ'}{dJ} = 2(1 + J')^2 - 2\alpha(1 + J')^{3/2} + (1 + J'). \tag{5.8.23}$$

Now, putting $N^2 = J' + 1$ and integrating, we arrive at the solution

$$J = C(a - N)^{(a^2 - 1)/a(a-b)} (N - b)^{(b^2 - 1)/b(b-a)} / N^{1/ab} \tag{5.8.24}$$

where a and $b = \dfrac{1}{2a}$ are roots of the quadratic

$$\lambda^2 - \alpha\lambda + \frac{1}{2} = 0, \tag{5.8.25}$$

and C is an arbitrary constant. Since $\left(\dfrac{dJ}{dY}\right) = N^2 - 1$, we may use (5.8.24) to find the solution as

$$Y = C_1 \int (a - N)^{-1/(2a^2-1)} \left(N - \frac{1}{2}a^{-1}\right)^{2a^2/(2a^2-1)} \frac{dN}{N^3} \qquad (5.8.26)$$

where C_1 is another constant. We can write (5.8.26) in terms of incomplete beta functions. Imposing the BCs (5.8.22) on this form, we arrive at the solution

$$Y = \frac{\beta\left(\frac{4a^2-1}{2a^2-1}, \frac{2(a^2-1)}{2a^2-1}, \frac{U-a^{-1}}{2a-a^{-1}}\right)}{\beta\left(\frac{4a^2-1}{2a^2-1}, \frac{2(a^2-1)}{2a^2-1}, 1\right)}. \qquad (5.8.27)$$

Here, β denotes the incomplete beta function. The solution (5.8.27) satisfies the conditions (5.8.22). To see that, we note that

$$N^2 = \frac{dI}{dY} = \frac{gh}{(u-c)^2}$$

so that

$$u = (2a - N^{-1})(gh)^{1/2}. \qquad (5.8.28)$$

Writing $U = \dfrac{u}{(gh)^{1/2}}$ and noting (5.8.19), (5.8.24), and (5.8.25), as well as the relation $J = I - Y$, we find that

$$U = \frac{1}{a} \text{ at } Y = 0 \qquad (5.8.29)$$
$$U = 2a \text{ at } Y = 1. \qquad (5.8.30)$$

The conditions (5.8.29) and (5.8.30) are clearly satisfied by the solution (5.8.27). We also note that the (normalised) wave propagation speed

$$\bar{c} = \frac{c}{(gh)^{1/2}} = 2\alpha = (a + 2a^{-1}). \qquad (5.8.31)$$

For $a > 1$, $\bar{c} > 2a > \bar{u}_{\max}$. Thus $u \neq c$ anywhere in the range of u, and so there is no critical point in the flow. We observe that (5.8.31) gives the wave speed in terms of the free surface $h(x, t)$, which itself remains arbitrary.

Now we derive the above solution by (exact) linearisation. For that purpose, we seek the solution of (5.8.10), (5.8.11), (5.8.12$_1$), and (5.8.13) directly in the self-similar form

$$u = y^{1/2} P(Y) \qquad (5.8.32)$$
$$w_1 = y^{1/2} Q(Y) \qquad (5.8.33)$$
$$c = C_0 h^{1/2}, \quad Y = y/h. \qquad (5.8.34)$$

The form (5.8.32)-(5.8.34) is easily guessed by writing the power law similarity forms (see Sachdev and Philip (1986) for derivation by the group theoretic method). Equations (5.8.10)-(5.8.11) transform to

$$2Y^2 P' - 2YQ' - Q = 0 \tag{5.8.35}$$

$$\left(\frac{1}{2}P + YP'\right)Q - Y^2 PP' + 2\alpha g^{1/2} Y^{3/2} P' + g = 0 \tag{5.8.36}$$

where we choose $c_0 = 2\alpha g^{1/2}$ to conform to the earlier form of the solution; the prime denotes derivative with respect to Y. From (5.8.36), we have

$$Q = \frac{Y^2 PP' - 2\alpha g^{1/2} Y^{3/2} P' - g}{\frac{1}{2}P + YP'}. \tag{5.8.37}$$

Differentiating (5.8.37), we have

$$Q' = \frac{1}{\left(\frac{1}{2}P + YP'\right)^2}\left[\left(\frac{1}{2}Y^2 P^2 - \alpha g^{1/2} Y^{3/2} P + gY\right)P''\right.$$
$$\left. + (YP^2 - \frac{3}{2}\alpha g^{\frac{1}{2}} Y^{\frac{1}{2}} P + \frac{3}{2}g)P' + Y^2 PP'^2 + Y^3 P'^3\right]. \tag{5.8.38}$$

Eliminating Q and Q' from (5.8.35) with the help of (5.8.37) and (5.8.38), we have

$$(Y^3 P^2 - 2\alpha g^{1/2} Y^{5/2} P + 2gY^2)P''$$
$$+ (2gY + 2Y^2 P^2 - 4\alpha g^{1/2} Y^{3/2} P)P'$$
$$+ (Y^3 P - 2\alpha g^{1/2} Y^{5/2})P'^2 - \frac{1}{2}gP = 0. \tag{5.8.39}$$

Equation (5.8.39) simplifies considerably if we introduce the normalised variable

$$U = u/(gh)^{1/2} = g^{-1/2} Y^{1/2} P(Y); \tag{5.8.40}$$

we obtain

$$(U^2 - 2\alpha U + 2)\frac{d^2 U}{dY^2} + (U - 2\alpha)\left(\frac{dU}{dY}\right)^2 = 0. \tag{5.8.41}$$

If we interchange the dependent and independent variables in (5.8.41) (a hodograph transformation) we get the linear ODE

$$(U^2 - 2\alpha U + 2)\frac{d^2 Y}{dU^2} + (2\alpha - U)\frac{dY}{dU} = 0 \tag{5.8.42}$$

with the solution

$$Y = A_1 + B_1 \beta \left(\frac{4 - \lambda_1^2}{2 - \lambda_1^2}, \frac{2 - 2\lambda_1^2}{2 - \lambda_1^2}, \frac{U - \lambda_1}{2\lambda_1^{-1} - \lambda_1} \right), \tag{5.8.43}$$

where λ_1 is the smaller root of $\lambda^2 - 2\alpha\lambda + 2 = 0$. The function Q in (5.8.37) in terms of U is

$$Q = \frac{g^{1/2}Y(U - 2\alpha)U' - \frac{1}{2}g^{1/2}(U^2 - 2\alpha U + 2)}{Y^{1/2}U'}. \tag{5.8.44}$$

The boundary condition (5.8.13) reduces to $Y^{1/2}Q = 0$ on $Y = 0$, and in view of (5.8.44) becomes $U = \lambda_1$ at $Y = 0$. The condition $(5.8.12_1)$ in view of (5.8.44) becomes $U = \lambda_2$ at $Y = 1$. With these conditions, the solution (5.8.43) assumes the form

$$Y = \beta \left(\frac{4 - \lambda_1^2}{2 - \lambda_1^2}, \frac{2 - 2\lambda_1^2}{2 - \lambda_1^2}, \frac{U - \lambda_1}{2\lambda_1^{-1} - \lambda_1} \right) \Big/ \beta \left(\frac{4 - \lambda_1^2}{2 - \lambda_1^2}, \frac{2 - 2\lambda_1^2}{2 - \lambda_1^2}, 1 \right). \tag{5.8.45}$$

The above derivation requires that $(6 - 2\lambda_1^2)/(2 - \lambda_1^2)$ must not be a negative integer or zero. We have thus recovered (5.8.27) with $\lambda_1 = a^{-1}$.

5.9   *C*-Integrable Nonlinear PDEs

Calogero (1991) called those nonlinear PDEs which can be linearised by an appropriate change of variables "*C*-integrable:" this is in contrast to *S*-integrable PDEs which require spectral transform techniques. A class of *C*-integrable equations was obtained via the change of dependent variable

$$v(x, t) = u(x, t) \exp \left\{ \int_{a(t)}^{x} dx' F[u(x', t)] \right\} \tag{5.9.1}$$

which associates a nonlinear evolution PDE for $u(x, t)$ to a linear evolution PDE satisfied by $v(x, t)$. Here, the function $F(u)$ may be chosen arbitrarily. The transformation (5.9.1) is not the most general but generates an interesting class of equations of evolutionary type. It follows from (5.9.1) that

$$u[a(t), t] = v[a(t), t] \tag{5.9.2}$$

and

$$\frac{u_x(x, t)}{u(x, t)} + F[u(x, t)] = \frac{v_x(x, t)}{v(x, t)}. \tag{5.9.3}$$

If $F(u)$ and $a(t)$ are known, then (5.9.1) gives v in terms of u. On the other hand, if $v(x, t)$, the solution of a linear PDE, is known, to find $u(x, t)$ one must solve (5.9.3), a nonlinear nonautonomous first-order PDE with

(5.9.2) as the boundary condition. t, in the above, is a parameter. The special choices $F(u) = Au^\lambda$, $A \ln(u)$, where A and λ are constants, lead to simple transformations

$$u(x,t) = v(x,t) \left\{ 1 + \lambda A \int_{a(t)}^{x} dx' [v(x',t)]^\lambda \right\}^{-1/\lambda} \tag{5.9.4}$$

and

$$u(x,t) = v(x,t) \exp \left\{ -A \exp(-Ax) \int_{a(t)}^{x} \exp(Ax') \ln[v(x',t)] dx' \right\}, \tag{5.9.5}$$

respectively.

As the simplest example, if v satisfies the heat equation

$$v_t - v_{xx} = 0, \tag{5.9.6}$$

then it can be easily shown that $u(x,t)$ satisfies the generalized Burgers equation

$$u_t - u_{xx} + fu = 2u_x F(u) \tag{5.9.7}$$
$$f_x + uF'(u)f = -u_x^2 F''(u), \tag{5.9.8}$$

where $u = u(x,t), f = f(x,t)$, and the prime denotes differentiation with respect to the argument. For consistency of (5.9.1), (5.9.6), (5.9.7), and (5.9.8), $a(t)$ must satisfy the ODE

$$\{ F\dot{a}(t) + F^2 + F'u_x + f \}|_{x=a(t)} = 0 \tag{5.9.9}$$

or, in view of (5.9.2) and (5.9.3),

$$\{ F\dot{a}(t) + F^2 + F'(v_x - Fv) + f \}|_{x=a(t)} = 0. \tag{5.9.10}$$

Here, dot denotes $\dfrac{d}{dt}$ and F, etc., are evaluated at $x = a(t)$, that is, F stands for $F\{u[a(t),t]\}$.

Calogero gives a large number of simple second-order linear equations with constant coefficients of both parabolic and hyperbolic types which, with the help of (5.9.1), generate nonlinear PDEs of C-integrable type. Some of the latter are solved explicitly.

The transformation (5.9.1) is clearly a generalisation of the Cole-Hopf transformation and generates a large class of nonlinear PDEs whose potential must be carefully examined. A rather similar approach was adopted much earlier by Sachdev (1978). A generalized Cole-Hopf transformation was introduced into linear parabolic and hyperbolic equations with variable coefficients to find what class of nonlinear PDEs would be generated, and hence find solutions of the latter via the known solutions of the former. Here we give details relating to the parabolic case and summarise the results for the hyperbolic one.

We begin with the linear parabolic equation

$$\phi_t + b\phi_x + c\phi + f = \epsilon a \phi_{xx} \tag{5.9.11}$$

where $a, b,$ and c are functions of x and t, and ϵ is a parameter. We introduce the generalised Cole-Hopf transformation

$$F(u) = k(x, t)(\ln \phi)_x \tag{5.9.12}$$

into (5.9.11). The derivatives are easily found:

$$k\frac{\phi_x}{\phi} = F, \ k\frac{\phi_{xx}}{\phi} = F'(u)u_x - \frac{k_x}{k}F + \frac{F^2}{k}$$

$$k\frac{\phi_t}{\phi} = \epsilon a\left(F'u_x - \frac{k_x}{k}F + \frac{F^2}{k}\right) - bF - ck - \frac{fk}{\phi}$$

$$k\frac{\phi_{xt}}{\phi} = F'u_t + \epsilon a\frac{FF'}{k}u_x$$

$$\qquad + F\left(-\frac{k_t}{k} + \frac{\epsilon a}{k}\left(-\frac{k_x}{k}F + \frac{F^2}{k}\right) - \frac{b}{k}F - c - \frac{f}{\phi}\right)$$

$$k\frac{\phi_{xxx}}{\phi} = \left(\frac{F}{k} - \frac{k_x}{k}\right)\left(F'u_x - \frac{k_x}{k}F + \frac{F^2}{k}\right)$$

$$\qquad + \left(F''u_x^2 + F'u_{xx} - \left(\frac{k_x}{k}\right)_x F - \frac{k_x}{k}F'u_x\right.$$

$$\qquad\left. + \frac{2FF'}{k}u_x - \frac{F^2}{k^2}k_x\right). \tag{5.9.13}$$

Differentiating (5.9.11) with respect to x, we have

$$k\frac{\phi_{xt}}{\phi} + (b - \epsilon a_x)k\frac{\phi_{xx}}{\phi} + (b_x + c)\frac{k\phi_x}{\phi} + kc_x + \frac{kf_x}{\phi} = \epsilon ak\frac{\phi_{xxx}}{\phi}. \tag{5.9.14}$$

Substituting the ϕ derivatives from (5.9.13) into (5.9.14), we get the nonlinear PDE for u alone provided that $f = 0$:

$$u_t + \left(b - \epsilon a_x + 2\epsilon a\frac{k_x}{k} - 2\epsilon a\frac{F}{k}\right)u_x - \epsilon a\frac{F''}{F'}u_x^2$$

$$\qquad + \frac{F}{F'}\left[\frac{\epsilon a}{k^2}(kk_{xx} - 2k_x^2) - \frac{k_t}{k} - \frac{bk_x}{k} + \epsilon a_x\frac{k_x}{k}\right.$$

$$\qquad\left. + b_x + \frac{kc_x}{F} + F\left(\frac{2\epsilon ak_x}{k^2} - \frac{\epsilon a_x}{k}\right)\right] = \epsilon au_{xx}. \tag{5.9.15}$$

We consider some special cases of (5.9.15).

(i) The quadratic term u_x^2 is absent so that (5.9.15) is a generalised Burgers equation if $F''(u) = 0$, that is, if F is a linear function of u. In particular, if $F = u$ and a is a constant, (5.9.15) can be written as an exchange process (Murray (1968), (1970)):

$$u_t + v_x + \left(\epsilon a \frac{k_x}{k^2} u - \epsilon a \frac{k_{xx}}{k} - \frac{k_t}{k} - \frac{bk_x}{k} \right) u + kc_x = 0 \qquad (5.9.16)$$

$$v = \left(b + 2\epsilon a \frac{k_x}{k} \right) u - \frac{\epsilon a}{k} u^2 - \epsilon a u_x. \qquad (5.9.17)$$

The system (5.9.16)-(5.9.17) generalises the Burgers equation expressed as a system

$$u_t + v_x = 0 \qquad (5.9.18)$$

$$v = -\epsilon a u_x + \frac{u^2}{2} \qquad (5.9.19)$$

and gives a class of nonlinear parabolic equations with constant viscous coefficient, which may be transformed into a linear parabolic equation via (5.9.12).

(ii) If we choose $k = -2a$ so that the coefficient in the transformation (5.9.12) conforms to that in the standard Cole-Hopf transformation, then (5.9.15) reduces to

$$u_t + (b + \epsilon a_x + \epsilon F)u_x - \epsilon a \frac{F''}{F'} u_x^2 + \frac{F}{F'} \left[\epsilon \left(a_{xx} - \frac{a_x^2}{a} \right) \right.$$
$$\left. - \frac{\epsilon F}{2} \frac{a_x}{a} - \frac{a_t}{a} - \frac{ba_x}{a} + b_x - \frac{2ac_x}{F} \right] = \epsilon a u_{xx}. \qquad (5.9.20)$$

With $F = u$, (5.9.20) is a GBE with damping and variable viscosity. The term $b + \epsilon a_x$ in the coefficient of u_x, even when assumed to be constant, plays a nontrivial role in the solution of (5.9.20), (Murray (1970)). If we put $b + \epsilon a_x = 0$ and $F = u$, (5.9.20) simplifies to

$$u_t + \epsilon u u_x - u \left(\epsilon u \frac{a_x}{a} + \frac{a_t}{a} \right) - 2ac_x = \epsilon a u_{xx} \qquad (5.9.21)$$

so that we may have a damping term which is either a constant or a linear function or a quadratic function of u, depending on the choice of a and c. The variable a could be interpreted as a variable coefficient of viscosity. In (5.9.21), if we further let a and (hence) k to be functions of t alone, we arrive at the equation

$$u_t + \epsilon u u_x - \frac{a_t}{a} u = \epsilon a(t) u_{xx}. \qquad (5.9.22)$$

Equation (5.9.22) includes Burgers model for turbulence, but with a variable viscosity (see Case and Chu (1969); Murray (1973)).

(iii) If we let a, k, and c in (5.9.20) be constant, but allow b to be a function of x and t, then it becomes

$$u_t + (b + \epsilon F)u_x - \epsilon a \frac{F''}{F'} u_x^2 + \frac{F}{F'} b_x = \epsilon a u_{xx}, \qquad (5.9.23)$$

a useful equation with constant diffusivity, which, for $F = u$, represents a GBE with a linear damping term which may also depend on x and t through b_x.

For a discussion of the general solution of the linear parabolic equation (5.9.11), one may refer to Friedman (1964) and Colton (1976). For similarity solution of (5.9.11), when a, b, and c are functions of x alone, one may refer to Lehnigk (1976) and Sachdev (1978). For the fundamental solution of (5.9.11), Swan (1977) may be consulted.

In a subsequent study, Nimmo and Crighton (1982) sought, in a systematic way, Bäcklund transformations (BTs) for the nonlinear parabolic equations of the form

$$u_t + u_{xx} + H(u_x, u, x, t) \equiv q + r + H(p, u, x, t) = 0 \qquad (5.9.24)$$

with the usual notation

$$q = u_t, \ p = u_x, \ r = u_{xx}, \qquad (5.9.25)$$

where $H(u_x, u, x, t)$ is an arbitrary function of the indicated variables. Equation (5.9.24) includes BE with $H = 2uu_x$ and an inhomogeneous BE with $H = 2uu_x - f(x, t)$ as special cases (after allowing for minor changes in the sign of t and some scaling). Consider the transformation

$$p' = f(t, x, u, u', p, q)$$
$$\qquad\qquad\qquad\qquad\qquad\qquad\qquad (5.9.26)$$
$$q' = \psi(t, x, u, u', p, q)$$

where $p' = \dfrac{\partial u'}{\partial x}, q' = \dfrac{\partial u'}{\partial t}$. The functions f and ψ are sought such that the pair of equations (5.9.26) is integrable if and only if u and u' satisfy the equations

$$r + q + H(p, u, x, t) = 0$$
$$\qquad\qquad\qquad\qquad\qquad\qquad\qquad (5.9.27)$$
$$r' + q' + G(p', u', x, t) = 0.$$

When this is the case, Equations (5.9.26) define a BT between the two equations in (5.9.27).

The main conclusion of Nimmo and Crighton (1982) regarding (5.9.24) is that the only nonlinear equations in this class that admit BTs are a slight generalisation of the Burgers equation, namely the inhomogeneous Burgers equation. They specifically show that the BTs of the above type do not exist for modified BE with cubic nonlinearity, for the spherical and cylindrical Burgers equations, or, indeed, for any other equation of this class (see also Sachdev (1987)).

Sachdev (1978) also attempted to generate nonlinear hyperbolic equations via (5.9.12) from the corresponding linear one, namely

$$\phi_t + b\phi_x + c\phi + f = \epsilon a\phi_{xt}, \qquad (5.9.28)$$

where a, b, c, and f are functions of x and t only, and ϵ is a parameter. Since the steps are entirely analogous to those for the parabolic case, we skip the details here. Several exchange processes were identified, including those treated by Goldstein and Murray (1959).

Chapter 6

Nonlinearisation and Embedding of Special Solutions

6.1 Introduction

It is straightforward to first linearise a given nonlinear PDE, seek to solve the resulting simpler PDE, and use it later to build up nonlinear effects either approximately or (more rarely) exactly. Many a time the linear solution is the large-time asymptotic solution describing the situation when the nonlinear effects have become negligible. The improvised nonlinear solution describes an earlier in time (or space) regime when the nonlinearity is still important. Before we take up some well-known model equations that use these ideas, we illustrate the approach with the help of some simple examples (Whitham (1974)). This concept was used by Landau (1945) and Whitham (1950), (1952) in the context of shock waves at large distances from their place of origin. From an explosion, for example.

Starting with the first-order wave equation

$$c_t + cc_x = 0, \tag{6.1.1}$$

we consider the situation when the disturbances upon the unperturbed state c_0 are small so that $\dfrac{c - c_0}{c_0} \ll 1$ and (6.1.1) may be replaced by

$$c_t + c_0 c_x = 0 \tag{6.1.2}$$

with the solution

$$c - c_0 = f(x - c_0 t). \tag{6.1.3}$$

The solution (6.1.3) completely misses out the nonlinear effects of build-up of an initial profile, however smooth initially, into a shock (See Section 2.2). Suppose we attempt the simple regular perturbation approach

$$c = c_0 + \epsilon c_1(x,t) + \epsilon^2 c_2(x,t) + \cdots \qquad (6.1.4)$$

where ϵ is the maximum initial value of $\dfrac{c - c_0}{c_0}$. Substituting (6.1.4) into (6.1.1) and equating coefficients of ϵ^n to zero, we have

$$c_{1t} + c_0 c_{1x} = 0 \qquad (6.1.5)$$

$$c_{2t} + c_0 c_{2x} = -c_1 c_{1x} \qquad (6.1.6)$$

$$c_{3t} + c_0 c_{3x} = -c_2 c_{1x} - c_1 c_{2x}, \quad \text{etc.} \qquad (6.1.7)$$

Each of these equations has the form

$$\phi_t + c_0 \phi_x = \Phi(x,t). \qquad (6.1.8)$$

Introducing the variable $y = x - c_0 t$, (6.1.8) becomes

$$\left(\frac{\partial \phi}{\partial t}\right)_{y=\text{const}} = \Phi(y + c_0 t, t) \qquad (6.1.9)$$

with the solution

$$\phi = \int_0^t \Phi(y + c_0 \tau, \tau) d\tau + \Psi(y). \qquad (6.1.10)$$

If we take specific IC

$$c(x,0) = c_0 + \epsilon P(x), \qquad (6.1.11)$$

the functions c_n satisfy

$$c_1 = P(x), \quad c_n = 0 \quad (n > 1) \quad \text{at } t = 0. \qquad (6.1.12)$$

The conditions (6.1.12) imply that the complementary functions $\Psi(y)$ are zero for solutions $c_n, n > 1$. It is now easy to find first few c_n:

$$\begin{aligned} c_1 &= P(y) \\ c_2 &= -tP(y)P'(y) \\ c_3 &= \frac{t^2}{2}(P^2 P')', \quad \text{etc.} \end{aligned} \qquad (6.1.13)$$

The general term in (6.1.13) will contain a term of the form $t^{n-1} R_n(y)$; the general order of the terms in the expansion (6.1.4) will be $\epsilon^n t^{n-1}$, showing that the series is not uniformly valid for $t \to \infty$.

As we have discussed in detail in Section 2.2, the exact solution of (6.1.1) subject to (6.1.12) has an implicit form, which may be written recursively as

$$c = c_0 + \epsilon P(x - ct)$$
$$= c_0 + \epsilon P(x - [c_0 + \epsilon P]t), \text{ etc.} \tag{6.1.14}$$

The naive form of the perturbation solution (6.1.4), as Whitham (1974) points out, may be recovered from (6.1.14) by an inadvisable expansion of the latter.

Whitham's approximate nonlinearisation technique starts with the linear solution, say, for spherical and cylindrical waves. Assuming that the nonlinear effects are relatively weak, the prescription is to introduce the new characteristic variable such that it satisfies the characteristic condition exactly; the amplitude factor is left uncorrected. We refer the reader to Chapter 9 of Whitham (1974) for this simple and effective (approximate) method of nonlinearisation, with applications to sonic boom theory and supersonic flow past a body. Here we illustrate Whitham's technique with a simple example (Whitham (1974), p.322).

Consider the nonlinear PDE

$$\phi_t + (c_0 + c_1\phi)\phi_x + \frac{\beta c_0}{x}\phi = 0. \tag{6.1.15}$$

The linearised form of (6.1.15) is

$$\phi_t + c_0\phi_x + \frac{\beta c_0}{x}\phi = 0. \tag{6.1.16}$$

Using Lagrange's method, we can write the solution of (6.1.16) as

$$\phi = \frac{f(t - x/c_0)}{x^\beta}. \tag{6.1.17}$$

The original PDE has the characteristic form

$$(c_0 + c_1\phi)\frac{d\phi}{dx} = -\frac{\beta c_0}{x}\phi \tag{6.1.18}$$

$$\frac{dt}{dx} = \frac{1}{c_0 + c_1\phi}. \tag{6.1.19}$$

Equation (6.1.18) has the exact solution

$$\phi e^{c_1\phi/c_0} = \frac{f(\tau)}{x^\beta} \tag{6.1.20}$$

where τ is the characteristic variable to be determined from (6.1.19). If we assume that ϕ is small, then

$$\phi = \frac{f(\tau)}{x^\beta} \tag{6.1.21}$$

is a uniformly valid approximation to (6.1.20). Comparing (6.1.17) and (6.1.21), we verify Whitham's first step, namely the change of a linearised solution such that the characteristic variable is required to satisfy the exact characteristic equation. To determine τ, we expand (6.1.19) and (6.1.20) in powers of φ, the expansions being convergent for $|\varphi| < c_0/c_1$. We thus obtain

$$\frac{dt}{dx} = \frac{1}{c_0} + \frac{\gamma_1 f(\tau)}{x^\beta} + \frac{\gamma_2 f^2(\tau)}{x^{2\beta}} + \cdots \tag{6.1.22}$$

where the coefficients γ_n are functions of c_0, c_1; in particular, $\gamma_1 = -c_1/c_0^2$. Integrating (6.1.22), we have

$$t = T(\tau) + \frac{x}{c_0} + \frac{\gamma_1 f(\tau)}{1-\beta} x^{1-\beta} + \frac{\gamma_2 f^2(\tau)}{1-2\beta} x^{1-2\beta} + \cdots \tag{6.1.23}$$

where we assume $\beta \neq 1, \frac{1}{2}$, for which logarithmic terms would appear in (6.1.23). The first uniformly valid approximation is simply

$$t = T(\tau) + \frac{x}{c_0} + \frac{\gamma_1 f(\tau)}{1-\beta} x^{1-\beta}, \tag{6.1.24}$$

and this agrees exactly with the result obtained from (6.1.21) and

$$\frac{dt}{dx} = \frac{1}{c_0} - \frac{c_1}{c_0^2}\phi. \tag{6.1.25}$$

Equations (6.1.21) and (6.1.24) give the (first) uniformly valid approximation and are consistent with Whitham's nonlinearisation procedure.

6.2   Exact Nonlinearisation of N Wave Solutions for Generalised Burgers Equations

Before we take up the discussion of GBEs, we quickly review the results for the IVP for the plane Burgers equation, which can be obtained exactly via (exact) linearisation through the Cole-Hopf transformation.

We consider the discontinuous N wave IC for the Burgers equation

$$u_t + u u_x = \frac{\delta}{2} u_{xx} \tag{6.2.1}$$

$$u(x,0) = \begin{cases} x & \text{if } |x| < l_0 \\ 0 & \text{otherwise.} \end{cases} \tag{6.2.2}$$

Here, δ is a (small) coefficient of viscosity.

Introducing the Cole-Hopf transformation

$$u = -\frac{\delta Q_x}{Q} \tag{6.2.3}$$

into (6.2.1), integrating once with respect to x and ignoring the function of integration, we arrive at the heat equation

$$Q_t = \frac{\delta}{2} Q_{xx}. \tag{6.2.4}$$

The IC (6.2.2) in view of (6.2.3) changes to

$$Q(x,0) = \begin{cases} e^{-x^2/2\delta} & \text{if } |x| < l_0 \\ e^{-l_0^2/2\delta} & \text{if } |x| \geq l_0. \end{cases} \tag{6.2.5}$$

The solution of the heat equation (6.2.4) subject to IC (6.2.5) can be written as

$$\begin{aligned} Q(x,t) = \frac{1}{(2\pi\delta t)^{1/2}} & \left[\int_{-l_0}^{l_0} e^{-[y^2+(x-y)^2/t]/2\delta} dy \right. \\ & + \left. e^{-l_0^2/2\delta} \left\{ \int_{-\infty}^{-l_0} e^{-(x-y)^2/2\delta t} dy + \int_{l_0}^{\infty} e^{-(x-y)^2/2\delta t} dt \right\} \right]. \end{aligned} \tag{6.2.6}$$

Making use of the properties of error function and its asymptotics for large t, we can write (6.2.6) as

$$\begin{aligned} Q(x,t) = e^{-l_0^2/2\delta} + \frac{e^{-x^2/2\delta t}}{(\pi t)^{1/2}} & \left[2 \int_0^{l_0/(2\delta)^{1/2}} e^{-z^2} dz \right. \\ & \left. - \left(\frac{2}{\delta} \right)^{1/2} l_0 e^{-l_0^2/2\delta} \right] + O\left(\frac{1}{t} \right) \end{aligned} \tag{6.2.7}$$

and

$$\begin{aligned} Q_x(x,t) = -\frac{1}{\delta} \frac{x}{t^{1/2}} \frac{e^{-x^2/2\delta t}}{t} & \left[\frac{2}{\pi^{1/2}} \int_0^{l_0/(2\delta)^{1/2}} e^{-z^2} dz \right. \\ & \left. - \left(\frac{2}{\pi\delta} \right)^{1/2} l_0 e^{-l_0^2/2\delta} \right] + O\left(\frac{1}{t^2} \right) \end{aligned} \tag{6.2.8}$$

as $t \to \infty$, uniformly with respect to the variable $\xi = x/(2\delta t)^{1/2}$.

Equations (6.2.3), (6.2.7) and (6.2.8) give

$$\begin{aligned} u(x,t) = \frac{\frac{x}{t^{1/2}} \frac{e^{-x^2/2\delta t}}{t} \left[\frac{2}{\pi^{1/2}} \int_0^{l_0/(2\delta)^{1/2}} e^{-z^2} dz - \left(\frac{2}{\pi\delta} \right)^{1/2} l_0 e^{-l_0^2/2\delta} \right]}{e^{-l_0^2/2\delta} + \frac{e^{-x^2/2\delta t}}{t^{1/2}} \left[\frac{2}{\pi^{1/2}} \int_0^{l_0/(2\delta)^{1/2}} e^{-z^2} dz - \left(\frac{2}{\pi\delta} \right)^{1/2} l_0 e^{-l_0^2/2\delta} \right]} \\ + O\left(\frac{1}{t} \right). \end{aligned} \tag{6.2.9}$$

Defining

$$c_0 = \frac{2}{\pi^{1/2}} e^{l_0^2/2\delta} \int_0^{l_0/(2\delta)^{1/2}} e^{-z^2} dz - \left(\frac{2}{\pi\delta}\right)^{1/2} l_0, \tag{6.2.10}$$

(6.2.9) simplifies to

$$u(x,t) = \frac{(2\delta)^{1/2}\xi}{t^{1/2}\left[1 + \frac{t^{1/2}}{c_0} e^{\xi^2}\right]} + O\left(\frac{1}{t}\right). \tag{6.2.11}$$

It is easy to check that the first term in (6.2.11) is, in fact, an exact solution of the Burgers equation; we refer to it as

$$u^\infty(x,t) = \frac{x/t^{1/2}}{t^{1/2}\left[1 + \frac{t^{1/2}}{c_0} e^{x^2/2\delta t}\right]}. \tag{6.2.12}$$

The decay of the solution (6.2.12) is described by the lobe Reynolds number of the antisymmetric N wave defined by

$$R(t) = \frac{\int_0^\infty u^\infty(x,t)dx}{\delta}$$
$$= \log\left(1 + \frac{c_0}{t^{1/2}}\right). \tag{6.2.13}$$

It is obtained on substituting for $u^\infty(x,t)$ in (6.2.13) from (6.2.12), and integrating with respect to x, etc. The N wave solution (6.2.12) skips the early embryonic part of the evolution of IC (6.2.2), but describes it from the point the shocks have assumed their Taylor structures, all the way to the linear (old age) regime when the nonlinear term in (6.2.1) is small and the solution is given simply by that of the heat equation, easily derived from (6.2.12) in the large t limit:

$$u \sim \frac{c_0 x}{t^{3/2}} e^{-x^2/2\delta t}. \tag{6.2.14}$$

Let the Reynolds number at some initial time t_0 be defined by

$$R(t_0) = \log\left(1 + \frac{c_0}{t_0^{1/2}}\right) \equiv R_0, \text{ say;} \tag{6.2.15}$$

then, we can write

$$c_0 = (e^{R_0} - 1)t_0^{1/2}. \tag{6.2.16}$$

The solution (6.2.12) can now be written as

$$u^\infty(x,t) = \frac{x}{t}\left[1 + \left(\frac{t}{t_0}\right)^{1/2} \frac{e^{x^2/2\delta t}}{e^{R_0} - 1}\right]^{-1} \tag{6.2.17}$$

which, for large R_0, can be approximated by

$$u^\infty(x,t) \approx \frac{x}{t}\left[1 + \left(\frac{t}{t_0}\right)^{1/2} e^{R_0(x^2/2R_0\delta t - 1)}\right]^{-1} \qquad (6.2.18)$$

for all x and t. If we let $R_0 \to \infty$ in (6.2.18), we have

$$u^\infty(x,t) \approx \begin{cases} x/t & -(2R_0\delta t)^{1/2} < x < (2R_0\delta t)^{1/2} \\ 0 & |x| > (2R_0\delta t)^{1/2}. \end{cases} \qquad (6.2.19)$$

The limiting form (6.2.19) is the exact solution of the inviscid form of the Burgers equation (6.2.1), namely

$$u_t + uu_x = 0. \qquad (6.2.20)$$

We rewrite the solution (6.2.12) in the form

$$u^\infty(x,t) = (2\delta)^{1/2}\xi/V(\eta, T) \qquad (6.2.21)$$

in terms of the "canonical" variables (belonging to the heat equation)

$$T = t^{1/2}, \quad \eta = \xi^2 = x^2/2\delta t, \quad c_0 = a^{-1}, \quad \text{say}; \qquad (6.2.22)$$

here, the function

$$V(\eta, T) = \sum_{i=0}^\infty f_i(T)\frac{\eta^i}{i!} \qquad (6.2.23)$$

where

$$f_0(T) = T + aT^2 \qquad (6.2.24)$$

and

$$f_i(T) = aT^2 \quad \text{for all } i \geq 1. \qquad (6.2.25)$$

The first term in (6.2.24) is the contribution from the inviscid solution (6.2.19) while the second term arises from the old-age solution (6.2.14). $f_i(T)$, $i \geq 1$ in (6.2.25) have contribution only from the old-age solution (6.2.14). The form (6.2.21) motivates the N wave solutions of generalised Burgers equations for which, in general, we have no Cole-Hopf type of transformation.

Thus, the Burgers equation and its (exact) solutions for different ICs provide invaluable insight into the structure of the solutions and guide the search for solutions of generalised Burgers equations (GBEs).

One could look upon $f_i(T)$ in (6.2.23)-(6.2.25) in the following way. Since $f_i(T)$ appear in the denominator in (6.2.21), the most dominant contribution T^2 for large time comes from the old-age (linear) solution. If the idea is to extend the old-age solution backward in time, one could start with the contribution from the old age as the dominant term in $f_i(T)$ and write out a series in descending powers of T to include other influences such as nonlinearity, damping, or geometrical spreading. This approach delivers the exact solution u^∞ of the Burgers equation. For other equations one ends up with infinite series for the functions $f_i(T)(i = 0, 1, \ldots)$ in descending powers of T, which are found by solving an infinite system of coupled nonlinear ODEs.

This was the approach adopted by Sachdev, Joseph, and Mayil Vaganan (1996) for the GBE

$$u_t + u^n u_x + \left(\alpha + \frac{j}{2t}\right)u + \left(\beta + \frac{\gamma}{x}\right)u^{n+1} = \frac{\delta}{2}u_{xx} \qquad (6.2.26)$$

where j, α, β, and γ are nonnegative constants and n is a positive integer. Equation (6.2.26) includes many important models of physical problems as special cases. These include Burgers equation $(n = 1, \alpha = \beta = \gamma = j = 0)$, nonplanar Burgers equation $(n = 1, \alpha = \beta = \gamma = 0)$, modified Burgers equation $(n = 2$ or 4, $\alpha = j = \beta = \gamma = 0)$, damped Burgers equation $(n = 1, j = \beta = \gamma = 0)$, and Burgers-Fisher equation $(n = 1, j = \gamma = 0)$.

We first give the general approach and then consider one of the cases in some detail.

We seek exact N wave solutions of (6.2.26) for the cases (i) $\beta = 0$, (ii) $\alpha > 0$, $\beta > 0$, which tend in the limit $t \to \infty$ to the old-age solution of the respective equations. We observe that

$$u(x, t) = c\frac{x/t^{1/2}}{t^{1+j/2}}e^{-\alpha t}e^{-x^2/2\delta t} \qquad (6.2.27)$$

is solution of the linearised form of (6.2.26), namely

$$u_t + \left(\alpha + \frac{j}{2t}\right)u = \frac{\delta}{2}u_{xx} \qquad (6.2.28)$$

where c is the so called old age constant, which depends in a complicated way on the initial conditions.

Motivated by the form (6.2.21) of the Burgers equation, we introduce the variables

$$u(x, t) = \frac{x/t^{1/2}}{[V(\xi, \tau)]^{1/n}}, \quad \xi(x, t) = \frac{x}{(2\delta t)^{1/2}}, \quad \tau = t^{1/2} \qquad (6.2.29)$$

into (6.2.26) and obtain the following PDE for $V(\xi, \tau)$:

$$n\,\xi\,VV_{\xi\xi} - (n+1)\xi V_\xi^2 + 2nVV_\xi + 2n\xi^2 VV_\xi - 2n\tau\xi VV_\tau$$

$$+2n^2(j-1)\xi V^2 + 4n^2\alpha\tau^2\xi V^2 + 4n^2(2\delta)^{(n-1)/2}\tau\xi^n V$$

$$-4n(2\delta)^{(n-1)/2}\tau\xi^{n+1}V_\xi + 4n^2\beta(2\delta)^{n/2}\tau^2\xi^{n+1}V$$

$$+4n^2\gamma(2\delta)^{(n-1)/2}\tau\xi^n V = 0. \qquad (6.2.30)$$

The planar solution (6.2.21) suggests that we write

$$V(\xi, \tau) = \sum_{i=0}^{\infty} f_i(\tau) \frac{\xi^i}{i!}. \tag{6.2.31}$$

Inserting (6.2.31) into (6.2.30) and equating coefficients of $\xi^i, i \geq 0$ to zero lead to the following system of coupled nonlinear ODEs for f_i's:

$$2n \sum_{k=0}^{i} \frac{f_k}{k!} \frac{f_{i+1-k}}{(i-k)!} + n \sum_{k=0}^{i-1} \frac{f_k}{k!} \frac{f_{i-k}}{(i-1-k)!} - (n+1) \sum_{k=0}^{i-1} \frac{f_{k+1}}{k!} \frac{f_{i-k}}{(i-1-k)!}$$

$$+ (2n^2(j-1) + 4n^2\alpha\tau^2) \sum_{k=0}^{i-1} \frac{f_k}{k!} \frac{f_{i-1-k}}{(i-1-k)!} \tag{6.2.32}$$

$$- 2n\tau \sum_{k=0}^{i-1} \frac{f_k'}{k!} \frac{f_{i-1-k}}{(i-1-k)!} + 2n \sum_{k=0}^{i-2} \frac{f_k}{k!} \frac{f_{i-1-k}}{(i-2-k)!} +$$

$$+ 4n^2(2\delta)^{(n-1)/2}(1+\gamma)\tau \frac{f_{i-n}}{(i-n)!}$$

$$+ 4n^2\beta(2\delta)^{n/2}\tau^2 \frac{f_{i-n-1}}{(i-n-1)!}$$

$$- 4n(2\delta)^{(n-1)/2}\tau \frac{f_{i-n}}{(i-n-1)!} = 0.$$

We construct solution (6.2.29) of (6.2.26) where $V(\xi, \tau)$ is given by (6.2.31). The functions $f_i(\tau)$ governed by (6.2.32) must be such that (6.2.29) tends to the linear solution (6.2.27) in the limit $t \to \infty$. This requires that

$$\frac{c^n i! f_{2i}(\tau)}{n^i (2i)! \tau^{nj+2n} e^{n\alpha\tau^2}} \to 1 \text{ as } \tau \to \infty \tag{6.2.33}$$

and

$$\frac{f_{2i+1}(\tau)}{\tau^{nj+2n} e^{n\alpha\tau^2}} \to 0 \tag{6.2.34}$$

uniformly in $i = 0, 1, 2, \ldots$.

For $i = 0$, (6.2.32) is simply $2n f_0(\tau) f_1(\tau) = 0$. Since $f_0(\tau) \neq 0$, we must have

$$f_1(\tau) = 0. \tag{6.2.35}$$

Using (6.2.35) in (6.2.32), we may show that, when n is even,

$$f_k(\tau) = 0, \quad k = 1, 3, 5, \ldots, n-1 \tag{6.2.36}$$

and

$$f_{n+1}(\tau) = -\frac{4}{3}(n+1)! n(2\delta)^{(n-1)/2}(1+\gamma)\tau;$$

for odd n and $\beta \neq 0$,

$$f_k(\tau) = 0, k = 1, 3, 5, \ldots n, \tag{6.2.37}$$

and

$$f_{n+2}(\tau) = -\frac{4}{3}n(n+2)!\beta(2\delta)^{n/2}\tau^2.$$

If $\beta = 0$ and n is an odd integer, we have

$$f_k = 0, \quad k = 1, 3, 5, \ldots \tag{6.2.38}$$

The case $n = 1$ in (6.2.26), as indicated earlier, has many important GBEs as its special cases. We shall therefore consider the cases $n = 1$ and $n \neq 1$ separately.

a. $n \neq 1$.

In this case, (6.2.32) with $i = 1$ and $f_1(\tau) = 0$ (see (6.2.35)) becomes

$$3f_2(\tau) + 2n(j - 1 + 2\alpha\tau^2)f_0(\tau) - 2\tau f_0'(\tau) = 0. \tag{6.2.39}$$

Since f_{2i} are required to satisfy (6.2.33) for $i = 0, 1, 2, \ldots$, an appropriate choice of f_0 is

$$f_0(\tau) = a\tau^{nj+2n}e^{n\alpha\tau^2} \tag{6.2.40}$$

where

$$a = \frac{1}{c^n}. \tag{6.2.41}$$

Insertion of (6.2.40) into (6.2.39) gives

$$f_2 = 2an\tau^{nj+2n}e^{n\alpha\tau^2} \tag{6.2.42}$$

which satisfies (6.2.33) with $i = 1$. Since we know f_0 and f_2 from (6.2.40) and (6.2.42) and $f_1 = f_3 = 0$ (see (6.2.36)), we can find $f_4, f_6, \ldots$, etc. from (6.2.32) rewritten as

$$\frac{n(i+2)}{i!}f_0f_{i+1}$$

$$= (n+1)\sum_{k=0}^{i-1}\frac{f_{k+1}}{k!}\frac{f_{i-k}}{(i-1-k)!} + 2n\tau\sum_{k=0}^{i-1}\frac{f_k'}{k!}\frac{f_{i-1-k}}{(i-1-k)!}$$

$$-2n\sum_{k=1}^{i}\frac{f_k}{k!}\frac{f_{i+1-k}}{(i-k)!} - n\sum_{k=1}^{i-1}\frac{f_k}{k!}\frac{f_{i+1-k}}{(i-1-k)!}$$

$$-(2n^2(j-1) + 4n^2\alpha\tau^2)\sum_{k=0}^{i-1}\frac{f_k}{k!}\frac{f_{i-1-k}}{(i-1-k)!}$$

$$-2n\sum_{k=0}^{i-2}\frac{f_k}{k!}\frac{f_{i-1-k}}{(i-2-k)!} - 4n^2(2\delta)^{(n-1)/2}(1+\gamma)\tau\frac{f_{i-n}}{i-n}$$

$$-4n^2\beta(2\delta)^{n/2}\tau^2\frac{f_{i-n-1}}{(i-n-1)!}$$

$$+4n(2\delta)^{(n-1)/2}\tau\frac{f_{i-n}}{(i-n-1)!}. \tag{6.2.43}$$

To complete the construction of the N wave solutions we must prove that (6.2.33) and (6.2.34) hold generally. We do this by induction argument. Let this be true for $k = 0, 1, 2, \ldots i$. We first assume that i is odd. Then using (6.2.33)-(6.2.34) for $k = 0, 1, \ldots i$ in (6.2.43) and dividing throughout by

$$\frac{n(i+2)(i+1)!n^{(i+1)/2}a^2\tau^{2(nj+2n)}e^{2n\alpha\tau^2}}{i!((i+1)/2)!},$$

we get, after some simplification,

$$\frac{f_{i+1}((i+1)/2)!}{n^{(i+1)/2}(i+1)!a\tau^{nj+2n}e^{n\alpha\tau^2}} \simeq \frac{i!((i+1)/2)!}{n^{(i+1)/2}n(i+2)!}$$

$$\left[nn^{(i+1)/2}\left\{\sum_{l=0}^{(i-1)/2}\frac{(4l-i+1)(i+1-2l)}{l!((i+1-2l)/2)!}+\frac{(i+1)(i+2)}{((i+1)/2)!}\right\}\right]=1$$

since

$$\sum_{l=0}^{(i-1)/2}\frac{(4l-i+1)(i+1-2l)}{l!((i+1-2l)/2)!}=0$$

is an identity for each positive integer i. The proof of (6.2.34) for i an even integer is similar. Thus, we have constructed N wave solution of the GBE (6.2.26) which has the correct old-age asymptotic behaviour (6.2.27) for large t.

We conclude this section with an example of (6.2.26) with $n = 1, j = \gamma = 0$, corresponding to the Burgers-Fisher equation

$$u_t + uu_x + \alpha u + \beta u^2 = \frac{\delta}{2}u_{xx}. \tag{6.2.44}$$

In this case (6.2.32) becomes

$$2\sum_{k=0}^{i}\frac{f_k}{k!}\frac{f_{i+1-k}}{(i-k)!}+\sum_{k=0}^{i-1}\frac{f_k}{k!}\frac{f_{i+1-k}}{(i-1-k)!}$$

$$-2\sum_{k=0}^{i-1}\frac{f_{k+1}}{k!}\frac{f_{i-k}}{(i-1-k)!}+2(2\alpha\tau^2-1)\sum_{k=0}^{i-1}\frac{f_k}{k!}\frac{f_{i-1-k}}{(i-1-k)!}$$

$$-2\tau\sum_{k=0}^{i-1}\frac{f_k'}{k!}\frac{f_{i-1-k}}{(i-1-k)!}+2\sum_{k=0}^{i-2}\frac{f_k}{k!}\frac{f_{i-1-k}}{(i-2-k)!}$$

$$+4(2\delta)^{n/2}\beta\tau^2\frac{f_{i-2}}{(i-2)!} + \frac{4\tau(2-i)}{(i-1)!}f_{i-1}(\tau) = 0, \quad i = 0,1,2,\ldots$$

$$(6.2.45)$$

For $i = 0$, (6.2.45) gives simply $2f_0(\tau)f_1(\tau) = 0$. Since $f_0(\tau) \neq 0$, we have $f_1(\tau) = 0$. For $i = 1$, (6.2.45) becomes

$$3f_2(\tau) + 2(-1 + 2\alpha\tau^2)f_0(\tau) - 2\tau f_0'(\tau) + 4\tau = 0. \tag{6.2.46}$$

We choose $f_0(\tau)$ in conformity with the old-age solution according to (6.2.33) with $j = 0$, $n = 1$:

$$f_0(\tau) = a\tau^2 e^{\alpha\tau^2}. \tag{6.2.47}$$

Therefore, (6.2.46) gives

$$f_2(\tau) = 2a\tau^2 e^{\alpha\tau^2} - \frac{4}{3}\tau \tag{6.2.48}$$

and, from (6.2.45), we have

$$f_3(\tau) = -2\sqrt{2}\beta\delta^{1/2}\tau^2$$

$$f_4(\tau) = 12a\tau^2 e^{\alpha\tau^2} + \frac{32}{15a}e^{-\alpha\tau^2} - \frac{16}{5}\tau + \frac{16}{5}\alpha\tau^3 \quad \text{etc.}$$

$$(6.2.49)$$

It may be easily verified that, as for the general case,

$$f_{2i}(\tau) \approx \frac{a(2i)!}{i!}\tau^2 e^{\alpha\tau^2} \tag{6.2.50}$$

$$f_{2i+1}(\tau) \approx O(\tau^2) \tag{6.2.51}$$

as $\tau \to \infty$ uniformly in i, proving that the series (6.2.31) converges for large t and the solution (6.2.29) so constructed has the appropriate (old age) behaviour given by (6.2.27) in the limit $t \to \infty$.

In the present analysis we chose $f_0(\tau)$ simply to be the contribution arising from the old-age behaviour. A slightly different approach which expresses this contribution to f_0 in descending powers of τ (as also for other f_i's) will be discussed in Sections 6.4 and 6.5, where we discuss N wave solutions of two important GBEs – the nonplanar Burgers equation and the modified Burgers equation. The solution thus obtained will also be compared with the numerical solution of certain IVPs to demonstrate their asymptotic character.

6.3   Burgers Equation in Cylindrical Coordinates with Axisymmetry

Apart from the Burgers equation, there is another interesting GBE which admits Cole-Hopf transformation and, hence, exact solution of IVPs from the linearised form (see Section 5.3 for multidimensional Burgers equation and its linearisation). This is the Burgers equation in cylindrical coordinates with axisymmetry, namely

$$
\begin{aligned}
u_t + uu_r &= \nu \left[\frac{1}{r}(ru_r)_r - \frac{u}{r^2} \right] \\
&= \nu \left(\frac{1}{r}(ru)_r \right)_r .
\end{aligned}
\tag{6.3.1}
$$

Here, r is the radial coordinate and ν is the coefficient of viscosity. The solution of this equation also motivates an approach to seeking solution of GBEs for which no Cole-Hopf transformation is available.

The transformation (Nerney, Schmahl, and Musielak (1996))

$$
u = \frac{-2\nu}{\theta} \frac{\partial \theta}{\partial r}
\tag{6.3.2}
$$

changes (6.3.1) to the linear diffusion equation

$$
\theta_t = \frac{\nu}{r}(r\theta_r)_r.
\tag{6.3.3}
$$

Equation (6.3.2) may also be written as

$$
\theta(r, t) = k(t) \exp \left(-\frac{1}{2\nu} \int_0^r u(\omega, t)d\omega \right)
\tag{6.3.4}
$$

where

$$
k(t) = \theta(0, t).
\tag{6.3.5}
$$

The initial values $\theta(r, 0)$ and $u(r, 0)$ are connected by

$$
\theta(r, 0) = \theta_0(r) = k(0) \exp \left(-\frac{1}{2\nu} \int_0^r u_0(\omega)d\omega \right)
\tag{6.3.6}
$$

for $r_0 \leq r \leq R$.

The general solution of (6.3.3) may formally be written as

$$
\theta(r, t) = \int_0^\infty \int_0^\infty \theta_0(r') J_0(kr') J_0(kr) e^{-\nu k^2 t} kr' dk dr'.
\tag{6.3.7}
$$

The integral with respect to k in (6.3.7) can be immediately evaluated by using the formula

$$
\int_0^\infty e^{-\nu k^2 t} J_0(kr) J_0(kr') k dk = \frac{1}{2\nu t} \exp \left(-\frac{r^2 + r'^2}{4\nu t} \right) I_0 \left(\frac{rr'}{2\nu t} \right)
\tag{6.3.8}
$$

(Watson (1962)) where the modified function $I_0(\alpha r') = J_0(i\alpha r')$ is real. Equation (6.3.7) may now be written as

$$\theta(r,t) = \frac{1}{2\nu t} \exp\left(-\frac{r^2}{4\nu t}\right) \int_0^\infty \theta_0(r') \exp\left(-\frac{r'^2}{4\nu t}\right) J_0\left(\frac{ir'r}{2\nu t}\right) r'\,dr'. \tag{6.3.9}$$

The initial condition for (6.3.1) is assumed by Nerney et al. (1996) as

$$u(r,0) = u_0\frac{r_0}{r}, \quad r_0 \le r \le R \tag{6.3.10}$$

to simulate an incompressible flow where the source of water is assumed to emanate from the origin of coordinates and spread out in a circularly symmetric pattern. The initial flow covers the domain $r_0 < r < R$, where $r_0 \simeq 0$ and R is longer than any other length scale in the problem; subsequently, we let $R \to \infty$.

Transforming (6.3.10) according to (6.3.6) we have

$$\theta_0(r) = k_0\left(\frac{r_0}{r}\right)^a \tag{6.3.11}$$

where

$$a = \frac{u_0 r_0}{2\nu} \tag{6.3.12}$$

is a Reynolds number and is assumed to be finite. With $\theta_0(r)$ given by (6.3.11), we may explicitly evaluate (6.3.9) using Hankel's generalisation of (6.3.8) (Watson (1962)). Thus, we obtain

$$\theta = k_0 r_0^a \Gamma\left(1 - \frac{a}{2}\right)(4\nu t)^{-a/2} \sum_{n=0}^\infty \frac{\left(\frac{a}{2}\right)_n}{(n!)^2}\left(-\frac{r^2}{4\nu t}\right)^n \tag{6.3.13}$$

where $\left(\frac{a}{2}\right)_n$ is Pochhammer notation

$$\left(\frac{a}{2}\right)_n = \frac{a}{2}\left(\frac{a}{2}+1\right)\cdots\left(\frac{a}{2}+n-1\right) \tag{6.3.14}$$

and

$$\left(\frac{a}{2}\right)_0 = 1. \tag{6.3.15}$$

Now we use (6.3.2) to write the solution in terms of confluent hypergeometric functions

$$u(r,t) = \frac{a}{2}\frac{r}{t}\frac{M(a/2+1, 2, -r^2/(4\nu t))}{M(a/2, 1, -r^2/(4\nu t))}$$

$$= \frac{r}{t}\frac{\sum_{n=1}^\infty \left(\frac{a}{2}\right)_n \frac{n}{(n!)^2}\left(-\frac{r^2}{4\nu t}\right)^{n-1}}{\sum_{n=0}^\infty \frac{1}{(n!)^2}\left(\frac{a}{2}\right)_n\left(-\frac{r^2}{4\nu t}\right)^n}. \tag{6.3.16}$$

The important observation is that the solution (6.3.16) can be written as the products of the inviscid solution and an infinite sum of the similarity variable $r^2/4\nu t$. The effect of nonlinearity is manifested by the factor n in the numerator multiplying the powers of the similarity variable. The form (6.3.16) is similar to that for the Burgers equation (see (6.2.17) of Section 6.2) that we attempt to simulate for other GBEs in the following section, and succeed in some cases in obtaining exact solutions. How various stages of decay come about in the solution (6.3.16) — from early nonlinear steepening to later diffusion domination leading to old-age and, hence, decay to zero — is entirely analogous to that for the Burgers equation (see Sachdev (1987)). Nerney et al. depict the solution (6.3.16) as it evolves in time.

Another simple example derives from the initial condition

$$u(r,0) = u_0 \frac{r}{r_0} = \frac{r}{t_0}, \quad 0 \leq r \leq r_0; \quad t = 0 \tag{6.3.17}$$

where

$$t_0 = \frac{r_0}{u_0}. \tag{6.3.18}$$

Here, the condition (6.3.17) describes a wind emanating from the origin with a linearly increasing velocity out to some radius r_0, larger than any other length scale in the problem.

Equations (6.3.4) and (6.3.17) give

$$\theta_0(r) = k_0 e^{-br^2} \tag{6.3.19}$$

where

$$b = \frac{u_0}{4\nu r_0}. \tag{6.3.20}$$

The solution of (6.3.3) subject to (6.3.19) is easily found to be

$$\theta(r,t) = \frac{k_0}{1 + t/t_0} \exp\left[-\frac{r^2}{4\nu t} \frac{1}{1 + t_0/t}\right] \tag{6.3.21}$$

and, hence, we get via (6.3.2) the solution

$$u = \frac{r}{t + t_0}. \tag{6.3.22}$$

In the present case the solution (6.3.22) approaches the inviscid solution $u = r/t$ as $t \to \infty$, and not the linear diffusive (old age) solution.

Some of these features will come out in the discussion of other GBEs in Sections 6.4 and 6.5.

6.4 Nonplanar Burgers Equation – A Composite Solution

The equation

$$u_t + u u_x + \frac{1}{2}\frac{d}{dt}(\ln A)u = \frac{\delta}{2}u_{xx} \qquad (6.4.1)$$

describes the propagation of weakly nonlinear longitudinal waves in a gas or liquid, subject not only to the diffusive effects associated with viscosity and thermal conductivity represented by the right-hand side of (6.4.1), but also to the geometrical effects of change of ray tube area $A(t)$ represented by the last term on the left (see Lighthill (1956) and Leibovich and Seebass (1974)). Assuming the form of ray tube area to be $A = A_0\, t^j$, where j is a positive or negative constant (the latter for a contracting ray tube) and A_0 is another (positive) constant, (6.4.1) reduces to

$$u_t + u u_x + \frac{1}{2}\frac{ju}{t} = \frac{\delta}{2}u_{xx}. \qquad (6.4.2)$$

Equation (6.4.2) does not admit Cole-Hopf-like transformation for exact linearisation. Taking a cue from the analytic form of N wave solution for the plane Burgers equation (and also from the solution for the axis-symmetric Burgers equation discussed in Section 6.3), we attempt to simulate the N wave solutions of (6.4.2) which, unlike in Section 6.2, do not merely extend the (old-age) linear solution back in time, but also include the nonlinear effects as represented by the inviscid solution. This happens in a simple way for the Burgers equation – the contributions from the inviscid and old-age solutions to $f_0(T)$ in (6.2.24) in the representation (6.2.21) of Section 6.2 simply add up after appropriate peeling off of some factors from the inviscid and old-age solutions. We investigate what happens when we attempt to mimic the solution (6.2.21) of Section 6.2 for the nonplanar equation (6.4.2) (Sachdev, Joseph, and Nair (1994)).

Following the discussion in Section 6.2, we first assume that $j = m/n$, $(-1 < j < 2)$, a rational number, with $m < n$ or $m > n$ (this restriction will become clear later). We first consider the case $j > 0$ such that m and n are positive integers with no common divisor. In this case the inviscid form of (6.4.2)

$$u_t + u u_x + \frac{1}{2}mu/nt = 0 \qquad (6.4.3)$$

has an exact N-wave solution

$$u(x,t) = \frac{(2\delta)^{1/2}}{2n/(2n-m)}\frac{x}{(2\delta t)^{1/2}}\frac{1}{t^{1/2}}. \qquad (6.4.4)$$

The old-age (linear) form of (6.4.2) is

$$u_t + \frac{1}{2}\frac{mu}{nt} = \frac{\delta}{2}u_{xx} \qquad (6.4.5)$$

and has the antisymmetric solution

$$u(x,t) = C\frac{x}{t^{1/2}}\frac{1}{t^{(2n+m)/2n}}e^{-x^2/2\delta t} \tag{6.4.6}$$

where C is an arbitrary constant. The idea is to attempt to construct a composite solution of (6.4.2) such that it has the inviscid form (6.4.4) in the limit of large initial Reynolds number and tends to the old-age solution (6.4.6) as $t \to \infty$. This is what happens for the plane Burgers equation (see Section 6.2). Motivated by the form (6.2.11) or (6.2.12) of Section 6.2 for the Burgers equation, we write

$$u(x,t) = (2\delta)^{1/2}\xi/V(\eta,T) \tag{6.4.7}$$

where

$$\xi = x/(2\delta t)^{1/2}, \quad \eta = \xi^2, \quad T = t^{1/2} \tag{6.4.8}$$

and substitute it in (6.4.2) to get the PDE governing $V(\eta, T)$:

$$V(jV - TV_T) + (2T - V)(V - 2\eta V_\eta) + 3VV_\eta + 2\eta VV_{\eta\eta} - 4\eta V_\eta^2 = 0. \tag{6.4.9}$$

We seek a representation of the solution of (6.4.9) in the form

$$V(\eta,T) = \sum_{i=0}^{\infty} f_i(T)\frac{\eta^i}{i!} \tag{6.4.10}$$

(see (6.2.23) of Section 6.2). Substituting (6.4.10) into (6.4.9) and equating coefficients of like powers of η to zero, we get an infinite system of coupled nonlinear ODEs for $f_i(T)(i = 1, 2, 3, \ldots)$:

$$3f_1 + 2T - Tf_0' + (j-1)f_0 = 0 \tag{6.4.11}$$

$$5f_0f_2 + 2jf_0f_1 - f_1^2 - T(f_0f_1)' - 2Tf_1 = 0 \tag{6.4.12}$$

$$7f_0f_3 + [2(j+1)f_0 - Tf_0' - 3f_1 - 6T]f_2$$
$$-Tf_0f_2' + 2f_1[(j+1)f_1 - Tf_1'] = 0 \tag{6.4.13}$$

$$(j-1)\sum_{k=0}^{i}\frac{f_k}{k!}\frac{f_{i-k}}{(i-k)!} - T\sum_{k=0}^{i}\frac{f_k}{k!}\frac{f_{i-k}'}{(i-k)!} + 2T\frac{f_i}{i!} + 3\sum_{k=0}^{i}\frac{f_k}{k!}\frac{f_{i-k+1}}{(i-k)!}$$

$$-4T\frac{f_i}{(i-1)!} + 2\sum_{k=0}^{i-1}\frac{f_k}{k!}\frac{f_{i-k}}{(i-k-1)!} + 2\sum_{k=0}^{i-1}\frac{f_k}{k!}\frac{f_{i-k+1}}{(i-k-1)!}$$

$$-4\sum_{k=0}^{i-1}\frac{f_{k+1}}{k!}\frac{f_{i-k}}{(i-k-1)!} = 0, i = 3, 4, 5, \ldots \tag{6.4.14}$$

The structure of system (6.4.11)-(6.4.14) is such that once f_0, f_1, and f_2 can be found, all other $f_i, i \geq 3$ are obtained by algebraic operations alone. We determine $f_0, f_1,$ and f_2 from (6.4.11) and (6.4.12) by making use of the inviscid and old-age behaviours (6.4.4) and (6.4.6), respectively. Introducing the variable

$$\tau = T^{1/n} = t^{1/2n} \tag{6.4.15}$$

in (6.4.11) and (6.4.12), we get

$$3f_1 + 2\tau^n - (\tau/n)f_0' + ((m-n)/n)f_0 = 0 \tag{6.4.16}$$

$$5f_0 f_2 + (2m/n)f_0 f_1 - f_1^2 - (\tau/n)(f_0 f_1)' - 2\tau^n f_1 = 0. \tag{6.4.17}$$

The inviscid and old-age forms of solutions (6.4.4) and (6.4.6) of (6.4.2) suggest (in the light of the solution of Burgers equation) that the first term in the expansions for $f_i(\tau)(i = 0, 1, 2)$ is proportional to $t^{1/2} = \tau^n$ while the last term should be proportional to $t^{(2n+m)/2n} = \tau^{2n+m}$. Therefore, we seek solutions for f_i in the form

$$f_0(\tau) = \tau^n \sum_{i=0}^{n+m} a_i \tau^i$$

$$f_1(\tau) = \tau^n \sum_{i=0}^{n+m} b_i \tau^i \tag{6.4.18}$$

$$f_2(\tau) = \tau^n \sum_{i=0}^{n+m} c_i \tau^i.$$

Substituting (6.4.18) into (6.4.16) and (6.4.17) and equating coefficients of like powers of τ to zero, we get the following system for the coefficients $a_i, b_i,$ and c_i appearing in (6.4.18):

$$
\begin{aligned}
a_0 &= 2n/(2n-m) \\
b_0 &= 0, \; c_0 = 0 \\
a_{n+m} &= b_{n+m} = c_{n+m} \\
b_i &= [(2n+i-m)/3n]a_i, \quad i = 1, 2, \ldots (n+m)
\end{aligned} \tag{6.4.19}
$$

and

$$5\alpha_i + [(2m-2n-i)/n]\beta_i - 2b_i - \gamma_i = 0 \quad i = 1, 2, \ldots (n+m) \tag{6.4.20}$$

$$5\alpha_i + [(2m-2n-i)/n]\beta_i - \gamma_i = 0,$$
$$i = n+m+1, \ldots 2(n+m) - 1 \tag{6.4.21}$$

where $\alpha_i, \beta_i,$ and γ_i are defined by

$$\alpha_i = \begin{cases} \displaystyle\sum_{q=0}^{i} a_q c_{i-q} & i = 0, 1, \ldots (n+m) \\[4ex] \displaystyle\sum_{q=i-(n+m)}^{n+m} a_q c_{i-q} & i = (n+m+1), \ldots 2(n+m) \end{cases} \tag{6.4.22}$$

$$\beta_i = \begin{cases} \displaystyle\sum_{q=0}^{i} a_q b_{i-q} & i = 0, 1, \ldots (n+m) \\ \displaystyle\sum_{q=i-(n+m)}^{n+m} a_q b_{i-q} & i = (n+m+1), \ldots 2(n+m) \end{cases} \tag{6.4.23}$$

$$\gamma_i = \begin{cases} \displaystyle\sum_{q=0}^{i} b_q b_{i-q} & i = 0, 1, \ldots (n+m) \\ \displaystyle\sum_{q=i-(n+m)}^{n+m} b_q b_{i-q} & i = (n+m+1), \ldots 2(n+m) \end{cases} \tag{6.4.24}$$

Using the expressions for b_i from (6.4.19) into (6.4.21), we get the following system of linear algebraic equations for $c_1, c_2, \ldots c_{n+m-1}$:

$$5 \sum_{q=i-(n+m)}^{n+m} a_q c_{i-q} + \left(\frac{2m - 2n - i}{n}\right) \sum_{q=i-(n+m)}^{n+m}$$

$$\left(\frac{2n + i - q - m}{3n}\right) a_q a_{i-q} - \sum_{q=i-(n+m)}^{n+m}$$

$$\left(\frac{2n + i - q - m}{3n}\right)\left(\frac{2n + q - m}{3n}\right) a_q a_{i-q} = 0$$

$$i = n + m + 1, \ldots, 2(n+m) - 1. \tag{6.4.25}_i$$

We begin evaluation of c_i starting with $(6.4.25)_i$ for $i = 2(n+m) - 1$. Since $c_{n+m} = a_{n+m}$ (see (6.4.19)), we can solve the c_{n+m-1} in terms of a_i. Then from $(6.4.25)_{2(n+m)-2}$, we can get c_{n+m-2}. Proceeding in this manner, we obtain $c_{n+m-1}, c_{n+m-2}, \ldots c_1$ in terms of $a_i, i = 1, 2, \ldots (n+m)$. Substituting these in (6.4.20), we get $(n+m)$ algebraic equations for $(n+m-1)$ unknowns $a_1, a_2, \ldots a_{n+m-1}$; these can be solved in terms of the $(a_{n+m})^{-1}$, the old-age constant, which itself must be obtained from fitting the asymptotic linear form with the numerical solution. To show that the solution (6.4.7) has the correct old-age behaviour (6.4.6), we first observe that $f_i/\tau^{2n+m} \to a_{n+m}$ as $\tau \to \infty$ for $i = 0, 1, 2$ by construction. It is then possible to show by an induction argument using (6.4.13) and (6.4.14) that $f_i/\tau^{2n+m} \to a_{n+m}$ as $\tau \to \infty$ for $i \geq 3$. We conclude that the solution (6.4.7) has the right old-age behaviour as $\tau \to \infty$.

To determine the decay of N wave we define the lobe Reynolds number

$$R(t) = \frac{1}{\delta} \int_0^\infty u(x, t) dx \tag{6.4.25}$$

which is nondimensional area under one lobe of the N wave. Integrating (6.4.2) with respect to x from 0 to ∞ and using (6.4.7), (6.4.10), (6.4.15), and (6.4.18$_1$), we get the ODE

$$\frac{dR}{dt} + \frac{mR}{2nt} = -\frac{1}{2t} \frac{1}{\displaystyle\sum_{i=0}^{n+m} a_i t^{i/2n}}. \tag{6.4.27}$$

Multiplying (6.4.27) by the integrating factor $t^{m/2n}$ and integrating, we obtain

$$R(t) = \frac{1}{2t^{m/2n}} \int_t^\infty \frac{ds}{s^{(2n-m)/2n} \left[\displaystyle\sum_{i=0}^{n+m} a_i s^{i/2n}\right]}. \tag{6.4.28}$$

Here, we have imposed the condition that $R(t = \infty) = 0$, requiring that $Rt^{m/2n} \to 0$ as $t \to \infty$.

Now we consider some special cases which admit explicit solution. The nonlinear algebraic system of equations for the unknown coefficients in f_i often admit more than one solution. A unique choice must be made by rejecting those solutions which give singularities; the numerical solution also aids this choice.

(i) The simplest case is one with cylindrical symmetry corresponding to

$$j = 1(m = 1, n = 1) \tag{6.4.29}$$

in (6.4.2). Here,

$$\tau = T = t^{1/2} \tag{6.4.30}$$

and (6.4.11)-(6.4.12) become

$$3f_1 - \tau f_0' + 2\tau = 0 \tag{6.4.31}$$

$$5f_0 f_2 - f_1^2 + 2f_0 f_1 - 2\tau f_1 - \tau(f_0 f_1)' = 0. \tag{6.4.32}$$

We seek solution of (6.4.31)-(6.4.32) in the form

$$f_0 = a_0\tau + a_1\tau^2 + a_2\tau^3 \tag{6.4.33}$$
$$f_1 = b_0\tau + b_1\tau^2 + b_2\tau^3 \tag{6.4.34}$$
$$f_2 = c_0\tau + c_1\tau^2 + c_2\tau^3. \tag{6.4.35}$$

Substituting (6.4.33)-(6.4.35) into (6.4.31) and equating coefficients of like powers of τ to zero, we get

$$b_0 = \frac{1}{3}(a_0 - 2), b_1 = 2a_1/3, b_2 = a_2. \tag{6.4.36}$$

Substituting (6.4.33)-(6.4.35) into (6.4.32) etc., we find that

$$5a_0 c_0 - b_0^2 - 2b_0 = 0 \tag{6.4.37}$$

$$5a_0c_1 + 5a_1c_0 - 2b_0b_1 - a_0b_1 - a_1b_0 - 2b_1 = 0 \qquad (6.4.38)$$

$$5a_0c_2 + 5a_1c_1 + 5a_2c_0 - b_1^2 - 2b_0b_2 - 2a_0b_2 - 2a_1b_1 - 2b_0a_2 - 2b_2 = 0 \quad (6.4.39)$$

$$5a_1c_2 + 5a_2c_1 - 2b_1b_2 - 3a_2b_1 - 3a_1b_2 = 0 \qquad (6.4.40)$$

$$5a_2c_2 - b_2^2 - 4a_2b_2 = 0. \qquad (6.4.41)$$

From (6.4.36) and (6.4.41) we have

$$c_2 = a_2. \qquad (6.4.42)$$

Equation (6.4.40), with the help of (6.4.36) and (6.4.42), gives

$$c_1 = \frac{4}{15}a_1. \qquad (6.4.43)$$

From (6.4.36) and (6.4.37) we get

$$c_0 = \frac{1}{45a_0}(a_0^2 + 2a_0 - 8). \qquad (6.4.44)$$

Substituting for $c_i(i = 0, 1, 2)$ from (6.4.42)-(6.4.44) and $b_i(i = 0, 1, 2)$ from (6.4.36) into (6.4.38), we get either $a_1 = 2$ or $a_1 = 0$. The case $a_1 = 2$ gives the following solution for the coefficients:

$$
\begin{aligned}
a_0 &= 2, & a_1 &= \pm 3a_2^{1/2} \\
b_0 &= 0, & b_1 &= \pm 2a_2^{1/2}, b_2 = a_2 \\
c_0 &= 0, & c_1 &= \pm \frac{4}{5}a_2^{1/2}, c_2 = a_2.
\end{aligned}
\qquad (6.4.45)
$$

Rejecting the solution with a negative sign (remember that the old-age constant a_2^{-1} is positive), since it gives a singularity in the expression for Reynolds number at a finite time, we have the following forms for f_i,

$$
\begin{aligned}
f_0(t) &= 2t^{1/2} + 3a_2^{1/2}t + a_2t^{3/2} \\
f_1(t) &= 2a_2^{1/2}t + a_2t^{3/2} \\
f_2(t) &= \frac{4}{5}a_2^{1/2}t + a_2t^{3/2},
\end{aligned}
\qquad (6.4.46)
$$

in terms of the old-age constant a_2^{-1}. Consider the second alternative $a_1 = 0$. Using these values of a_1, b_i, and $c_i(i = 0, 1, 2)$ from (6.4.36) and (6.4.42)-(6.4.44), and after some simplification, we arrive at the quadratic

$$2a_0^2 + a_0 - 1 = 0 \qquad (6.4.47)$$

for a_0. This gives $a_0 = -1$ or $1/2$. Rejecting the negative value and putting together b_i and c_i found earlier in terms of a_i's etc., we have

$$a_0 = 1/2, \quad a_1 = 0$$

$$b_0 = -1/2, \quad b_1 = 0, \quad b_2 = a_2 \tag{6.4.48}$$
$$c_0 = -3/10, \quad c_1 = 0, \quad c_2 = a_2$$

in terms of a_2^{-1} alone. The functions $f_i(t)$ now become

$$f_0(t) = t^{1/2} \left(\frac{1}{2} + a_2 t \right)$$
$$f_1(t) = t^{1/2} \left(-\frac{1}{2} + a_2 t \right) \tag{6.4.49}$$
$$f_2(t) = t^{1/2} \left(-\frac{3}{10} + a_2 t \right).$$

The numerical solution of the problem shows that of the two possible choices (6.4.46) and (6.4.49), the latter gives better results. With $f_0(t)$ from (6.4.49), we get the Reynolds number from (6.4.7), (6.4.10), and (6.4.26) as

$$R(t) = t^{-1/2}(R_0 t_0^{1/2} - h(t) + h(t_0)) \tag{6.4.50}$$

where

$$h(s) = \sqrt{2} \arctan \left(\sqrt{2a_2 s} \right)/\sqrt{a_2} \tag{6.4.51}$$

and a_2^{-1} is the old-age constant.

$j = \dfrac{1}{2}.$

In this case, $\tau = T^{1/2} = t^{1/4}$ and (6.4.11)-(6.4.12) become

$$3f_1 - \frac{1}{2}f_0 - \frac{\tau}{2}f_0' + 2\tau^2 = 0 \tag{6.4.52}$$

$$5f_0 f_2 + f_0 f_1 - f_1^2 - 2\tau^2 f_1 - \frac{1}{2}\tau(f_0 f_1)' = 0. \tag{6.4.53}$$

The appropriate forms for f_0, f_1, and f_2 here are

$$f_0(\tau) = \tau^2[a_0 + a_1\tau + a_2\tau^2 + a_3\tau^3]$$
$$f_1(\tau) = \tau^2[b_0 + b_1\tau + b_2\tau^2 + b_3\tau^3] \tag{6.4.54}$$
$$f_2(\tau) = \tau^2[c_0 + c_1\tau + c_2\tau^2 + c_3\tau^3].$$

Substituting (6.4.54) into (6.4.52) and equating coefficients of like powers of τ to zero, we get

$$b_0 = \frac{a_0}{2} - \frac{2}{3}, b_1 = \frac{2a_1}{3}, b_2 = \frac{5a_2}{6}, b_3 = a_3 \tag{6.4.55}$$

while the same process for (6.4.53) yields

$$c_0 = \frac{1}{5a_0} \left(\frac{3a_0^2}{4} - \frac{a_0}{3} - \frac{8}{9} \right)$$

$$c_1 = \frac{1}{5a_0}\left(\frac{5a_0a_1}{3} - \frac{2a_1}{9} + \frac{8a_1}{9a_0}\right) \tag{6.4.56}$$

$$c_2 = \frac{37}{60}a_2$$

$$c_3 = a_3$$

where a_1, a_2 and a_3 satisfy

$$a_0a_2(3a_0^2 + 4a_0 - 8) - a_1^2(a_0^2 + 2a_0 - 8) = 0 \tag{6.4.57}$$

$$8a_1a_2 - 2a_0a_1a_2 - a_0^2a_1a_2 - 8a_0a_3 + 6a_0^2a_3 + 9a_0^3a_3 = 0 \tag{6.4.58}$$

$$-a_0^2a_2^2 + 8a_1a_3 - 2a_0a_1a_3 + 3a_0^2a_1a_3 = 0. \tag{6.4.59}$$

If we choose the coefficient of a_2 in (6.4.57) to be zero, we get

$$a_0 = 0, \quad \frac{-4 + 4\sqrt{7}}{6}, \quad \frac{-4 - 4\sqrt{7}}{6}. \tag{6.4.60}$$

Equation (6.4.57) implies that $a_1 = 0$. If we put $a_0 = 0$, $a_1 = 0$ in (6.4.58), we get $a_3 = 0$; this is impossible since a_3^{-1} is the old-age constant. If we assume that none of the values of a's in (6.4.60) holds, then (6.4.57) gives

$$a_2 = \frac{a_1^2}{a_0}\frac{a_0^2 + 2a_0 - 8}{3a_0^2 + 4a_0 - 8}. \tag{6.4.61}$$

Putting (6.4.61) into (6.4.58), we get

$$a_1^3 = \frac{9a_0^3(3a_0^2 + 4a_0 - 8)}{(a_0^2 + 2a_0 - 8)^2}\left(a_0a_3 + \frac{2}{3}a_3 - \frac{8a_3}{9a_0}\right). \tag{6.4.62}$$

Substituting (6.4.61)-(6.4.62) into (6.4.59) we get, after some simplification,

$$a_1\frac{16a_3(3a_0 - 4)}{9a_0^2(3a_0^2 + 4a_0 - 8)} = 0 \tag{6.4.63}$$

implying that either $a_0 = \frac{4}{3}$ or $a_1 = 0$. The choice $a_0 = \frac{4}{3}$ leads to an infinite slope $u_x(0, t)$ at a finite time t. Rejecting this possibility, we choose $a_1 = 0$, and, hence, $a_2 = 0$, in view of (6.4.57). Equation (6.4.58) then gives

$$-8a_0a_3 + 6a_0^2a_3 + 9a_0^3a_3 = 0. \tag{6.4.64}$$

Since $a_3 \neq 0$, we have either $a_0 = -4/3$ or $2/3$. Rejecting, again, the negative root, we arrive at the unique choice $a_0 = 2/3$. Therefore,

$$a_0 = \frac{2}{3}, \quad a_1 = 0, \quad a_2 = 0$$

$$b_0 = -\frac{1}{3}, \quad b_1 = b_2 = 0, \quad b_3 = a_3 \tag{6.4.65}$$

$$c_0 = -\frac{7}{30}, \quad c_1 = c_2 = 0, \quad c_3 = a_3,$$

determining all the coefficients in terms of the old-age constant a_3^{-1}.

The functions $f_i(i = 1, 2, 3)$ in (6.4.54) are thus found to be

$$f_0(\tau) = \tau^2 \left(\frac{2}{3} + a_3 \tau^3 \right)$$

$$f_1(\tau) = \tau^2 \left(-\frac{1}{3} + a_3 \tau^3 \right) \tag{6.4.66}$$

$$f_2(\tau) = \tau^2 \left(-\frac{7}{30} + a_3 \tau^3 \right),$$

where $\tau = t^{1/4}$ and a_3^{-1} is the old-age constant. Using $f_0(\tau)$ from (6.4.66) along with (6.4.7), (6.4.10), and (6.4.26), we arrive at the Reynolds number

$$R(t) = R(t_0) \left(\frac{t_0}{t} \right)^{1/4} - \frac{1}{2} t^{-1/4} (h(t) - h(t_0)) \tag{6.4.67}$$

where

$$h(s) = \frac{4 \arctan((-a_0^{1/3} + 2a_3^{1/3} s^{1/4})/(\sqrt{3} a_0^{1/3}))}{\sqrt{3} a_0^{2/3} a_3^{1/3}}$$

$$+ \frac{4 \log \left(a_0^{1/3} + a_3^{1/3} s^{1/4} \right)}{3 a_0^{2/3} a_3^{1/3}}$$

$$- \frac{2 \log \left(a_0^{2/3} - a_0^{1/3} a_3^{1/3} s^{1/4} + a_3^{2/3} s^{1/2} \right)}{3 a_0^{2/3} a_3^{1/3}} \tag{6.4.68}$$

where $a_0 = \frac{2}{3}$, as found earlier, and a_3^{-1} is the old-age constant.

We give the final results for $j = \frac{1}{3}, \frac{1}{4}$ since the process of evaluation of coefficients, etc. is similar.

$$j = \frac{1}{3}$$

$$f_0(\tau) = \frac{3}{4} \tau^3 + a_1 \tau^7$$

$$f_1(\tau) = -\frac{1}{4} \tau^3 + a_1 \tau^7 \tag{6.4.69}$$

$$f_2(\tau) = -\frac{11}{60} \tau^3 + a_1 \tau^7$$

where $\tau = t^{1/6}$ and

$$R(t) = R(t_0) \left(\frac{t_0}{t}\right)^{1/6} - \left(\frac{1}{2}\right) t^{-1/6}(h(t) - h(t_0)) \tag{6.4.70}$$

where

$$
\begin{aligned}
h(s) &= \frac{3 \arctan ((-\sqrt{2}a_0^{1/4} + 2a_1^{1/4}s^{1/6})/(\sqrt{2}a_0^{1/4}))}{\sqrt{2}a_0^{3/4}a_1^{1/4}} \\
&+ \frac{3 \arctan (\sqrt{2}a_0^{1/4} + 2a_1^{1/4}s^{1/6})/\sqrt{2}a_0^{1/4})}{\sqrt{2}a_0^{3/4}a_1^{1/4}} \\
&- \frac{3 \log(\sqrt{a_0} - \sqrt{2}a_0^{1/4}a_1^{1/4}s^{1/6} + \sqrt{a_1}s^{1/3})}{2\sqrt{2}a_0^{3/4}a_1^{1/4}} \\
&+ \frac{3 \log \left(\sqrt{a_0} + \sqrt{2}a_0^{1/4}a_1^{1/4}s^{1/6}\sqrt{a_1}s^{1/3}\right)}{2\sqrt{2}a_0^{3/4}a_1^{1/4}}
\end{aligned} \tag{6.4.71}
$$

with $a_0 = \dfrac{3}{4}$ and a_1^{-1} as the old-age constant.

$j = \dfrac{1}{4}$

$$
\begin{aligned}
f_0 &= \frac{4}{5}\tau^4 + a_1\tau^9 \\
f_1 &= -\frac{\tau^4}{5} + a_1\tau^9 \\
f_2 &= -\frac{3\tau^4}{20} + a_1\tau^9
\end{aligned} \tag{6.4.72}
$$

where $\tau = t^{1/8}$ and a_1^{-1} is the old-age constant.

$$R(t) = R(t_0) \left(\frac{t_0}{t}\right)^{1/8} - \frac{1}{2}t^{-1/8}(h(t) - h(t_0)) \tag{6.4.73}$$

where

$$
\begin{aligned}
h(s) &= \frac{8 \log(a_0^{1/5} + a_1^{1/5}s^{1/8})}{5a_0^{4/5}a_1^{1/5}} \\
&- \frac{8 \cos(\pi/5) \log \left(a_0^{2/5} + a_1^{2/5}s^{1/4} - 2a_0^{1/5}a_1^{1/5} \cos(\pi/5)s^{1/8}\right)}{5a_0^{4/5}a_1^{1/5}} \\
&- \frac{8 \cos(3\pi/5) \log \left(a_0^{2/5} + a_1^{2/5}s^{1/4} - 2a_0^{1/5}a_1^{1/5} \cos(3\pi/5)s^{1/8}\right)}{5a_0^{4/5}a_1^{1/5}}
\end{aligned}
$$

$$+ \frac{16 \arctan\left[\left\{\left(a_1^{1/5} s^{1/8} - a_0^{1/5} \cos(\pi/5)\right) \csc(\pi/5)\right\} / a_0^{1/5}\right] \sin(\pi/5)}{5 a_0^{4/5} a_1^{1/5}}$$

$$+ \frac{16 \arctan\left[\left\{\left(a_1^{1/5} s^{1/8} - a_0^{1/5} \cos(3\pi/5)\right) \csc(3\pi/5)\right\} / a_0^{1/5}\right] \sin(3\pi/5)}{5 a_0^{4/5} a_1^{1/5}}$$

$$(6.4.74)$$

where $a_0 = \frac{4}{5}$ and a_1^{-1} is the old-age constant.

For $j = \frac{m}{n}, -1 < j < 0$, the above procedure does not lead to a non-singular choice of coefficients in the function $f_0(t)$. Instead, what was attempted was a rather ad hoc choice of the function $f_0(t)$, which is simply a sum of contributions arising from the inviscid and old-age solutions, in analogy with the case for the plane Burgers equation for which this is exactly true. Surprisingly, the numerical solution shows that this ad hoc choice gives excellent results for the Reynolds number $R(t)$ and, hence, the description of the rate of decay of N wave with time. Such a choice was first used for nonplanar N waves by Sachdev and Seebass (1973), for which we now have given an exact treatment. For details of the case for $-1 < j < 0$, see Sachdev, Srinivasa Rao, and Joseph (1999).

The analytic results for $0 < j < 2$ given above agree very well with the numerical solution, except in the very early stages of the evolution of discontinuous N wave when it adjusts itself via an embryonic shock to the Taylor shock. Indeed, these solutions have the validity in the same sense as the solution (6.4.12) of the plane Burgers equation in Section 6.2 (see again Sachdev, Srinivasa Rao, and Joseph (1999)). Evolution of a typical initial profile and decrease in the corresponding Reynolds number for $j = 1$ are shown in Figure 6.1 and Table 6.1, respectively.

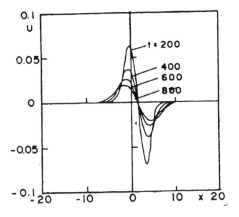

Figure 6.1. Evolution of a typical initial N wave profile for Equation (6.4.2) with $j = 1$ and $\delta = 0.001$.

Table 6.1. Comparison of the Reynolds numbers obtained from numerical and composite solutions (see Equation (6.4.50)) for the IC $u(x, t_i) = (1-j/2)x/t_i, |x| <$ d_0, $u(x, t_i) = 0$ otherwise; the parameters were chosen to be $t_i = 0.5$, $d_0 = 0.205$, $j = 1$, $\delta = 0.001$.

t	numerical	composite
100.0	1.0691	1.3227
200.0	0.6375	0.7240
300.0	0.4597	0.5001
400.0	0.3603	0.3824
500.0	0.2957	0.3097
600.0	0.2501	0.2603
700.0	0.2177	0.2245
800.0	0.1927	0.1974
900.0	0.1729	0.1761
1000.0	0.1568	0.1590
1100.0	0.1434	0.1449
1200.0	0.1322	0.1331
1300.0	0.1225	0.1231
1400.0	0.1142	0.1145
1500.0	0.1069	0.1071
1600.0	0.1005	0.1005
1700.0	0.0948	0.0947
1800.0	0.0897	0.0896
1900.0	0.0851	0.0849
2000.0	0.0809	0.0808
2100.0	0.0771	0.0770
2200.0	0.0736	0.0735
2300.0	0.0704	0.0704
2400.0	0.0674	0.0675
2500.0	0.0647	0.0649
2600.0	0.0621	0.0624
2700.0	0.0597	0.0601
2800.0	0.0574	0.0580
2900.0	0.0553	0.0560
3000.0	0.0533	0.0542

6.5 Modified Burgers Equation

The modified Burgers equation (MBE)

$$u_t + u^2 u_x = \frac{\delta}{2} u_{xx} \qquad (6.5.1)$$

is the counterpart of the modified K-dV equation

$$u_t + u^2 u_x + \delta u_{xxx} = 0. \qquad (6.5.2)$$

While the analysis of (6.5.2) turns out simpler than that for the K-dV equation, which has a quadratic nonlinearity instead of a cube nonlinearity and facilitates the solution of the latter via Miura's transformation, Whitham (1974), (6.5.1) is more difficult than the Burgers equation (BE), which has a quadratic nonlinearity instead of cubic nonlinearity. Unlike BE, (6.5.1) does not admit exact linearisation since no Cole-Hopf-like transformation seems to exist for the purpose. Moreover, (6.5.1) does not enjoy the symmetry $u(-x, t) = -u(x, t)$. For N wave initial conditions, lack of this symmetry makes the analysis hard (see Sachdev and Srinivasa Rao (1999)). Indeed, the matched asymptotic expansion technique which was so successful in handling nonplanar Burgers equation (Crighton and Scott (1979)) fails to provide information about the shock wave displacement due to diffusion and other details such as the old-age constant (Lee-Bapty and Crighton (1987)). This analysis basically provides the evolution of (6.5.1) subject to IC

$$u(x, 0) = \begin{cases} -x & |x| \le 1 \\ 0 & \text{otherwise,} \end{cases} \tag{6.5.3}$$

only in the very early stages of the evolution of the N wave for very small δ, when (6.5.1) behaves essentially like the hyperbolic equation

$$u_t + u^2 u_x = 0 \tag{6.5.4}$$

and the method of characteristics proves useful. Even this turns out to be quite complicated, displaying the peculiar situation in which the steady Taylor shocks which separate the lossless portions of the wave develop an internal singularity at and beyond some finite time t.

As we pointed out earlier, (6.5.1) does not enjoy the antisymmetry property, and so the two lobes of the N wave in (6.5.3) evolve nonsymmetrically to each other; the node of the wave (unlike for the planar or nonplanar BEs) itself moves as the wave evolves. Numerical solution of (6.5.1) and (6.5.3) indicates, however, that after sufficiently long time when the wave is still in its nonlinear evolution and is far from its old-age regime, the node becomes essentially stationary. So, with this basic assumption, we find out the asymptotic solution of the N wave using the exact nonlinearisation technique. We closely follow the work of Sachdev and Srinivasa Rao (1999).

We consider a more general form of MBE, namely

$$u_t + u^n u_x = \frac{\delta}{2} u_{xx} \tag{6.5.5}$$

where n is an even integer. The special case $n = 2$ and the more general case $n \ge 2$ were treated by matched asymptotic expansions by Lee-Bapty and Crighton (1987) and Harris (1996), respectively.

The linear form of (6.5.5) is

$$u_t = \frac{\delta}{2} u_{xx}. \tag{6.5.6}$$

It has the antisymmetric solution

$$u = c \frac{(x - x_0)}{t^{3/2}} e^{-(x-x_0)^2/2\delta t} \qquad (6.5.7)$$

where x_0 is an (arbitrary) point of antisymmetry and c is the old-age constant. We choose x_0 to be the point where the node of the evolving N wave finally comes to (near) rest.

Introducing the independent variables

$$\xi = \frac{(x - x_0)}{(2\delta t)^{1/2}}, \quad \tau = t^{1/2} \qquad (6.5.8)$$

and the dependent variable

$$v = [(2\delta)^{1/2}\xi u^{-1}]^n \qquad (6.5.9)$$

into (6.5.5), we get

$$2nvv_\xi + \xi(nvv_{\xi\xi} - 2n\tau vv_\tau - 2n^2 v^2)$$
$$+2n\xi^2 vv_\xi + 4n^2(2\delta)^{(n-1)/2}\tau\xi^n v - 4n(2\delta)^{(n-1)/2}\tau\xi^{n+1}v_\xi$$
$$-(n+1)\xi v_\xi^2 = 0. \qquad (6.5.10)$$

Observe the negative power of u in (6.5.9); the transformation (6.5.9) is motivated by the N-wave solution of the plane Burgers equation (see Section 6.2). We attempt the following form for v wherein the old-age (linear) solution (6.5.7) is the dominant term and lower-order terms take into account the effects of nonlinearity:

$$v = \sum_{i=0}^{\infty} f_i(\tau) \frac{\xi^i}{i!}. \qquad (6.5.11)$$

Substituting (6.5.11) into (6.5.10) and equating coefficients of various powers of ξ to zero, we get

$$2n \sum_{k=0}^{i} \frac{f_k}{k!} \frac{f_{i+1-k}}{(i-k)!} + n \sum_{k=0}^{i-1} \frac{f_k}{k!} \frac{f_{i+1-k}}{(i-1-k)!}$$

$$-(n+1) \sum_{k=0}^{i-1} \frac{f_{k+1}}{k!} \frac{f_{i-k}}{(i-1-k)!} - 2n^2 \sum_{k=0}^{i-1} \frac{f_k}{k!} \frac{f_{i-1-k}}{(i-1-k)!}$$

$$-2n\tau \sum_{k=0}^{i-1} \frac{f_k'}{k!} \frac{f_{i-1-k}}{(i-1-k)!} + 2n \sum_{k=0}^{i-2} \frac{f_k}{k!} \frac{f_{i-1-k}}{(i-2-k)!}$$

$$+4n^2(2\delta)^{(n-1)/2}\tau \frac{f_{i-n}}{(i-n)!}$$

$$-4n(2\delta)^{(n-1)/2}\tau\frac{f_{i-n}}{(i-n-1)!} = 0 \quad i \geq 0. \tag{6.5.12}$$

It is found that $f_1 = 0$. For $i = 1, 3$, Equation (6.5.12) gives

$$3f_2 - 2nf_0 - 2\tau f_0' = 0 \tag{6.5.13}$$

$$5nf_0f_4 + 3(n-2)f_2^2 + 12n(1-n)f_0f_2 - 6n\tau(f_0f_2)' = 0. \tag{6.5.14}$$

We seek solution of (6.5.13) and (6.5.14) in the form

$$f_0(\tau) = \tau^{2n}\left(a_0 + \frac{a_1}{\tau} + \cdots \frac{a_p}{\tau^p}\right) \tag{6.5.15}$$

$$f_2(\tau) = \tau^{2n}\left(b_0 + \frac{b_1}{\tau} + \cdots \frac{b_p}{\tau^p}\right) \tag{6.5.16}$$

$$f_4(\tau) = \tau^{2n}\left(c_0 + \frac{c_1}{\tau} + \cdots \frac{c_p}{\tau^p}\right) \tag{6.5.17}$$

where the leading terms come from the contributions of the linear solution (6.5.7). Substituting (6.5.15)-(6.5.17) into (6.5.13) and (6.5.14) and equating the coefficients of like powers of τ to zero, we get algebraic equations relating a_i, b_i, and c_i. Solving these equations, we get the coefficients b_i's and c_i's in terms of the first and last coefficients in (6.5.15), namely a_0 and a_p. We give explicit solution of the algebraic systems for $p = 2, 3, 4$.

$p = 2$

$$
\begin{aligned}
a_1 &= 2(a_0a_2)^{1/2} \\
b_0 &= 2na_0, b_1 = 4(n - 1/3)(a_0a_2)^{1/2}, b_2 = 2(n - 2/3)a_2 \\
c_0 &= 12n^2a_0, c_1 = 8(3n^2 - 2n - 1/5)(a_0a_2)^{1/2} \\
c_2 &= (180n^3 - 240n^2 + 32n + 32)a_2/15n
\end{aligned} \tag{6.5.18}
$$

$p = 3$

$$
\begin{aligned}
a_1 &= 3a_0^{2/3}a_3^{1/3}, a_2 = 3a_0^{1/3}a_3^{2/3} \\
b_0 &= 2na_0, b_1 = 2(3n - 1)a_0^{2/3}a_3^{1/3} \\
b_2 &= 2(3n - 2)a_0^{1/3}a_3^{2/3}, b_3 = 2(n - 1)a_3 \\
c_0 &= 12n^2a_0, c_1 = 12(15n^2 - 10n - 1)a_0^{2/3}a_3^{1/3}/5 \\
c_2 &= 12(15n^3 - 20n^2 + 2n + 2)a_0^{1/3}a_3^{2/3}/5n \\
c_3 &= 12(5n^3 - 10n^2 + 3n + 2)a_3/5n
\end{aligned} \tag{6.5.19}
$$

$p = 4$

$$
\begin{aligned}
a_1 &= 4a_0^{3/4}a_4^{1/4}, a_2 = 6a_0^{1/2}a_4^{1/2} \\
a_3 &= 4a_0^{1/4}a_4^{3/4} \\
b_0 &= 2na_0, b_1 = (-8/3 + 8n)a_4^{1/4}a_0^{3/4}
\end{aligned}
$$

$$b_2 = (-8 + 12n)a_0^{1/2}a_4^{1/2}$$
$$b_3 = (-8 + 8n)a_0^{1/4}a_4^{3/4} \qquad (6.5.20)$$
$$b_4 = (-8/3 + 2n)a_4$$
$$c_0 = 12a_0 n^2$$
$$c_1 = (-16/5 - 32n + 48n^2)a_0^{3/4}a_4^{1/4}$$
$$c_2 = (128/15 + 128/15n - 96n + 72n^2)a_0^{1/2}a_4^{1/2}$$
$$c_3 = (80/3 + 256/15n - 96n + 48n^2)a_0^{1/4}a_4^{3/4}$$
$$c_4 = (224/15 + 128/15n - 32n + 12n^2)a_4$$

In fact, if we let $f_0(\tau)$ have the form

$$f_0(\tau) = \tau^{2n}(a_0^{1/p} + a_p^{1/p}/\tau)^p \qquad (6.5.21)$$

for all $p = 2, 3, 4, \ldots$ (this is clearly true for $p = 2, 3,$ and 4), then the solution of (6.5.13) and (6.5.14), respectively, gives

$$f_2(\tau) = -\frac{2}{3}\tau^{2n}(a_0^{1/p} + a_p^{1/p}/\tau)^{p-1}(-3a_0^{1/p}n + a_p^{1/p}(-3n + p)/\tau)$$

$$f_4(\tau) = -\frac{4}{15n}\tau^{2n}(a_0^{1/p} + a_p^{1/p}/\tau)^{p-2}\left[-45a_0^{2/p}n^3\right. \qquad (6.5.22)$$

$$+a_0^{1/p}a_p^{1/p}(-90n^3 + 30n^2p + 3pn)/\tau$$

$$+a_p^{2/p}(-45n^3 + 30n^2p - 5np^2 + 6pn - 2p^2)/\tau^2\Big]$$

in decreasing powers of τ for all $p = 2, 3, 4, \ldots$: Other $f_i(\tau), i > 4$, can be found from (6.5.12), recursively.

The simple form (6.5.21) for $f_0(\tau)$ makes the calculation of the Reynolds number for arbitrary p rather easy.

It is clear from the above that each representation has two arbitrary constants: the old-age constant a_0 and the last constant in f_0, namely a_p. It is also evident that the solutions that we have constructed have linear solution as their asymptotic in the limit $t \to \infty$. A rigorous argument for this can be given in a manner similar to that for the nonplanar equation in Section 6.4.

The Reynolds number of the N-wave solution of (6.5.5) is defined as

$$R(t) = \frac{1}{\delta}\int_{-\infty}^{x_0} u(x, t)dx. \qquad (6.5.23)$$

Integrating (6.5.5) with respect to x form $-\infty$ to x_0 and using the vanishing conditions at $-\infty$, we have

$$R'(t) = \frac{1}{2}u_x(x_0, t). \qquad (6.5.24)$$

From (6.5.8), (6.5.9), (6.5.11), and (6.5.21), we have

$$u_x(x_0, t) = -\frac{1}{t^{3/2}}\left(c_1 + \frac{c_2}{t^{1/2}}\right)^{-p/n} \tag{6.5.25}$$

where $c_1 = c^{-n/p}, c_2 = a_p^{1/p}$, and c is the old-age constant equal to $a_0^{-1/n}$ in the expansion for f_0, and a_p is the coefficient of the last term in the same. Using (6.5.25) in (6.5.24), integrating the latter with the condition $R = R(t_0)$ at $t = t_0$, we have

$$R(t) = R(t_0) + \frac{1}{c_2} \log\left[\left(c_1 + \frac{c_2}{t^{1/2}}\right) \Big/ \left(c_1 + \frac{c_2}{t_0^{1/2}}\right)\right], n = p$$

$$= R(t_0) - \frac{n}{c_2(p-n)}\left\{\left(c_1 + \frac{c_2}{t^{1/2}}\right)^{-(p-n)/n}\right. \tag{6.5.26}$$

$$\left. - \left(c_1 + \frac{c_2}{t_0^{1/2}}\right)^{-(p-n)/n}\right\}, n \neq p.$$

It may be noted that even though each lobe of the N wave evolves under (6.5.5) rather differently, their respective areas remain the same. Numerical solution of IVP (6.5.5) and (6.5.3) shows that the formula (6.5.26) works rather well, its accuracy increasing with increasing p (Sachdev and Srinivasa Rao (1999)). The old-age constant c and the other coefficient $c_2 = a_p^{1/p}$ appearing in (6.5.26) are obtained by matching the latter with the numerical solution. Evolution of the N wave for $n = 2$ and the corresponding Reynolds numbers at different times are shown in Figures 6.2, 6.3, and Table 6.2, respectively.

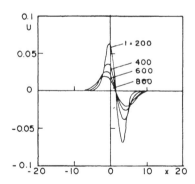

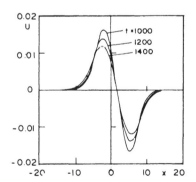

Figure 6.2. Evolution of N wave under Equation (6.5.1) with IC (6.5.3) and $\delta = 0.01$ at $t = 200, 400, 600, 800.$

Figure 6.3. Evolution of N wave under Equation (6.5.1) with IC (6.5.3) and $\delta = 0.01$ at $t = 1000, 1200, 1400.$

Table 6.2. Comparison of the Reynolds numbers obtained by the nonlinearisation method with $p = 4$ (see Equation (6.5.26)) and numerical solution. Here, $n = 2$, $\delta = 0.01$ in Equation (6.5.5). The initial profile is given by (6.5.3).

t	R_{num}	R_{anal}
100.0	22.5408	24.6575
200.0	17.7679	18.3592
300.0	15.1795	15.3699
400.0	13.4729	13.5248
500.0	12.2374	12.2374
600.0	11.2892	11.2719
700.0	10.5319	10.5125
800.0	9.9088	9.8944
900.0	9.3835	9.3785
1000.0	8.9346	8.9391
1100.0	8.5444	8.5590
1200.0	8.2012	8.2258
1300.0	7.8963	7.9306
1400.0	7.6231	7.6666
1500.0	7.3728	7.4286
1600.0	7.1490	7.2127
1700.0	6.9443	7.0155
1800.0	6.7562	6.8346
1900.0	6.5826	6.6677
2000.0	6.4218	6.5132
2100.0	6.2722	6.3696
2200.0	6.1325	6.2356
2300.0	6.0018	6.1103
2400.0	5.8791	5.9927
2500.0	5.7637	5.8820
2600.0	5.6547	5.7776
2700.0	5.5518	5.6790
2800.0	5.4542	5.5855
2900.0	5.3616	5.4969
3000.0	5.2736	5.4126

6.6 Embedding of Similarity Solution in a Larger Class

Before we discuss this matter for nonlinear PDEs, it is instructive to view the role of special exact solution of nonlinear ODEs in the larger family of asymptotic solutions. To illustrate this point we consider the Thomas-Fermi equation

$$y''(x) = x^{-1/2}[y(x)]^{3/2} \qquad (6.6.1)$$

and observe that it has a special exact solution vanishing at infinity, namely

$$y_s(x) = \frac{144}{x^3}. \tag{6.6.2}$$

The following results have been proved regarding the place of (6.6.2) in the larger family of solutions (Hille (1970), (1970a), Sachdev (1991)).

1. Equation (6.6.1) has a one-parameter family of solutions of the form

$$y_\infty(x, a) = x^{-3} \left[144 + \sum_{n=1}^{\infty} a_n x^{-n\sigma} \right] \tag{6.6.3}$$

   where $a_1 = a$ is arbitrary and other coefficients are uniquely determined in terms of a. The series (6.6.3) converges for large x. In particular, if $a = 0$, (6.6.3) reduces to the special solution (6.6.2). For σ in (6.6.3), see (6.6.8).

2. Equation (6.6.1) has a one-parameter family of solutions

$$y_0(x, c) = x^{-3} \left\{ 144 + \sum_{n=1}^{\infty} c_n x^{n\tau} \right\} \tag{6.6.4}$$

   where $c_1 = c$ is arbitrary, the other coefficients are uniquely determined, and the series converges for small values of x (see (6.6.8) for the definition of τ).

3. The boundary value problem for (6.6.1) with

$$y(a) = b, \quad \lim_{x \to \infty} y(x) = 0, \quad a \geq 0, \quad b > 0 \tag{6.6.5}$$

   has a unique solution. The singular solution $y_s(x) = 144x^{-3}$ satisfies (6.6.5) for any choice of (a, b) on its graph C. This curve separates the points of the first quadrant into two domains: D_- below C and D_+ above it. For $(a, b) \in D_-$, there is a unique value of $\beta > 0$ such that

$$y = \beta^{-3} y_\infty(\beta^{-1}x) \tag{6.6.6}$$

   defines the solution of (6.6.5). All these integral curves stay in D_- and are defined for $x \geq 0$. They can be extended to the left of (a, b) to the y-axis. If $(a, b) \in D_+$, then the corresponding solution of (6.6.5) stays in D_+. Its graph has a vertical asymptote.

4. The singular boundary value problem

$$\lim_{x \downarrow 0} y(x) = \infty, \quad y(a) = b, \quad a > 0, \quad b > 0$$

has a unique solution which is strictly decreasing. The function

$$\nu(x) = x^3 y(x) \tag{6.6.7}$$

is bounded and monotone with limit 144 as $x \downarrow 0$. If $(a, b) \in D_-$, then $y(x)$ stays in D_- and $\nu(x)$ increases to 144 as $x \downarrow 0$. If $(a, b) \in D_+$, then $y(x)$ stays in D_+, and $\nu(x)$ decreases to 144 as $x \downarrow 0$.

The exponents σ and τ in (6.6.3) and (6.6.4) are

$$\sigma = \frac{1}{2}\left[\sqrt{73} - 7\right], \quad \tau = \frac{1}{2}\left[\sqrt{73} + 7\right] \tag{6.6.8}$$

and arise naturally when the exact solution is sought to be extended by writing $\nu(x) = x^3 y(x)$ and then $\nu = 144 + u$, etc. (See Sachdev (1991)).

This kind of analysis is not so common in nonlinear PDEs. We conjecture that the similarity solutions have such a special role in demarcating different classes of solutions of PDEs. We consider below plane gasdynamic equations and show how their self-similar solutions describing piston-driven shocks of large or arbitrary strength may be extended so that the enlarged class contains the self-similar solution as a special case.

We follow the work of Sachdev, Gupta, and Ahluwalia (1992). We consider one of the simplest cases investigated in this work, namely when the special form of gasdynamic equations arises from the conservation of mass and momentum, and the piston-driven shocks are of infinite strength. The analysis in this case is relatively simpler; we refer the reader to the original work for solutions of other forms of gasdynamic equations and for shocks of arbitrary strength.

The gasdynamic equations governing plane, unsteady, anisentropic flows are

$$\rho_t + u\rho_x + \rho u_x = 0 \tag{6.6.9}$$
$$\rho(u_t + uu_x) + p_x = 0 \tag{6.6.10}$$
$$S_t + uS_x = 0 \tag{6.6.11}$$
$$p = \rho^\gamma \exp\left(\frac{S - S_0}{c_v}\right) \tag{6.6.12}$$

where ρ, u, p, and S are the density of the fluid, particle velocity, pressure, and entropy, respectively, at any point x and time t. The variables p, ρ, and S are related by the thermodynamic relation (6.6.12). γ and c_v are ratio of specific heats (c_p/c_v) and the specific heat at constant volume, respectively.

The system of nonlinear PDEs (6.6.9)-(6.6.11) may be rewritten in the conservation form as

$$\rho_t + (\rho u)_x = 0 \tag{6.6.13}$$

$$(\rho u)_t + (\rho u^2 + p)_x = 0 \tag{6.6.14}$$

$$\left(\frac{1}{2}\rho u^2 + \frac{p}{\gamma - 1}\right)_t + \left(\left(\frac{1}{2}\rho u^2 + \frac{p}{\gamma - 1}\right)u + pu\right)_x = 0. \tag{6.6.15}$$

The Rankine-Hugoniot conditions holding across a shock moving into a nonuniform medium with velocity $u_0 = 0$, pressure $p = p_0$ (a constant), and density $\rho = \rho_*(x)$ are

$$u = \left(\frac{2}{\rho_*}\right)^{1/2} (p - p_0)[(\gamma + 1)p + (\gamma - 1)p_0]^{-1/2} \tag{6.6.16}$$

$$\frac{\rho}{\rho_*} = \frac{(\gamma + 1)p + (\gamma - 1)p_0}{(\gamma - 1)p + (\gamma + 1)p_0} \tag{6.6.17}$$

$$U = (2\rho_*)^{-1/2}[(\gamma + 1)p + (\gamma - 1)p_0]^{1/2} \tag{6.6.18}$$

where U is the shock velocity; the trajectory of the shock is given by

$$\frac{dx}{dt} = U. \tag{6.6.19}$$

For strong shocks, the relations (6.6.16)-(6.6.18) simplify to

$$u = \left(\frac{2}{\gamma + 1}\frac{p}{\rho_*}\right)^{1/2} \tag{6.6.20}$$

$$\frac{\rho}{\rho_*} = \frac{\gamma + 1}{\gamma - 1} \tag{6.6.21}$$

$$U = \left(\frac{\gamma + 1}{2}\frac{p}{\rho_*}\right)^{1/2}. \tag{6.6.22}$$

Equations (6.6.13)-(6.6.14) suggest the introduction of the variables τ and ξ through the differential relations

$$d\tau = \rho dx - (\rho u)dt \tag{6.6.23}$$

$$d\xi = \rho u dx - (p + \rho u^2)dt \tag{6.6.24}$$

yielding

$$dx = -\frac{u}{p}d\xi + \frac{(p + \rho u^2)}{p\rho}d\tau \tag{6.6.25}$$

$$dt = -\frac{1}{p}d\xi + \frac{u}{p}d\tau. \tag{6.6.26}$$

The shock trajectory (6.6.19) in terms of τ and ξ becomes

$$d\xi + \phi(\tau)d\tau = 0 \tag{6.6.27}$$

where

$$\phi(\tau) = \frac{p_0}{\rho_*(x)} U^{-1}. \tag{6.6.28}$$

To facilitate the fitting of shock conditions (6.6.20)-(6.6.22), a new variable s is introduced so that

$$ds = d\xi + \phi(\tau)d\tau \tag{6.6.29}$$

which, in view of (6.6.27), states that s is constant along the shock. In terms of the variables s and τ, (6.6.25) and (6.6.26) become

$$dx = -\frac{u}{p}ds + \frac{(p + \rho u^2 + \rho u\phi)}{\rho p}d\tau \tag{6.6.30}$$

$$dt = -\frac{1}{p}ds + \frac{u + \phi}{p}d\tau. \tag{6.6.31}$$

For convenience we choose $s = 0$ as the level line corresponding to the shock path, since (6.6.29) is invariant under translation in s. The other level lines $\tau = $ constant give paths of constant entropy (see (6.6.23)) in (τ, s) plane. In view of this, we may drop Equation (6.6.11). Equations (6.6.9)-(6.6.10) with τ and s as independent variables become

$$u_\tau + (u + \phi)u_s - (f/\gamma)p^{-1/\gamma}p_s = 0 \tag{6.6.32}$$

$$p_\tau + (u + \phi)p_s - pu_s = 0 \tag{6.6.33}$$

where

$$f(\tau) = (p^{1/\gamma}\rho^{-1}). \tag{6.6.34}$$

The relation (6.6.34) follows from (6.6.12). The shock trajectory for the case of strong shock is easily found in the (s, τ) plane as

$$ds = d\xi = 0. \tag{6.6.35}$$

The relation $d\xi = 0$ derives from (6.6.27) and (6.6.28) since $\phi(\tau) \to 0$ as $p_0 \to 0$ in the strong shock limit. Thus, $s = \xi = 0$ give the shock locus. In this special case, the governing equations (6.6.32)-(6.6.34) simplify to

$$u_\tau + uu_s - \frac{\gamma - 1}{\gamma}fp^{-1/\gamma}p_s = 0 \tag{6.6.36}$$

$$p_\tau + up_s - pu_s = 0 \tag{6.6.37}$$

where

$$f(\tau) = \frac{1}{\rho(\gamma - 1)}p^{1/\gamma}. \tag{6.6.38}$$

We introduce the variables

$$W = \frac{u}{\phi}, \quad H = \left(\frac{p}{b_0}\right)^{(\gamma-1)/\gamma} \tag{6.6.39}$$

into (6.6.36)-(6.6.37), where $\phi(\tau)$ is an arbitrary function of τ and b_0 is an arbitrary constant with the dimension of pressure; we get the system

$$H_\tau + \phi W H_s - \frac{\gamma - 1}{\gamma}\phi H W_s = 0 \tag{6.6.40}$$

$$\phi(W_\tau + \phi W W_s) - b_0^{(\gamma-1)/\gamma} f H_s + W\frac{d\phi}{d\tau} = 0. \tag{6.6.41}$$

If we specify the function ϕ and the constant b_0 such that

$$\phi(\tau) = U(\tau) \tag{6.6.42}$$

and

$$\rho_* U^2 = b_0 = \frac{\rho_0 x_0^2}{t_0^2}, \tag{6.6.43}$$

then the strong shock conditions (6.6.20)-(6.6.22) in terms of the functions W and H simplify:

$$W_{s=0} = \frac{2}{\gamma + 1} \tag{6.6.44}$$

$$H_{s=0} = \left(\frac{2}{\gamma + 1}\right)^{(\gamma-1)/\gamma}. \tag{6.6.45}$$

We seek a solution of the system (6.6.40)-(6.6.41) subject to the conditions (6.6.44)-(6.6.45) in the form

$$W(s,\tau) = \sum_{j=0}^{\infty} W^{(j)}(\tau)\frac{s^j}{j!} \tag{6.6.46}$$

$$H(s,\tau) = \sum_{j=0}^{\infty} H^{(j)}(\tau)\frac{s^j}{j!}. \tag{6.6.47}$$

Putting (6.6.46)-(6.6.47) into (6.6.40) and (6.6.41) and equating coefficients of different powers of s to zero, we get the following algebraic relations for first- and higher- order coefficients:

$$W^{(n+1)}(\tau) = \frac{\gamma}{(\gamma - 1)\phi H^{(0)}}\left[\phi W^{(0)} H^{(n+1)} + \frac{dH^{(n)}}{d\tau}\right.$$

$$\left. +\phi T^{(n)}(\{W^{(i)}\}, \{H^{(i)}\})\right] \tag{6.6.48}$$

$$H^{(n+1)}(\tau) = \frac{1}{D_0}\left[W^{(n)}\frac{d\phi}{d\tau} + \phi\left(\frac{dW^{(n)}}{d\tau} + \frac{\gamma}{\gamma - 1}\frac{W^{(0)}}{H^{(0)}}\frac{dH^{(n)}}{d\tau}\right.\right.$$

$$\left.\left. +\phi S^{(n)}\left(\{W^i\}, \{H^{(j)}\}\right)\right)\right] \tag{6.6.49}$$

with the notation

$$T^{(n)}(\{W^{(i)}\}, \{H^{(i)}\}) = \sum_{k=1}^{n} \binom{n}{k} \left[W^{(k)} H^{(n+1-k)} - \frac{\gamma-1}{\gamma} H^{(k)} W^{(n+1-k)} \right]$$

(6.6.50)

$$S^{(n)}(\{W^{(i)}\}, \{H^{(i)}\}) = \sum_{k=1}^{n} \binom{n}{k} W^{(k)} W^{(n+1-k)}$$

$$+ \frac{\gamma}{\gamma-1} \frac{W^{(0)}}{H^{(0)}} T^{(n)}(\{W^i\}, \{H^{(i)}\}),$$

$$D_0 = b_0^{(\gamma-1)/\gamma} f - \frac{\gamma}{\gamma-1} \left(\frac{W^{(0)}}{H^{(0)}} \right)^2. \qquad (6.6.51)$$

It is interesting to note that the recursion relations (6.6.48)-(6.6.51) involve only algebraic operations; no differential equations need be solved. The zeroth order terms in (6.6.46)-(6.6.47) (corresponding to $s = 0$) are given by (6.6.44) and (6.6.45):

$$W^{(0)} = \frac{2}{\gamma+1} \qquad (6.6.52)$$

$$H^{(0)} = \left(\frac{2}{\gamma+1} \right)^{(\gamma-1)/\gamma}. \qquad (6.6.53)$$

In the following, the higher terms in the series solution (6.6.46)-(6.6.47) would be found by choosing the functions $\phi(\tau)$ and $f(\tau)$ appropriately. These will yield power law and exponential types of solutions.

1. Power law similarity solutions of the system (6.6.40)-(6.6.41)

We choose

$$\rho_* = \rho_0 \left(1 + \frac{a\tau}{\rho_0 x_0} \right)^{-2\alpha} \qquad (6.6.54)$$

$$\phi = \frac{x_0}{t_0} \left(1 + \frac{a\tau}{\rho_0 x_0} \right)^{\alpha} \qquad (6.6.55)$$

$$f = c_* \frac{x_0^2}{t_0^2} \left(1 + \frac{a\tau}{\rho_0 x_0} \right)^{2\alpha} \qquad (6.6.56)$$

where c_* is a parameter defined by

$$c_* = \frac{1}{2} \left(\frac{2}{\gamma+1} \right)^{(\gamma+1)/\gamma} b_0^{-(\gamma+1)/\gamma}. \qquad (6.6.57)$$

Here, $\alpha(\neq 0)$ is the similarity exponent, and x_0, t_0, and ρ_0 are arbitrary dimensional constants representing distance, time, and density, respectively. The constant a assumes values $+1$ or -1.

Substituting (6.6.54)-(6.6.56) into (6.6.48)-(6.6.49), we get

$$W^{(n+1)}(z) = W_*^{(n+1)} z^{n+1} \tag{6.6.58}$$

$$H^{(n+1)}(z) = H_*^{(n+1)} z^{n+1}, n = 0, 1, \ldots \tag{6.6.59}$$

where

$$z = \frac{t_0}{\rho_0 x_0^2} \left(1 + \frac{a\tau}{\rho_0 x_0} \right)^{-(\alpha+1)} \tag{6.6.60}$$

and

$$W_*^{(n+1)} = \frac{\gamma}{(\gamma-1)H^{(0)}} \left[W^{(0)} H_*^{(n+1)} - na(\alpha+1)H_*^n + T_*^{(n)} \right.$$
$$\left. \left(\{W_*^{(i)}\}, \{H_*^{(i)}\} \right) \right]$$

$$H_*^{(n+1)} = \frac{1}{D_*} \left[(\alpha - n(\alpha+1))a W_*^{(n)} - \frac{\gamma an(\alpha+1)W^{(0)}}{(\gamma-1)H^{(0)}} H_*^{(n)} \right.$$

$$\left. + S_*^{(n)} \left(\{W_*^{(i)}\}, \{H_*^{(i)}\} \right) \right] \tag{6.6.61}$$

are constants, and

$$T_*^{(n)}(\{W_*^{(i)}\}, \{H_*^{(i)}\}) = \sum_{k=1}^n \binom{n}{k} \left[W_*^{(k)} H_*^{(n+1-k)} - \frac{\gamma-1}{\gamma} H_*^{(k)} W_*^{(n+1-k)} \right]$$

$$S_*^{(n)}(\{W_*^{(i)}\}, \{H_*^{(i)}\}) = \sum_{k=1}^n \binom{n}{k} W_*^{(k)} W_*^{(n+1-k)}$$

$$+ \frac{\gamma W^{(0)}}{(\gamma-1)H^{(0)}} T_*^{(n)} \left(\{W_*^{(i)}\}, \{H_*^{(i)}\} \right) \tag{6.6.62}$$

$$D_* = b_0^{(\gamma-1)/\gamma} c_* - \frac{\gamma}{\gamma-1} \frac{W^{(0)^2}}{H^{(0)}} = -\frac{1}{\gamma-1} \left(\frac{2}{\gamma+1} \right)^{1/\gamma}. \tag{6.6.63}$$

We may now write the series solution (6.6.46)-(6.6.47) as

$$W = \sum_{j=0}^\infty W_*^{(j)} \frac{\sigma^j}{j!} \tag{6.6.64}$$

$$H = \sum_{j=0}^\infty H_*^{(j)} \frac{\sigma^j}{j!} \tag{6.6.65}$$

where

$$\sigma = \frac{t_0}{\rho_0 x_0^2} \left(1 + \frac{a\tau}{\rho_0 x_0}\right)^{-(\alpha+1)} s. \qquad (6.6.66)$$

The coefficients $W_*^{(j)}$ and $H_*^{(j)}$ in the series (6.6.64) and (6.6.65) for $j \geq 0$ are given by (6.6.52), (6.6.53), (6.6.61), and (6.6.62) with $j = n + 1, n = 0, 1, \ldots$. It may be checked that the solution (6.6.64)-(6.6.65) is a similarity solution, with σ defined by (6.6.66) as the similarity variable, and satisfies the system of ODEs

$$[-a(\alpha+1)\sigma + W]\frac{dH}{d\sigma} - \frac{\gamma-1}{\gamma}H\frac{dW}{d\sigma} = 0 \qquad (6.6.67)$$

$$[-a(\alpha+1)\sigma + W]\frac{dW}{d\sigma} - c_* b_0^{(\gamma-1)/\gamma}\frac{dH}{d\sigma} + a\alpha W = 0 \qquad (6.6.68)$$

with IC (6.6.44) and (6.6.45) at the shock. The parametric representation of the shock may now be found to be

$$\frac{x}{x_0} = \begin{cases} \dfrac{1}{a(1+2\alpha)}\left[\left(1 + \dfrac{a\tau}{\rho_0 x_0}\right)^{2\alpha+1} - 1\right], & \alpha \neq -\dfrac{1}{2} \\[4mm] \dfrac{1}{a}\ln\left(\dfrac{a\tau}{\rho_0 x_0}\right) & \alpha = -\dfrac{1}{2} \end{cases} \qquad (6.6.69)$$

$$\frac{t}{t_0} = \begin{cases} \dfrac{1}{a(\alpha+1)}\left[\left(1 + \dfrac{a\tau}{\rho_0 x_0}\right)^{\alpha+1} - 1\right], & \alpha \neq -1 \\[4mm] \dfrac{1}{a}\ln\left[1 + \dfrac{a\tau}{\rho_0 x_0}\right], & \alpha = -1. \end{cases} \qquad (6.6.70)$$

We have used the conditions that $x = t = 0$ when $\tau = 0$. Eliminating τ between (6.6.69) and (6.6.70), we get the shock locus as

$$\frac{x(t)}{x_0} = \begin{cases} \dfrac{1}{a(2\alpha+1)}\left[\left(1 + \dfrac{a(\alpha+1)}{t_0}t\right)^{(2\alpha+1)/(\alpha+1)} - 1\right], & \alpha \neq -1, -\frac{1}{2} \\[4mm] \dfrac{1}{a}\left[1 - \exp\left(-\dfrac{at}{t_0}\right)\right], & \alpha = -1 \\[4mm] \dfrac{2}{a}\ln\left[1 + \dfrac{at}{2t_0}\right], & \alpha = -\frac{1}{2} \end{cases}$$

$$\qquad (6.6.71)$$

and, hence, the shock velocity and the undisturbed density (using (6.6.54) and (6.6.69)):

$$U = \begin{cases} \dfrac{x_0}{t_0}\left(1 + \dfrac{a(\alpha+1)t}{t_0}\right)^{\alpha/(\alpha+1)}, & \alpha \neq -1 \\[2ex] \dfrac{x_0}{t_0}\exp\left(-\dfrac{at}{t_0}\right), & \alpha = -1 \end{cases} \tag{6.6.72}$$

$$\rho_* = \begin{cases} \rho_0\left[1 + \dfrac{a(2\alpha+1)x}{x_0}\right]^{-2\alpha/(2\alpha+1)}, & \alpha \neq -\tfrac{1}{2} \\[2ex] \rho_0\exp\left(\dfrac{ax}{x_0}\right), & \alpha = -\tfrac{1}{2}. \end{cases} \tag{6.6.73}$$

Making use of (6.6.30)-(6.6.31) and remembering that $w = w(\sigma), H = H(\sigma)$ where σ is given by (6.6.66), we get

$$\frac{x(H,\tau)}{x_0} = \begin{cases} \dfrac{\left[1 + \frac{a\tau}{\rho_0 x_0}\right]^{2\alpha+1}}{a\alpha}\displaystyle\int_{H_*}^{H}\left[\frac{\gamma-1}{\gamma}H\left(\frac{dW}{dH}\right)^2 - c\right]H^{-\gamma/(\gamma-1)}dH \\[2ex] \qquad + \dfrac{1}{a(2\alpha+1)}\left[\left(1 + \dfrac{a\tau}{\rho_0 x_0}\right)^{2\alpha+1} - 1\right], \quad \alpha \neq -\dfrac{1}{2} \\[3ex] \dfrac{2}{a}\displaystyle\int_{H_*}^{H}\left[c - \dfrac{\gamma-1}{\gamma}H\left(\dfrac{dW}{dH}\right)^2\right]H^{-\gamma/(\gamma-1)}dH \\[2ex] \qquad + \dfrac{1}{a}\ln\left[1 + \dfrac{a\tau}{\rho_0 x_0}\right], \quad \alpha = -\dfrac{1}{2} \end{cases} \tag{6.6.74}$$

$$\frac{t(H,\tau)}{t_0} = \begin{cases} \dfrac{(1 + a\tau/\rho_0 x_0)^{\alpha+1}}{a\alpha}\displaystyle\int_{H_*}^{H}\left[\frac{\gamma-1}{\gamma}H\left(\frac{dW}{dH}\right)^2 - c\right]\frac{H^{-\gamma/(\gamma-1)}}{W}dH \\[2ex] \qquad + \dfrac{1}{a(\alpha+1)}\left[\left(1 + \dfrac{a\tau}{\rho_0 x_0}\right)^{\alpha+1} - 1\right], \quad \alpha \neq -1 \\[3ex] \dfrac{1}{a}\displaystyle\int_{H_*}^{H}\left[c - \dfrac{\gamma-1}{\gamma}H\left(\dfrac{dW}{dH}\right)^2\right]\frac{H^{-\gamma/(\gamma-1)}}{W}dH \\[2ex] \qquad + \dfrac{1}{a}\ln\left[1 + \dfrac{a\tau}{\rho_0 x_0}\right], \quad \alpha = -1 \end{cases} \tag{6.6.75}$$

where $c = \frac{1}{2}\left(\frac{2}{\gamma+1}\right)^{(\gamma+1)/\gamma}$, $H_* \leq H < \infty$ if $a\alpha > 0$, and $1 < H \leq H_*$ if $a\alpha < 0$.

Setting $\tau = $ constant in (6.6.74) and (6.6.75), we get the lines of constant entropy or paths of gas particles in a parametric form. For $\tau = 0$, these reduce to the parametric representation of the law of piston motion:

$$\frac{x(H)}{x_o} = \frac{1}{a\alpha} \int_{H_*}^{H} \left[\frac{\gamma-1}{\gamma} H \left(\frac{dW}{dH} \right)^2 - c \right] H^{-\gamma/(\gamma-1)} dH \quad (6.6.76)$$

$$\frac{t(H)}{t_o} = \frac{1}{a\alpha} \int_{H_*}^{H} \left[\frac{\gamma-1}{\gamma} H \left(\frac{dW}{dH} \right)^2 - c \right] \frac{H^{-\gamma/(\gamma-1)}}{W} dH. \quad (6.6.77)$$

The solution (6.6.52), (6.6.53), (6.6.64)-(6.6.66), (6.6.61), (6.6.62), (6.6.76), and (6.6.77) of a certain piston problem involves four arbitrary constants ρ_0, x_0, t_0, and α. The series (6.6.64)-(6.6.66), alongside (6.6.52)-(6.6.53), can be summed up only for $\alpha = -1, -1/2$. For $\alpha = -1$, we have a first integral of the ODE system (6.6.67)-(6.6.68), namely $W = H^{\gamma/(\gamma-1)}$, and then the similarity solution simplifies to give the following.

$\alpha = -1$.

$$u = UH^{\gamma/(\gamma-1)}, \rho_* = \rho_0 \left(1 - \frac{ax}{x_0} \right)^{-2}$$

$$f = c_* \frac{x_0^2}{t_0^2} \left(1 + \frac{a\tau}{\rho_0 x_0} \right)^{-2}, \quad U = \frac{x_0}{t_0} \exp \left(-\frac{at}{t_0} \right)$$

Shock locus:

$$\frac{x_s(t)}{x_0} = \frac{1}{a} \left[1 - \exp \left(-\frac{at}{t_0} \right) \right] \quad (6.6.78)$$

Piston path :

$$\frac{x(H)}{x_0} = \frac{1}{a} \int_{H_*}^{H} \left[c - \frac{\gamma-1}{\gamma} H \left(\frac{dW}{dH} \right)^2 \right] H^{-\gamma/(\gamma-1)} dH$$

$$\frac{t(H)}{t_0} = \frac{1}{a} \int_{H_*}^{H} \left[c - \frac{\gamma-1}{\gamma} H \left(\frac{dW}{dH} \right)^2 \right] \frac{H^{-\gamma/(\gamma-1)}}{W} dH$$

where $H_* \leq H < \infty$ if $a = -1$ or $1 < H \leq H_*$ if $a = 1$. Again, $H_* = H^{(0)}$. This solution involves only three arbitrary constants, ρ_0, x_0, and t_0.

$\alpha = -1/2$.

$$u = UW, \rho_* = \rho_0 \exp \left(\frac{ax}{x_0} \right)$$

$$U = \frac{x_0}{t_0} \left(1 + \frac{at}{2t_0} \right)^{-1}, \quad f = c_* \frac{x_0^2}{t_0^2} \left(1 + \frac{a\tau}{\rho_0 x_0} \right)^{-1}$$

Shock locus:

$$\frac{x_s(t)}{x_0} = \frac{2}{a} \ln\left[1 + \frac{at}{2t_0}\right] \tag{6.6.79}$$

Piston path:

$$\frac{x(H)}{x_0} = \frac{2}{a} \int_{H_*}^{H} \left[c - \frac{\gamma - 1}{\gamma} H \left(\frac{dW}{dH}\right)^2\right] H^{-\gamma/(\gamma-1)} dH$$

$$\frac{t(H)}{t_0} = \frac{2}{a} \int_{H_*}^{H} \left[c - \frac{\gamma - 1}{\gamma} H \left(\frac{dW}{dH}\right)^2\right] \frac{H^{-\gamma/(\gamma-1)}}{W} dH$$

where

$$W = [-2\gamma cH + 2H^{\gamma/(\gamma-1)}]^{1/2}.$$

The choice

$$\rho_* = \rho_0 \exp\left(-\frac{2a\tau}{\rho_0 x_0}\right)$$

$$U = \frac{x_0}{t_0} \exp\left(\frac{a\tau}{\rho_0 x_0}\right) \tag{6.6.80}$$

$$f = c_* \frac{x_0^2}{t_0^2} \exp\left(\frac{2a\tau}{\rho_0 x_0}\right)$$

leads to exponential types of solutions of the system (6.6.40)-(6.6.41). Since the analysis is entirely similar to that for the power law case, we refer the reader to the original paper of Sachdev et al. (1992).

Sachdev et al. (1992) carried out a detailed study for shocks of arbitrary strength; the analysis in this case becomes even more arduous. Previous similarity solutions of the transformed system (6.6.32)-(6.6.34) found earlier by Sachdev and Reddy (1982) were recovered as special cases. The analysis was also carried out for another transformed system deriving from the conservation form of the equations of continuity and energy instead of equations of continuity and momentum. Some of the series solutions for certain sets of parameters were either summed or computed; the convergence in the latter case was checked numerically.

Chapter 7

Asymptotic Solutions by Balancing Arguments

7.1 Asymptotic Solution by Balancing Arguments – Examples from ODEs

For nonlinear problems for which exact solution is not tenable, one may look for asymptotic solutions for large time or distance. For this purpose, one may sift from the full equation(s) those terms which balance in terms of power of the limiting variable, the remaining terms being smaller in comparison in the limit. The effect of the neglected terms may be subsequently incorporated in the next order. This approach was first made popular in the context of nonlinear ODEs by Bender and Orszag (1978), but has since been used to great effect by Grundy and his collaborators (Grundy, Sachdev, and Dawson (1994); Grundy, Van Duijn, and Dawson (1994); Dawson, Van Duijn, and Grundy (1996); Escobedo and Grundy (1996); Van Duijn, Grundy, and Dawson (1997)).

We shall, in this section, discuss some examples from ODEs (Bender and Orszag (1978); Sachdev (1991)) for which the balancing argument has been used profitably.

Let us investigate the behaviour of

$$y^2 y''' = -\frac{1}{3} \qquad (7.1.1)$$

as the independent variable $x \to \infty$. One obvious possibility for y is a quadratic in x and a small correction term which may include the effect of the term $-\frac{1}{3}$ on the right-hand side of (7.1.1). If we substitute

$$y(x) \sim ax^2 + bx + c + \epsilon(x) \qquad (7.1.2)$$

into (7.1.1) and assume that $\epsilon(x)$ is small as $x \to \infty$, we have

$$a^2 x^4 \epsilon''' \sim -\frac{1}{3} \quad \text{as} \quad x \to \infty. \tag{7.1.3}$$

On integration, (7.1.3) gives

$$\epsilon \sim \frac{1}{18a^2 x} \quad \text{as} \quad x \to \infty \tag{7.1.4}$$

which, indeed, vanishes as $x \to \infty$. With the correction term (7.1.4) put in (7.1.2), one may justifiably generalise the latter and assume that

$$y(x) \sim ax^2 + bx + c + \frac{d}{x} + \frac{e}{x^2} + \frac{f}{x^3} + \dots \tag{7.1.5}$$

Substituting (7.1.5) into (7.1.1) and comparing like powers of x on both sides, it is found that

$$d = \frac{1}{18a^2}, \quad e = -\frac{b}{36a^3}, \quad f = \frac{3b^2 - 2ac}{180a^4}, \quad \text{etc.} \tag{7.1.6}$$

and, therefore, the solution takes the form

$$y(x) \sim ax^2 + bx + c + \frac{1}{18a^2 x} - \frac{b}{36a^3 x^2} + \frac{3b^2 - 2ac}{180a^4 x^3} + \dots, \quad x \to \infty. \tag{7.1.7}$$

The behaviour of the series (7.1.7) for large x may be investigated analytically or numerically. This series is clearly singular for $a = 0$ (b and c are other arbitrary constants), suggesting that there is another asymptotic behaviour which covers this case. If we attempt

$$y(x) \sim Ax^\alpha, \quad x \to \infty \tag{7.1.8}$$

for (7.1.1), we find that

$$A^3 \alpha(\alpha - 1)(\alpha - 2)x^{3\alpha - 3} \sim -\frac{1}{3}, \quad x \to \infty \tag{7.1.9}$$

giving $\alpha = 1$, which, however, leads to a contradiction. In such circumstances it is usual to attempt

$$y(x) \sim Ax(\ln x)^\alpha \tag{7.1.10}$$

which may correct the choice Ax. Putting (7.1.10) into (7.1.1), we find that

$$A^2 x^2 (\ln x)^{2\alpha} \left[-A\alpha \frac{(\ln x)^{\alpha - 1}}{x^2} + \frac{A\alpha(\alpha - 1)(\alpha - 2)(\ln x)^{\alpha - 3}}{x^2} \right] \sim -\frac{1}{3}. \tag{7.1.11}$$

Neglecting $(\ln x)^{\alpha-3}$ in comparison with $(\ln x)^{\alpha-1}$ as $x \to \infty$, we find that $-A^3\alpha(\ln x)^{3\alpha-1} \sim -\dfrac{1}{3}$ as $x \to \infty$. We thus have the choice $\alpha = \dfrac{1}{3}$ and $A = 1$. To lowest order, we have

$$y(x) \sim x(\ln x)^{1/3}, \quad x \to \infty. \tag{7.1.12}$$

The next step is to attempt to improve upon (7.1.12). One "plausible" choice is to write a descending power series in $\ln x$:

$$y(x) \sim x(\ln x)^{1/3} \left[1 + A/(\ln x) + B/(\ln x)^2 + C/(\ln x)^3 + \ldots \right]. \tag{7.1.13}$$

If we substitute (7.1.13) into (7.1.1) and equate various powers of $(\ln x)$ on both sides, it turns out that A is arbitrary and $B = -\dfrac{10}{27} - A^2$, $C = \dfrac{50}{27} A + \dfrac{5}{3}A^3$, etc. Thus, (7.1.13) becomes

$$y(x) \sim x(\ln x)^{1/3} \left[1 + \frac{A}{\ln x} - \frac{A^2 + \frac{10}{27}}{(\ln x)^2} + \frac{\frac{5}{3}A^3 + \frac{50}{27}A}{(\ln x)^3} + \ldots \right], x \to \infty. \tag{7.1.14}$$

Actually, Equation (7.1.1) can be solved in a closed form following a sequence of (perfectly logical) transformations, but the final solution is so implicit that it has little practical use. On the other hand, the behaviours (7.1.7) and (7.1.14) for $x \to \infty$ when combined with appropriate numerical solution can provide useful information about the structure of the solution for large x (see Bender and Orszag (1978)).

More interesting work on asymptotics of nonlinear ODE by balancing argument is due to Levinson (1969), who also gave an estimate of error in the approximate solution. We consider an example due to Levinson (1969), namely

$$xx'' + x' + tx = t^2 \tag{7.1.15}$$

where $x = x(t)$ and $x' = (dx/dt)$, etc. If we attempt the solution $x = Ct^a$, we find from (7.1.15) that

$$C^2 a(a - 1)t^{2a-2} + aCt^{a-1} + Ct^{a+1} = t^2. \tag{7.1.16}$$

All the terms in powers of t clearly do not balance. Instead, we have partial balances:

   (i) The first and third terms on the left-hand side balance so that $2a - 2 = a + 1$ or $a = 3$; the remaining terms are $O(t^2)$ and are therefore less important as $t \to \infty$. Moreover, if we choose $C = -1/6$, the dominant terms with exponent 4 exactly cancel and, therefore, $-t^3/6$ is a possible approximate solution.

(ii) If the last term on the left of (7.1.16) balances with t^2 on the right, we have $a + 1 = 2$ or $a = 1$; other terms are clearly of lesser order as $t \to \infty$. In this case, $x = t$ is a plausible approximate solution.

Since t^2 appears explicitly in (7.1.15), a power of t is an obvious guess function; other functions such as exponential may not be appropriate. A simple substitution of this kind would easily convince the reader.

We attempt to improve upon the approximate solution (i) of (7.1.15) by writing

$$x = -\frac{t^3}{6} + u. \tag{7.1.17}$$

On substituting (7.1.17) into (7.1.15) we get

$$u'' - \frac{6}{t^3}\left(1 - \frac{6u}{t^3}\right)^{-1} u' + \frac{54u}{t^4}\left(1 - \frac{6u}{t^3}\right)^{-1} u' = -\frac{9}{t}. \tag{7.1.18}$$

Expanding the inverse functions for large t, keeping only the linear terms on the left and pushing all other terms to right, we have the linear operator on the left as

$$\tilde{L}u = u'' - \frac{6u'}{t^3} + \frac{54u}{t^4}. \tag{7.1.19}$$

If u behaves like a power of t less than 3 (as it should), then $u/t^4, u'/t^3$ are smaller than u'' by a factor of t^{-2} for large t. We may therefore drop the former two and keep only the simple operator $Lu = u''$ on the left. The two linearly independent solutions of

$$Lu = u'' = 0$$

are $\psi_1 = 1, \psi_2 = t$. Now, using the variation of parameters, we can write the general solution of (7.1.18) in the form of an integral equation

$$u(t) = C_1 t + C_2 - 9t \log t + 6 \int_t^\infty (s - t)\left(1 - \frac{6u(s)}{s^3}\right)^{-1} (su'(s) - 9u(s))\frac{ds}{s^4}. \tag{7.1.20}$$

Here, the factor $(s - t)$ arises from

$$\frac{\psi_1(t)\psi_2(s) - \psi_1(s)\psi_2(t)}{\psi_1'(s)\psi_2(s) - \psi_1(s)\psi_2'(s)} = t - s \tag{7.1.21}$$

in the solution obtained by the method of variation of parameters and the term $-9t \log t$ comes from integration of $(t-s)(-9/s)$ with respect to s. The constants C_1 and C_2 in (7.1.20) are arbitrary. To prove that the integral equation (7.1.20) possesses a solution, we first differentiate it with respect to t and multiply $u'(t)$ so obtained by t to have another term of the same order as $u(t)$:

$$tu'(t) = C_1 t - 9t \log t - 9t - 6t \int_t^\infty \left(1 - \frac{6u(s)}{s^3}\right)^{-1} [su'(s) - 9u(s)]\frac{ds}{s^4}. \tag{7.1.22}$$

The system (7.1.20) and (7.1.22) is in a form to which the following theorem due to Levinson (1969) can be applied.

Theorem. Let $h(t), K_1(t, s)$, and $K_2(t, s)$ be continuous vector functions, and let $g(y, t)$ be a continuous scalar function. Consider the integral equation

$$y(t) = h(t) + \int_a^t K_1(t, s)g(y(s), s)ds + \int_t^\infty K_2(t, s)g(y(s), s)ds. \quad (7.1.23)$$

Let $g(0, t) = 0$. Let there be a continuous scalar function $H(t)$ such that

$$|h(t)| \le H(t). \quad (7.1.24)$$

Moreover, for y and $\tilde{y}$ satisfying

$$|y| \le 2H(t), \quad |\tilde{y}| \le 2H(t), \quad (7.1.25)$$

let there be a scalar function $\gamma(t)$ for which

$$|g(y, t) - g(\tilde{y}, t)| \le \gamma(t)|y - \tilde{y}| \quad (7.1.26)$$

and

$$\int_a^t |K_1(t, s)|\gamma(s)H(s)ds \le \frac{1}{4}H(t) \quad (7.1.27)$$

$$\int_t^\infty |K_2(t, s)|\gamma(s)H(s)ds \le \frac{1}{4}H(t). \quad (7.1.28)$$

Then, (7.1.23) has a solution which can be obtained by successive approximations starting with $y = 0$ as the zeroth iterate on the right-hand side. The processes will converge uniformly in a norm involving $H(t)$. The solution $y(t)$ will satisfy $|y(t)| \le 2H(t)$ and will be unique. If $h(t)$ depends continuously on the parameters, then y will also do so.

We shall apply this theorem to the system (7.1.20) and (7.1.22). Here,

$$h(t) = \begin{pmatrix} C_1 t + C_2 - 9t \log t \\ (C_1 - 9)t - 9t \log t \end{pmatrix} \quad (7.1.29)$$

We can therefore choose

$$H(t) = At \log t \quad (7.1.30)$$

where A is some large number such that $(2|C_1| + |C_2| + 27) \le A$. Also, we have

$$g(u(s), su'(s), s) = \left(1 - \frac{6u(s)}{s^3}\right)^{-1}(su'(s) - 9u(s))\frac{1}{s^4}. \quad (7.1.31)$$

So, for $|y(t)| \leq 2At \log t$ and for t sufficiently large (depending on A), one chooses

$$\gamma(t) = \frac{20}{t^4}. \tag{7.1.32}$$

To check (7.1.28), we have the left-hand side as

$$12 \int_t^\infty s \frac{20}{s^4} As \log s \, ds \leq 240A \left(\frac{1 + \log t}{t} \right) \tag{7.1.33}$$

$$\leq H(t) \frac{240(1 + \log t)}{t^2 \log t}$$

$$\leq \frac{1}{4} H(t) \tag{7.1.34}$$

for large enough t. Hence, the theorem applies and (7.1.20) and (7.1.22) have a solution $u(t), tu'(t)$ depending continuously on C_1 and C_2. Moreover,

$$|u(t)| + |tu'(t)| \leq 2H(t) = 2A \log t. \tag{7.1.35}$$

Using (7.1.35) in the integrals in (7.1.20) and (7.1.22), we get

$$u(t) = C_1 t + C_2 - 9t \log t + O \left(\frac{A \log t}{t} \right) \tag{7.1.36}$$

$$u'(t) = (C_1 - 9) - 9 \log t + O \left(\frac{A \log t}{t^2} \right) \tag{7.1.37}$$

where $A \geq (2|C_1| + |C_2| + 27)$. That (7.1.36)-(7.1.37) satisfy (7.1.18) can be checked by direct substitution. We have thus proved the existence of a two-parameter family of solutions $x(t) = -\frac{t^3}{6} + u(t)$ for Equation (7.1.1).

For this nonlinear equation there exists another family of asymptotic solutions corresponding to the second guess $x(t) = t + u$. We do not give the details here and refer the reader to Levinson (1969). The solution is found to be

$$x(t) = t + (C_1 \sin t + C_2 \cos t) t^{-1/2} + O(1/t) \tag{7.1.38}$$

where C_1 and C_2 are arbitrary constants.

It is possible to further refine these approximate solutions by iteration.

In the context of nonlinear PDEs, the balancing argument can be easily extended. One first introduces some cononical independent variables, the (relevant) similarity variable, and time, say, and then uses the balancing argument in the time variable. The lowest-order equation is still a PDE, but simpler than the original one and can often be solved. The solution thus obtained can be improved upon but the error analysis, similar to that in the work of Levinson (1969) for ODEs, seems difficult. What is done instead is to use the numerical solution of the original PDE subject to appropriate

IC/BC to determine the missing information in the asymptotic solution; the latter is oftentimes an improvised form of the solution of the linearised equation, which takes into account the effects of nonlinearity. This approach has been used to great advantage by Grundy and his collaborators (Grundy, Sachdev, and Dawson (1994); Grundy, Van Duijn, and Dawson (1994); Escobedo and Grundy (1996); Dawson, Van Duijn, and Grundy (1996); Van Duijn, Grundy, and Dawson (1997)).

7.2 Asymptotic Solution of Nonplanar Burgers Equation with N Wave Initial Conditions

We seek large-time solution to the nonplanar Burgers equation

$$u_t + uu_x + \frac{ju}{2t} = \frac{\delta}{2}u_{xx} \tag{7.2.1}$$

where $j > -1$ and δ is small, subject to discontinuous N wave initial conditions

$$u(x, t_i) = (1 - j/2)x/t_i, \quad |x| < d_0$$
$$= 0, \quad \text{otherwise} \tag{7.2.2}$$

where d_0 is the initial length of one lobe of the antisymmetric N wave.

We introduce the new variables

$$t, \eta = xt^a, u = t^c v(\eta, t) \tag{7.2.3}$$

into (7.2.1), where a and c are parameters to be chosen subsequently; we get

$$(c + j/2)v + a\eta v_\eta + tv_t = \frac{\delta}{2}t^{(2a+1)}v_{\eta\eta} - t^{(a+c+1)}vv_\eta. \tag{7.2.4}$$

We compare the relative importance, as $t \to \infty$, of the two terms explicitly involving t on the RHS of (7.2.4). If the second term on the right balances with all the terms in the left, we have

$$a + c + 1 = 0, \quad 2a + 1 < 0. \tag{7.2.5}$$

This approximation ignores the viscous dissipation effect represented by $v_{\eta\eta}$ term on the right of (7.2.4). The exact inviscid solution of (7.2.4) is $u = (1 - j/2)x/t$. Motivated by this solution, we choose

$$a = -1, \quad c = 0. \tag{7.2.6}$$

With a and c thus chosen, (7.2.4) becomes

$$\frac{j}{2}v - \eta v_\eta + tv_t = \frac{\delta}{2}t^{-1}v_{\eta\eta} - vv_\eta. \tag{7.2.7}$$

We now introduce the large-time asymptotic expansion

$$v(\eta, t) = v_0(\eta) + o(1) \tag{7.2.8}$$

as $t \to \infty$. Putting (7.2.8) into (7.2.7) and equating the leading order terms, we get

$$\frac{j}{2}v_0 - \eta v_0' \approx v_0 v_0' \tag{7.2.9}$$

with the general solution

$$v(\eta) \approx (1 - j/2)\eta + c_1. \tag{7.2.10}$$

If we require the solution to be antisymmetric with respect to $x = 0$, we must have $c_1 = 0$. Thus, (7.2.10) reduces to

$$v(\eta) = (1 - j/2)\eta + o(1) \tag{7.2.11}$$

as $t \to \infty$, or equivalently,

$$u(x, t) = (1 - j/2)\frac{x}{t} + o(1), \quad j \neq 2 \tag{7.2.12}$$

as $t \to \infty$. This is the zeroth order outer solution in the matched asymptotic expansion approach of Crighton and Scott (1979), who get the N-wave solution to first order by introducing appropriate shock layers at the front and the tail and matching them to the outer solution. Since this case has been dealt with in great detail by Crighton and Scott (1979), we refer the reader to their work. The other balance in (7.2.4) is between the first term on RHS with the LHS; here, viscous dissipation dominates nonlinear convection. Assuming that all the η derivatives are bounded, $t(\partial v/\partial t) = O(1)$ and $\eta = O(1)$ as $t \to \infty$, we have in this case

$$2a + 1 = 0, \quad a + c + 1 < 0 \tag{7.2.13}$$

or

$$a = -\frac{1}{2} \quad \text{and} \quad c < -\frac{1}{2}. \tag{7.2.14}$$

The (linear) old-age solution of (7.2.1) is

$$u(x, t) = C_1 \frac{x}{t^{(3+j)/2}} \exp(-x^2/2\delta t), \tag{7.2.15}$$

C_1 being the old-age constant; (7.2.15) motivates the choice

$$c = -(1 + j)/2 \tag{7.2.16}$$

(see (7.2.3)). Equation (7.2.4) now becomes

$$-v - \frac{\eta}{2}v_\eta + t v_t = \frac{\delta}{2}v_{\eta\eta} - t^{-(j+1)/2} v v_\eta. \tag{7.2.17}$$

We seek solution of (7.2.17) in the form

$$v(\eta, t) = v_0(\eta) + o(1). \tag{7.2.18}$$

Putting (7.2.18) into (7.2.17), we get, to the lowest order, the linear equation

$$\frac{\delta}{2}v_0'' + \frac{\eta}{2}v_0' + v_0 = 0, \tag{7.2.19}$$

which has the general solution

$$v_0(\eta) = A\eta \exp(-\eta^2/2\delta) + B\eta \exp(-\eta^2/2\delta) \int_0^\eta \eta^{-2} \exp(\eta^2/2\delta) d\eta. \tag{7.2.20}$$

Here, A and B are arbitrary constants. If we choose $B = 0$, the leading order solution of (7.2.17) is simply the "old-age" solution

$$v_0(\eta) = A\eta \exp(-\eta^2/2\delta). \tag{7.2.21}$$

To include the effect of the (neglected) convective term in (7.2.17), we write

$$v(\eta, t) = v_0(\eta, t) + \epsilon(\eta, t) \tag{7.2.22}$$

as $t \to \infty$. Substituting (7.2.22) into (7.2.17) and retaining the dominant terms in ϵ, we have

$$-\epsilon - \frac{\eta}{2}\epsilon_\eta + t\epsilon_t = \frac{\delta}{2}\epsilon_{\eta\eta} - t^{-(j+1)/2}\left\{A^2\eta(1 - \eta^2/\delta)\exp(-\eta^2/\delta)\right\}. \tag{7.2.23}$$

If we write

$$\epsilon = t^{-(j+1)/2}f(\eta) \tag{7.2.24}$$

in (7.2.23), we get

$$f'' + \frac{\eta}{\delta}f' + \frac{(3+j)}{\delta}f = \frac{2A^2}{\delta}\eta(1 - \eta^2/\delta)\exp(-\eta^2/\delta) \equiv R, \quad \text{say.} \tag{7.2.25}$$

The homogeneous part of (7.2.25) has the linearly independent solutions

$$f_1(\eta) = \exp(-\eta^2/2\delta) \, {}_1F_1\left(-\frac{j+2}{2}, \frac{1}{2}; \frac{\eta^2}{2\delta}\right) \tag{7.2.26}$$

$$f_2(\eta) = \frac{\eta}{(2\delta)^{1/2}} \exp(-\eta^2/2\delta) \, {}_1F_1\left(\frac{1-j}{2}, \frac{3}{2}; \frac{\eta^2}{2\delta}\right) \tag{7.2.27}$$

where

$${}_1F_1(a, c; z) = \sum_{k=0}^\infty A_k z^k$$

$$A_k = a(a+1)\ldots(a+k-1)/c(c+1)\ldots(c+k-1)k! \tag{7.2.28}$$

is the relevant solution of the confluent hypergeometric equation

$$zy'' + (c - z)y' - ay = 0. \tag{7.2.29}$$

A particular solution of (7.2.25) is obtained by the method of variation of parameters:

$$f_p(\eta) = f_1(\eta) \int_0^\eta -\frac{f_2(s)R(s)}{W(f_1, f_2)} ds + f_2(\eta) \int_0^\eta \frac{f_1(s)R(s)}{W(f_1, f_2)} ds \tag{7.2.30}$$

where R, f_1 and f_2 are as in (7.2.25), (7.2.26), and (7.2.27), respectively, and W is the Wronskian. The general solution of (7.2.25) is

$$f(\eta) = c_1 f_1(\eta) + c_2 f_2(\eta) + f_p(\eta). \tag{7.2.31}$$

Since the N-wave solution we seek is antisymmetric in x (and hence η) with respect to the origin, we must choose $c_1 = c_2 = 0$ in (7.2.31), and so

$$f(\eta) = f_p(\eta). \tag{7.2.32}$$

The solution (7.2.22) in view of (7.2.24) and (7.2.30) becomes

$$v(\eta) = A\eta \exp(-\eta^2/2\delta) + t^{-(j+1)/2} f(\eta) \tag{7.2.33}$$

as $t \to \infty$. In terms of original variables we have

$$u(x,t) = t^{-(1+j/2)} \left[Axt^{-1/2} \exp(-x^2/2\delta t) \right.$$
$$\left. + t^{-(j+1)/2} f(x/t^{1/2}) + O(t^{-(j+1)}) \right] \tag{7.2.34}$$

as $t \to \infty$.

As a quick check we derive the above results for Burgers equation ($j = 0$ in (7.2.1)) itself and verify their correctness with reference to the known exact solution. In this case (7.2.25) reduces to

$$f'' + \frac{\eta}{\delta} f' + \frac{3}{\delta} f = \frac{2A^?}{\delta} \eta \left(1 - \frac{\eta^?}{\delta} \right) \exp(-\eta^2/\delta). \tag{7.2.35}$$

A particular solution of (7.2.35) is

$$f_p(\eta) = -A^2 \eta \exp(-\eta^2/\delta). \tag{7.2.36}$$

The general solution of (7.2.35) is

$$f(\eta) = c_1 f_1(\eta) + c_2 f_2(\eta) - A^2 \eta \exp(-\eta^2/\delta) \tag{7.2.37}$$

where f_1 and f_2 are given by (7.2.26) and (7.2.27) with $j = 0$. Arguing as for the general j case, we may write the asymptotic solution in the present case as

$$u(x,t) = t^{-1} \left[A\frac{x}{t^{1/2}} \exp\left(-\frac{x^2}{2\delta t} \right) + t^{-1/2} f(x/t^{1/2}) + O(t^{-1}) \right] \tag{7.2.38}$$

as $t \to \infty$; here, f refers to (7.2.37) with $c_1 = c_2 = 0$. We now refer to the exact N-wave solution of plane Burgers equation (see Section 6.2)

$$u(x,t) = \frac{x/t}{1 + \left(\frac{t}{t_0}\right)^{1/2} \exp(x^2/2\delta t)} \tag{7.2.39}$$

$$= A\frac{x}{t^{3/2}} \exp\left(-\frac{x^2}{2\delta t}\right) \left\{1 - \frac{A}{t^{1/2}} \exp\left(-\frac{x^2}{2\delta t}\right) + O(t^{-1})\right\} \tag{7.2.40}$$

as $t \to \infty$; here $A = t_0^{1/2}$. The solution (7.2.40) is the same as (7.2.38), obtained by balancing argument with $f(\eta)$ given by (7.2.37).

We summarise here large-time behaviour of the nonnegative solutions of the generalised Burgers equation

$$u_t + (u^{\alpha+1})_x + \frac{Ju}{2t} = \delta u_{xx}, J > 0, \alpha > 0 \tag{7.2.41}$$

due to Grundy, Sachdev, and Dawson (1994), using the method of balances. Equation (7.2.41) is called subcylindrical if $J < 1$, and supercylindrical if $J > 1$. For $\alpha = 1/(J + 1)$, (7.2.41) possesses exact self-similar solution, found first by Sachdev and Nair (1987). Large-time nonnegative solutions of this equation were sought subject to the bounded initial data

$$u(x, 1) = u_1(x), \tag{7.2.42}$$

which have either finite support or vanish sufficiently quickly as $|x| \to \infty$. First, an integral invariance property was found by defining

$$M(t) = \int_{-\infty}^{\infty} u(x,t)dx. \tag{7.2.43}$$

Integrating (7.2.41) with respect to x from $-\infty$ to $+\infty$ and using vanishing conditions $u, u_x \to 0$ as $|x| \to \infty$, it may be checked that

$$\frac{dM}{dt} = -\frac{JM}{2t}$$

or

$$M(t) = M_1 t^{-J/2} \tag{7.2.44}$$

where M_1 is the constant of integration.

Introducing the variables

$$\eta = x/t^\delta, \quad u(x,t) = t^{-a}v(\eta,t), \quad a > 0, \delta > 0 \tag{7.2.45}$$

into (7.2.41), the balancing argument was used to identify three distinct cases. The resulting approximate equations were solved and improved upon, as illustrated for the N wave problem earlier.

The integral invariance property (7.2.44) was imposed in each case to fix the parameters.

1. $\alpha > J + 1$.

   In this case, diffusion dominates convection. The final solution has the form

   $$u(x,t) = \frac{M_1 t^{-(J+1)/2}}{2\sqrt{\pi}} e^{-x^2/4t} \{1 + o(1)\} \qquad (7.2.46)$$

   as $t \to \infty$, the result being uniform in $-\infty < x < \infty$. The asymptotic nature of (7.2.46) was verified with the numerical solution of (7.2.41) subject to the initial condition

   $$u_0 = \{H(x+1) - H(x-1)\}/2 \qquad (7.2.47)$$

   where H is the Heaviside function; here, the parameters were chosen to be $\alpha = 2/3, J = 2$. The convergence of the solution to the limiting profile (7.2.46) was clearly demonstrated.

2. $\alpha = 1/(J+1)$.

   Here, diffusion and convection both balance and the exact solution of Sachdev and Nair (1987), namely

   $$u = t^{-1/2\alpha} \frac{e^{-\eta^2/4}}{\left\{ B + \sqrt{\alpha\pi}\,\mathrm{erfc}\left(\frac{\eta\sqrt{\alpha}}{2}\right) \right\}^{1/\alpha}} , \qquad (7.2.48)$$

   was recovered. B is an arbitrary constant. For $\alpha = 1/3$ and $J = 2$, convergence of the solution with the initial condition (7.2.47) to the limiting form (7.2.48) was numerically demonstrated.

3. $\alpha < 1/(J+1)$.

   Here, convection dominates large-time solution when $\eta = O(1)$. In this case the analysis is relatively more complicated; it requires interposing of a shock layer at the leading edge and a trailing layer near the origin by matching the outer and inner solutions.

   The numerical scheme used to supplement the analytic results is due to Dawson (1991). This scheme has been found effective in the treatment of this class of nonlinear diffusion problems.

7.3 Asymptotic Profiles with Finite Mass in 1-D Contaminant Transport through Porous Media

Here, we give another example to illustrate the method of balancing to obtain large-time behaviour of solutions (Grundy, Van Duijn, and Dawson

(1994)). This concerns the large-time behaviour of positive solutions of

$$\frac{\partial}{\partial t}(u + u^p) = u_{xx} - u_x, p > 0 \qquad (7.3.1)$$

with $-\infty < x < \infty$ and $t \geq 0$ for pulse-type initial data. This equation, in suitably scaled variables, describes the one-dimensional flow of a solute through a porous medium with the solution undergoing absorption by the solid matrix of the media. Grundy et al. (1994) describe the physics in great detail and introduce suitable scaling, etc. to bring the model to the form (7.3.1). The nature of the asymptotic behaviour depends crucially on the value of the parameter p. The analysis shows that distinct behaviours come about for three ranges of p : $(1) 0 < p < 1, (2) 1 < p < 2, (3) p > 2$. The transformation of independent variables in (7.3.1) to

$$t, \xi = x - t \qquad (7.3.2)$$

is clearly suggested so that (7.3.1) takes the form

$$\frac{\partial}{\partial t}(u + u^p) = \frac{\partial^2 u}{\partial \xi^2} + \frac{\partial}{\partial \xi}(u^p). \qquad (7.3.3)$$

As for the example in Section 7.2, we introduce the similarity variable in (7.3.3), namely

$$\eta = \frac{\xi}{t^\delta} = \frac{x - t}{t^\delta}, \quad \delta > 0, \qquad (7.3.4)$$

together with the change of the dependent variable according to

$$u(\xi, t) = t^\alpha v(\eta, t) \qquad (7.3.5)$$

with $\alpha < 0$ to simulate solutions with temporal decay. Equation (7.3.3) becomes

$$\{t v_t + \alpha v - \delta \eta v_\eta\} + t^{\alpha(p-1)}\{t(v^p)_t + \alpha p v^p - \delta \eta (v^p)_\eta\}$$
$$= t^{1-2\delta} v_{\eta\eta} + t^{\alpha(p-1)+(1-\delta)}(v^p)_\eta, \qquad (7.3.6)$$

which, for convenience, may be referred to as $A + B = C + D$. Integrating (7.3.1) with respect to x from $x = -\infty$ to $x = +\infty$ and using vanishing condition for u and u_x at $\mp\infty$, we arrive at the mass conservation equality

$$\int_{-\infty}^\infty (u + u^p) dx = \int_\infty^\infty (u_0 + u_0^p) dx = M, \quad \text{say}, \qquad (7.3.7)$$

where u_0 refers to the initial value $u(x, 0) = u_0$. We shall use (7.3.7) as well as the balancing argument to find the parameters α and δ. Putting (7.3.4) and (7.3.5) into (7.3.7) we have

$$M = \int_{-\infty}^\infty (t^{\alpha+\delta} v + t^{\alpha p + \delta} v^p) d\eta. \qquad (7.3.8)$$

Since $\alpha < 0$, (7.3.8) implies that for $p > 1$, we have

$$M \sim O(t^{\alpha+\delta}) \quad \text{as} \quad t \to \infty. \tag{7.3.9}$$

Hence, for M to be invariant, we require that

$$\delta = -\alpha. \tag{7.3.10}$$

We seek solutions in the limit $t \to \infty$, $\eta = O(1)$ and assume that $v, v_\eta, v_{\eta\eta}$, as well as tv_t are bounded in this limit. With this assumption we check from (7.3.6) that for $p > 1$ and $\alpha < 0$, the term A on LHS of (7.3.6) dominates B. Now two possibilities arise with respect to RHS of (7.3.6): 1. A balances C, and D is less important. 2. A balances D, which itself dominates C. The asymptotic balance is really tantamount to asymptotic equivalence as $t \to \infty$.

1. In this case it is clear that

$$\delta = \frac{1}{2} \tag{7.3.11}$$

and

$$\alpha(p-1) + 1 - \delta < 0. \tag{7.3.12}$$

From (7.3.10) and (7.3.11) we have $\alpha = -\frac{1}{2}$, and so (7.3.12) becomes

$$p > 2. \tag{7.3.13}$$

Thus, the possibility $p > 2$ naturally arises from the analysis. With α, δ, and p thus determined, (7.3.6) becomes

$$tv_t - \frac{1}{2}(v + \eta v_\eta) + t^{-(p-1)/2}\{t(v^p)_t - \frac{p}{2}v^p - \frac{\eta}{2}(v^p)_\eta\}$$
$$= v_{\eta\eta} + t^{-(p-2)/2}(v^p)_\eta. \tag{7.3.14}$$

Now, writing

$$v(\eta, t) = v_0(\eta) + o(1) \tag{7.3.15}$$

with the assumptions that $tv_t = o(1)$ in the limit $t \to \infty$ and $\eta = O(1)$, we find from (7.3.14) that

$$v_0'' + \frac{1}{2}(\eta v_0' + v_0) = 0. \tag{7.3.16}$$

The general solution of (7.3.16) is

$$v_0 = k_1 e^{-\eta^2/4} - k_2 e^{-\eta^2/4} \int_\eta^\infty e^{s^2/4} ds = k_1 e^{-\eta^2/4} + O(\eta^{-1}) \tag{7.3.17}$$

as $\eta \to \infty$; k_1 and k_2 are arbitrary constants. Since the integral in (7.3.8) with respect to η over $(-\infty, \infty)$ must converge, we must put $k_2 = 0$ in (7.3.17) and thus have

$$v_0 = k_1 e^{-\eta^2/4}. \tag{7.3.18}$$

Substituting (7.3.15) and (7.3.18) into (7.3.8), we have

$$M = k_1 \int_{-\infty}^{\infty} e^{-\eta^2/4} d\eta \tag{7.3.19}$$

or

$$k_1 = \frac{M}{2\sqrt{\pi}}. \tag{7.3.20}$$

Thus, k_1 is found in terms of the initial mass M in (7.3.7). The solution for $p > 2$ is thus explicitly found to be

$$u(x, t) = \frac{M}{2\sqrt{\pi}} t^{-1/2} e^{-(x-t)^2/4t} \{1 + o(1)\} \tag{7.3.21}$$

as $t \to \infty$, $(x - t)/\sqrt{2t} = O(1)$. This result holds uniformly in x. To check the validity of (7.3.21), Grundy et al. (1994) solved (7.3.1) for $p = 3$ with IC

$$u_0(x) = H(x + 1) - H(x - 1) \tag{7.3.22}$$

where $H(x)$ is the Heaviside function. The area under the initial profile being $M = 4$, the asymptotic solution (7.3.21) becomes

$$t^{1/2} u(x, t) = \frac{2}{\sqrt{\pi}} e^{-\eta^2/4}. \tag{7.3.23}$$

The convergence of $t^{1/2} u(x, t)$ is slow, but evident. Grundy et al. (1994), in the appendix to their paper, obtain the next approximation

$$v(\eta, t) = v_0(\eta) + \frac{M^2}{4} \frac{t^{-1/2}}{\pi\sqrt{3}} \log t v_0'(\eta) + O(t^{-1/2}) \tag{7.3.24}$$

or, equivalently,

$$v(\eta, t) = v_0(\eta_1) + O(t^{-1/2}) \tag{7.3.25}$$

where

$$\eta_1 = \eta + \frac{M^2 t^{-1/2} \log t}{4\pi\sqrt{3}}. \tag{7.3.26}$$

The convergence of $t^{1/2} u(x, t)$, as obtained from the numerical solution to (7.3.25), is found to be much faster than it was to (7.3.21).

2. We now turn to the second possibility in (7.3.6) when the second term D on the RHS dominates the first, namely C. Under the same assumptions as in case 1, we have

$$\alpha(p - 1) + 1 - \delta = 0. \tag{7.3.27}$$

and

$$1 - 2\delta < 0. \tag{7.3.28}$$

Together with the condition (7.3.10), which holds in the present case, too, (7.3.27) and (7.3.28) give

$$\alpha = -\frac{1}{p}, \quad \delta = \frac{1}{p} \tag{7.3.29}$$

and

$$p < 2. \tag{7.3.30}$$

With values α and δ from (7.3.29), (7.3.6) becomes

$$tv_t - \frac{1}{p}(v + \eta v_\eta) + t^{-(p-1)/p}\left\{t(v^p)_t - v^p - \frac{\eta}{p}(v^p)_\eta\right\}$$
$$= t^{-(2-p)/p}v_{\eta\eta} + (v^p)_\eta. \tag{7.3.31}$$

We seek solution of (7.3.31) in the form

$$v = v_0(\eta) + O(1) \tag{7.3.32}$$

as $t \to \infty$, with $\eta = O(1)$ and $tv_t = o(1)$, which, as will become clear, may be referred to as the "outer limit." Putting (7.3.32) into (7.3.31) we have

$$(v_0 + \eta v_0') + p(v_0^p)' = 0 \tag{7.3.33}$$

to leading order. The general solution of the first-order ODE (7.3.33) is clearly

$$v_0^p + \frac{\eta}{p}v_0 = C \tag{7.3.34}$$

where C is the constant of integration. To fix C, we observe that, if $C < 0, v_0(\eta)$ is double-valued for

$$\eta \leq -p^2(-C)^{1/p}/(p-1)^{(p-1)/p}. \tag{7.3.35}$$

This solution is rejected. For $C > 0, v_0(\eta)$ is single-valued on $-\infty < \eta < \infty$, but as $\eta \to +\infty$ (7.3.34) shows that

$$v_0 \sim pC/\eta. \tag{7.3.36}$$

However, the mass invariance condition requiring $v_0(\eta)$ to be integrable over $-\infty < \eta < \infty$ would rule out (7.3.36) as well. Thus, we must have $C = 0$, and the solution (7.3.34) simply becomes

$$v_0 = \left(-\frac{\eta}{p}\right)_+^{1/(p-1)}. \tag{7.3.37}$$

v_0 in (7.3.37) is defined for all $\eta \in R$ but has infinite mass M (see (7.3.7)). To ensure a finite mass, we must choose (7.3.37) only over a finite interval $\eta_1 \le \eta \le 0$, say, and zero outside this interval. In this case, to leading order, we get

$$M = \int_{\eta_1}^{0} \left(-\frac{\eta}{p}\right)^{1/(p-1)} d\eta \qquad (7.3.38)$$

or

$$\eta_1 = -p \left(\frac{M}{p-1}\right)^{(p-1)/p} < 0. \qquad (7.3.39)$$

The solution (7.3.37) has a discontinuity at $\eta = \eta_1$. Thus, for this case with $1 < p < 2$, our solution, to lowest order, becomes

$$u(x,t) = t^{-1/p} \left(\frac{t-x}{pt^{1/p}}\right)^{1/(p-1)} \{1 + o(1)\} \qquad (7.3.40)$$

as $t \to \infty$, with $(x - t)/t^{1/p} = O(1)$. The condition $\eta_1 < \eta < 0$ implies that (7.3.40) holds for

$$\eta_1 t^{1/p} \le x - t < 0. \qquad (7.3.41)$$

The above solution, discontinuous at $\eta = \eta_1$, is an outer solution and must be supplemented by an outer layer. Before we turn to this matter we consider the case $p = 2$. In this case A, C, and D in (7.3.6) are asymptotically equivalent as $t \to \infty$, $\eta = O(1)$; therefore,

$$\alpha = -\delta = -\frac{1}{2}. \qquad (7.3.42)$$

We write

$$v = v_0(\eta) + o(1), \qquad (7.3.43)$$

and with the assumption that $tv_t = o(1)$, (7.3.6) gives

$$\alpha v_0 - \delta \eta v_0' = v_0'' + (v_0^2)', \qquad (7.3.44)$$

which has the exact solution

$$v_0(\eta) = \frac{e^{-\eta^2/4}}{k + \sqrt{\pi}\mathrm{erf}\,(\eta/2)} \qquad (7.3.45)$$

where k is the constant of integration and may be obtained from the mass-invariance condition

$$M = \int_{-\infty}^{\infty} v_0(\eta)d\eta. \qquad (7.3.46)$$

Thus, (7.3.45)-(7.3.46) give

$$k = \sqrt{\pi}(e^M + 1)/(e^M - 1). \qquad (7.3.47)$$

This solution (7.3.5) of (7.3.3) for $p = 2$ with $v = v_0(\eta)$ as in (7.3.45) and (7.3.47) is an exact similarity solution of the standard Burgers equation. Using the IC (7.3.22), Grundy et al. (1994) showed numerically that it converges, though slowly, to the above asymptotic solution. In an appendix to their paper, they improvised upon this solution to obtain

$$v(\eta, t) = v_0(\eta) - 0.0639t^{-1/2}(\log t)v_0'(\eta) + O(t^{-1/2})$$

$$= v_0(\eta_1) + O(t^{1/2}) \tag{7.3.48}$$

$$\eta_1 = \eta - 0.0639t^{-1/2}\log t.$$

The asymptotic form (7.3.48) was approached much faster as the numerical solution referred to above was compared with it at different times.

For $0 < p < 1$, Grundy et al. (1994), motivated by the numerical solution as well as properties of the PDE in the hyperbolic limit (when the viscous term is absent) sought the solution of (7.3.1) in the form

$$u = t^\beta v(\eta, t), \quad \beta < 0 \tag{7.3.49}$$

$$\eta = \frac{x}{t^\nu}, \quad \nu > 0. \tag{7.3.50}$$

Equation (7.3.1) in these variables becomes

$$t^{\beta(1-p)}\{tv_t + \beta v - \nu\eta v_\eta\} + \{t(v^p)_t + \beta p v^p - \nu\eta(v^p)_\eta\}$$
$$= t^{\beta(1-p)+1-2\nu}v_{\eta\eta} - t^{\beta(1-p)+1-\nu}v_\eta. \tag{7.3.51}$$

Writing (7.3.51) again as $A + B = C + D$, it is easy to check that, since $\beta < 0$ and $p < 1$, B dominates A as $t \to \infty$. Considering the possibilities on the RHS, it may be verified that the only consistent choice arises when D dominates C and asymptotically balances with B. This leads to

$$\beta(1 - p) + 1 - \nu = 0 \tag{7.3.52}$$

and

$$\beta(1 - p) + 1 - 2\nu < 0. \tag{7.3.53}$$

As in the previous cases, we get another constraint relating β and ν from the condition of mass-invariance (7.3.7):

$$M = \int_{-\infty}^{\infty} \{t^{\beta+\nu}v + t^{\beta p+\nu}v^p\}d\eta \tag{7.3.54}$$

and, for $p < 1$ and $\beta < 0$,

$$M \sim O(t^{\beta p+\nu}) \quad \text{as} \quad t \to \infty. \tag{7.3.55}$$

So, the invariance of M requires that

$$\nu = -\beta p. \tag{7.3.56}$$

Equations (7.3.52) and (7.3.56) then give

$$\beta = -1 \quad \text{and} \quad \nu = p. \tag{7.3.57}$$

Equation (7.3.51) now assumes the form

$$t(v^p)_t - p\{v^p + \eta(v^p)_\eta\} + t^{-(1-p)}\{tv_t - v - p\eta v_\eta\} = t^{-p}v_{\eta\eta} - v_\eta. \tag{7.3.58}$$

Writing

$$v(\eta, t) = v_0(\eta) + o(1), \tag{7.3.59}$$

assuming that $tv_t = o(1)$ as $t \to \infty$ and $\eta = O(1)$, and substituting it into (7.3.58), we get, to leading order,

$$p\{v_0^p + \eta(v_0^p)'\} = v_0'. \tag{7.3.60}$$

Integration of (7.3.60) gives

$$p\eta v_0^p - v_0 = C \tag{7.3.61}$$

where C is a constant. An argument similar to that for (7.3.34) shows that $C = 0$, and so

$$v_0 = (p\eta)_+^{1/(1-p)}. \tag{7.3.62}$$

The nontrivial part of the asymptotic solution (7.3.62) must, as before, be restricted to the interval $0 < \eta < \eta_2$. Using the mass-invariance condition (7.3.56), we get

$$M = \int_0^{\eta_2} (p\eta)^{p/(1-p)} d\eta, \tag{7.3.63}$$

yielding

$$\eta_2 = \left\{\frac{M}{1-p}\right\}^{1-p} p^{-p}. \tag{7.3.64}$$

In terms of x and t, the solution for $0 < p < 1$ may now be written as

$$u(x, t) = t^{-1}\left(\frac{px}{t^p}\right)^{1/(1-p)} \{1 + o(1)\} \tag{7.3.65}$$

as $t \to \infty$, $x/t^p = O(1)$. This solution holds over $0 < \eta < \eta_2$, that is, over

$$0 < x < \eta_2 t^p. \tag{7.3.66}$$

We call (7.3.65) the "outer solution," which may be written out as

$$u(x, t) = \begin{cases} 0 & x < 0 \\ t^{-1}(px/t^p)^{1/(1-p)} & 0 < x < \eta_2 t^p \\ 0 & x > \eta_2 t^p. \end{cases} \tag{7.3.67}$$

This solution is deficient in two ways:- it is discontinuous at $x = \eta_2 t^p$, and there is no moving interface to the left.

Both for $0 < p < 1$ and $1 < p < 2$, Grundy et al. (1994) introduce boundary layer solutions at the trailing edges $\eta = \eta_{1,2}$ and the leading edge $\eta = 0$. In the composite solution that they construct, two arbitrary constants remain undetermined in each of the boundary layer solutions. These must be determined in some ad hoc way by fitting with the numerical solutions; the composite solutions may help in reducing the excessive computing time required to calculate the slowly converging large-time solutions. The details of the boundary layers and their matching with the outer solutions may be found in the original paper of Grundy et al. (1994).

We shall now deal briefly with the same physical problem in two dimensions, namely contaminant transport in porous media for initial data with bounded support, and show how the analysis becomes more complicated. Although it is not feasible to construct the solution, as it was for one-dimensional, considerable qualitative details can be read off from the asymptotic analysis (Dawson, Van Duijn, and Grundy (1996)). We again start with the nondimensional form of the IVP

$$(u + u^p)_t + u_x = u_{xx} + u_{yy} \quad \text{for } (x, y, t) \in Q \tag{7.3.68}$$

$$u(x, y, 0) = u_0(x, y) \quad \text{for } (x, y) \in R^2, \tag{7.3.69}$$

where $Q = \{(x, y, t) : -\infty < x, y < \infty, t > 0\}$, and investigate the large-time behaviour of the nonnegative solutions of (7.3.68)-(7.3.69) ($u \geq 0, u$ being the redefined concentration) which satisfy mass conservation: we suppose that for all $t \geq 0, u(x, y, t) \to 0$ sufficiently fast as $|x|, |y| \to \infty$ so that $u + u^p$ is integrable for all $t \geq 0$. This implies that

$$\int\int_{R^2} (u + u^p) dx dy = \int\int_{R^2} (u_0 + u_0^p) dx dy := M \tag{7.3.70}$$

for all $t \geq 0$.

Intuitively, one expects that for $p \geq 1$, the solution u will become small for large time and, therefore, one may replace $u + u^p$ with u in (7.3.68) as a first approximation. This would lead to a linear convection-diffusion equation and, consequently, to a limit profile which is independent of the exponent p. This, however, was not what was observed numerically by Dawson et al. (1996).

As for the one-dimensional case, one introduces the moving coordinate system t, y and $\xi = t - x$ in (7.3.68) to find that

$$(u + u^p)_t + (u^p)_\xi = u_{\xi\xi} + u_{yy}. \tag{7.3.71}$$

Since $(u + u^p)_t = (1 + pu^{p-1})u_t$, one may view the term $(1 + pu^{p-1})$ as a concentration-dependent capacity. For $p \geq 1$, this capacity is bounded for all $u \geq 0$, while for $p < 1$ it blows up as $u \searrow 0$. In the former case, (7.3.71) is uniformly parabolic, and in the latter it is degenerate parabolic. In the theory of parabolic PDEs this has important consequences. If $u_0(x, y) = 0$ for all $x^2 + y^2 \geq R^2$, then for the case $p \geq 1$, $u(x, y, t) = 0$ outside a disc $D_{R(t)}(0)$ having a radius which expands in time ($r < R(t) < \infty$ and $R(t) \to \infty$ as $t \to \infty$). Hence, if $p < 1$ and, depending on the initial distribution, a free boundary may occur which separates the region where $u > 0$ from the region where $u = 0$. We shall treat these two cases separately.

1. $p > 1$.

We introduce the variables

$$\eta = \frac{\xi}{t^\beta} = \frac{t - x}{t^\beta}, \quad \zeta = \frac{y}{t^\delta} \tag{7.3.72}$$

and

$$u(\xi, y, t) = t^\alpha v(\eta, \zeta, t) \tag{7.3.73}$$

where $\beta, \delta \geq 0$ and $\alpha < 0$ to simulate temporal decay. In terms of the new variables, (7.3.71) becomes

$$\left(t \frac{\partial}{\partial t} + \alpha - \beta \eta \frac{\partial}{\partial \eta} - \delta \zeta \frac{\partial}{\partial \zeta} \right) (v + t^{\alpha(p-1)} v^p)$$
$$+ t^{1+\alpha(p-1)-\beta} \frac{\partial v^p}{\partial \eta} = t^{1-2\beta} \frac{\partial^2 v}{\partial \eta^2} + t^{1-2\delta} \frac{\partial^2 v}{\partial \zeta^2}. \tag{7.3.74}$$

Intuitively, we expect

$$v(\eta, \zeta, t) = v_0(\eta, \zeta) + o(1) \quad \text{as } t \to \infty \tag{7.3.75}$$

giving

$$u(x, y, t) = t^\alpha v_0 \left(\frac{t - x}{t^\beta}, \frac{y}{t^\delta} \right). \tag{7.3.76}$$

The parameters α, β, and δ must be found from (7.3.74) by the method of asymptotic balance as well as from the mass invariance condition (7.3.70). The latter, for $p > 1$ and in the limit $t \to \infty$, gives

$$\alpha + \beta + \delta = 0. \tag{7.3.77}$$

The balancing argument for (7.3.74) gives three possibilities:

(a) For $p > 3/2$, the time derivative and both the diffusion terms are dominant in (7.3.74), giving

$$\beta = \delta = \frac{1}{2} \quad \text{and} \quad \alpha = -1. \tag{7.3.78}$$

Since $v_0 = v_0(\eta, \zeta)$, the solution is simply the fundamental solution of the heat equation

$$v_0(\eta, \zeta) = \frac{M}{4\pi} e^{-(\eta^2 + \zeta^2)/4},$$

and hence

$$u(x, y, t) \to \frac{M}{4\pi t} e^{-\{(t-x)^2 + y^2\}/4t} \quad \text{as } t \to \infty. \qquad (7.3.79)$$

The solution (7.3.79) is radially symmetric with respect to the moving coordinates $(x = t, y = 0)$. Dawson et al. (1996), starting with a pulse initial condition, $u_0 = 1$, on the square of side 2 centered at the origin and $u_0 = 0$ elsewhere, arrived at this behaviour numerically for large time ~ 100. The exponent p here was chosen to be 3.

(b) For $p = \dfrac{3}{2}$, there is a balance between the time derivative, convection, and both diffusion terms. The parameters in this case also come out to be $\alpha = -1$, $\beta = \delta = \dfrac{1}{2}$, and v_0 satisfies the equation

$$v_0 + \frac{\eta}{2}(v_0)_\eta + \frac{\zeta}{2}(v_0)_\zeta - (v_0^{3/2})_\eta + (v_0)_{\eta\eta} + (v_0)_{\zeta\zeta} = 0 \text{ for } (\eta, \zeta) \in \Re^2$$
$$(7.3.80)$$

where

$$v_0 \geq 0 \quad \text{and} \quad \int\int_{\Re^2} v_0 d\eta d\zeta = M. \qquad (7.3.81)$$

Thus, for $p = 3/2$, we have the solution (7.3.76) with v_0 satisfying (7.3.80) and (7.3.81). This problem, though simpler than the original problem, cannot be solved in a closed form. Numerical solutions with $p = 3/2$ with the same IC as for the case (a) show that these solutions are symmetric in ζ but not in η (see Figure 7.1).

(c) $1 < p < \dfrac{3}{2}$. In this case, the time derivative term balances with ζ-diffusion and η-convection. It follows from (7.3.74) that, in this case,

$$\alpha = -\frac{3}{2p}, \quad \beta = \frac{3-p}{2p}, \quad \delta = \frac{1}{2}. \qquad (7.3.82)$$

Writing again $v = v_0(\eta, \zeta) + o(1)$, it follows from (7.3.74) that

$$\frac{3}{2p}v_0 + \frac{3-p}{2p}\eta(v_0)_\eta + \frac{\zeta}{2}(v_0)_\zeta - (v_0^p)_\eta + (v_0)_{\zeta\zeta} = 0 \qquad (7.3.83)$$

for $(\eta, \zeta) \in \Re^2$, where v_0 also must satisfy (7.3.81). The form of the asymptotic solution for $1 < p < \dfrac{3}{2}$, therefore, is

$$u(x, y, t) \to t^{-3/2p} v_0 \left(\frac{t-x}{t^{(3-p)/2p}}, \frac{y}{t^{1/2}} \right) \quad \text{as } t \to \infty \qquad (7.3.84)$$

where v_0 is the solution of (7.3.83) and (7.3.81). The numerical solution shows convergence to this form of the solution with appropriate initial conditions as before.

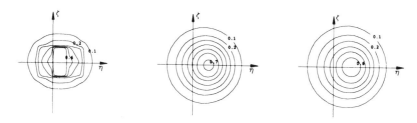

Figure 7.1. Numerical results of the IVP for the spreading of the contaminant (see Equation (7.3.71)) in the scaled variables for $p = 3/2$, $M = 8$. The results relate to $t = 1, 15, 100$, respectively.

As for the one-dimensional case, when $0 < p < 1$, one seeks solution in terms of the similarity variables

$$\eta = \frac{x}{t^b}, \zeta = \frac{y}{t^d} \tag{7.3.85}$$

where $b, d \geq 0$. The solution is assumed in the form

$$u(x, y, t) = t^a v(\eta, \zeta, t) \tag{7.3.86}$$

where $a < 0$ to consider temporal decay. Putting (7.3.85) and (7.3.86) into (7.3.68), we get

$$\left(t \frac{\partial}{\partial t} + ap - b\eta \frac{\partial}{\partial \zeta} - d\zeta \frac{\partial}{\partial \zeta} \right) (t^{a(1-p)} v + v^p)$$

$$+ t^{1+a(1-p)-b} \frac{\partial v}{\partial \eta} = t^{1+a(1-p)-2b} \frac{\partial^2 v}{\partial \eta^2} + t^{1+a(1-p)-2d} \frac{\partial^2 v}{\partial \zeta^2}. \tag{7.3.87}$$

The mass invariance condition (7.3.70) for $0 < p < 1$ gives

$$ap + b + d = 0. \tag{7.3.88}$$

Writing $v = v_0(\eta, \zeta) + O(1)$ and looking for different balances, one finds that the only sensible balance is between η-convection and ζ-diffusion; in this case we must have

$$a - \frac{-3}{3-p}, b = \frac{2p}{3-p}, \quad \text{and } d = \frac{p}{3-p}. \tag{7.3.89}$$

It is then governed by

$$\frac{3p}{3-p} v_0^p + \frac{2p}{3-p} \eta (v_0^p)_\eta + \frac{p}{3-p} \zeta (v_0^p)_\zeta - (v_0)_\eta + (v_0)_{\zeta\zeta} = 0 \quad \text{for } (\eta, \zeta) \in \Re^2. \tag{7.3.90}$$

It must also satisfy

$$v_0 \geq 0 \quad \text{and} \quad \int\int_{\Re^2} v_0^p d\eta d\zeta = M. \tag{7.3.91}$$

Equation (7.3.90) is too hard to solve explicitly. However, we may conclude that, for $0 < p < 1$,

$$u(x,y,t) \rightarrow t^{-3/(3-p)} v_0 \left(\frac{x}{t^{2p/(3-p)}}, \frac{y}{t^{p/(3-p)}} \right) \quad \text{as } t \rightarrow \infty. \tag{7.3.92}$$

This asymptotic form is confirmed by numerical solution of the IVP, posed for case (a) with $p > \frac{3}{2}$.

It is clear from the above discussion that the reduced equations for this two-dimensional problem, which hold for large t, do not generally admit explicit analysis. However, in conjunction with the numerical solution, they make the problem more tractable, reveal some qualitative features, and help one understand the structure of the solution in the asymptotic limit. We consider the case (c), $1 < p < 3/2$, in some detail to illustrate this point. In this case we have

$$\frac{3}{2p}v_0 + \frac{3-p}{2p}\eta\frac{\partial v_0}{\partial \eta} + \frac{\zeta}{2}\frac{\partial v_0}{\partial \zeta} - \frac{\partial v_0^p}{\partial \eta} + \frac{\partial^2 v_0}{\partial \zeta^2} = 0 \tag{7.3.93}$$

for $(\eta, \zeta) \in \Re^2$, and

$$v_0 \geq 0, \quad \int\int_{\Re^2} v_0 d\eta d\zeta = M. \tag{7.3.94}$$

We first observe that v_0 involves first derivative in η and second derivative in ζ, showing that the solutions are smoother in the transverse direction ζ than in the η direction, the direction of flow. The following points may now be noted.

(i) $$v_0(\eta, \zeta) = v_0(\eta, -\zeta) \quad \text{for all } (\eta, \zeta) \in \Re^2 \tag{7.3.95}$$

This follows immediately from the symmetry properties of (7.3.93) together with uniqueness of the solution (see Escobedo, Vazquez, and Zuazua (1993)).

(ii) There exists a constant $L > 0$ such that

$$v_0(\eta, \zeta) = 0 \quad \text{outside the strip } S = \{(\eta, \zeta) : 0 < \eta < L, -\infty < \zeta < \infty\}. \tag{7.3.96}$$

To show this we assume that v_0 decays sufficiently fast to zero as $|\eta|, |\zeta| \rightarrow \infty$. We write (7.3.90) in the divergence form

$$\frac{\partial}{\partial \eta}\left\{\frac{3-p}{2p}\eta v_0 - v_0^p\right\} + \frac{\partial}{\partial \zeta}\left\{\frac{1}{2}\zeta v_0 + \frac{\partial v_0}{\partial \zeta}\right\} = 0 \tag{7.3.97}$$

and introduce the transversal mass

$$M_t(\eta) = \int_{-\infty}^{\infty} v_0(\eta, \zeta)d\zeta \quad \text{for} \ -\infty < \eta < \infty. \tag{7.3.98}$$

Integration of (7.3.97) with respect to ζ gives

$$\frac{3-p}{2p} \frac{d}{d\eta}(\eta M_t) - \frac{d}{d\eta} \int_{-\infty}^{\infty} v_0^p(\eta, \zeta)d\zeta = 0. \tag{7.3.99}$$

Integration of (7.3.99), then, with respect to η yields

$$\frac{3-p}{2p}\eta M_t - \int_{-\infty}^{\infty} v_0^p(\eta, \zeta)d\zeta = C \quad \text{for} \ -\infty < \eta < \infty, \tag{7.3.100}$$

where C is the constant of integration. Letting $|\eta| \to \infty$ in (7.3.100) and using the rapid decay of v_0 at infinity yield $C = 0$, we have

$$\frac{3-p}{2p}\eta M_t = \int_{-\infty}^{\infty} v_0^p(\eta, \zeta)d\zeta \quad \text{for} \ -\infty < \eta < \infty. \tag{7.3.101}$$

Since $v_0 \geq 0$ and, hence, $M_t(\eta) \geq 0$, it follows from (7.3.101) that $v_0(\eta, \zeta) = 0$ for all $\eta \leq 0$. We also get from (7.3.101) that

$$\frac{3-p}{2p}\eta M_t \leq \sup_{\zeta \in \Re} v_0^{p-1}(\eta, \zeta)M_t$$

or

$$\left(\frac{3-p}{2p}\eta - \sup_{\zeta \in \Re} v_0^{p-1}(\eta, \zeta) \right) M_t \leq 0. \tag{7.3.102}$$

It follows from (7.3.102) that there exists a constant $L > 0$, depending on the maximum value of v_0 such that $M_t(\eta) = 0$ for all $\eta > L$. This proves the assertion (ii).

(iii) The inequality (7.3.102) implies that

$$\sup_{\zeta \in \Re} v_0^{p-1}(\eta, \zeta) \geq \frac{3-p}{2p}\eta \quad \text{for} \ 0 \leq \eta \leq L. \tag{7.3.103}$$

(iv) Having shown that the maximum value of v_0^{p-1} with respect to ζ is strictly positive for each $0 < \eta < L$, it follows from Escobedo, Vazquez, and Zuazua (1993) that

$$v_0 > 0 \text{ in } S. \tag{7.3.104}$$

(v)

$$v_0(\eta, \zeta) \leq \left(\frac{\eta}{p}\right)^{1/(p-1)} \qquad \text{for } (\eta, \zeta) \in S \tag{7.3.105}$$

(see (7.3.96) for definition of S). The upper bound in (7.3.105) is the one-dimensional solution (7.3.37); the result (7.3.105) follows from a straightforward application of comparison theorems.

(vi) The estimate (7.3.105) implies that $\partial/(\partial\eta)v_0^{(p-1)}(0, \zeta) \leq 1/p$ for all $-\infty < \zeta < \infty$. In fact,

$$\frac{\partial v_0^{p-1}}{\partial \eta} \leq \frac{1}{p} \text{ in } S. \tag{7.3.106}$$

To prove this, one first writes the equation for $w = v_0^{p-1}$ and then for the derivative $\dfrac{\partial w}{\partial \eta} = z$. The latter equation has the constant solution $z = 1/p$. A comparison argument then yields (7.3.106).

(vii) To find the behaviour of v_0 as $|\zeta| \to \infty$, let

$$M_L(\zeta) = \int_0^L v_0(\eta, \zeta)d\eta \tag{7.3.107}$$

denote the longitudinal mass. To find $M_L(\zeta)$, we integrate (7.3.97) with respect to η and put constant of integration equal to zero; we get

$$\frac{dM_L}{d\zeta} + \frac{1}{2}\zeta M_L = 0 \tag{7.3.108}$$

and, therefore,

$$M_L(\zeta) = M_L(0)e^{-\zeta^2/4}, -\infty < \zeta < \infty. \tag{7.3.109}$$

$M_L(0)$ is obtained from (7.3.94).

(viii) From (vi) and (vii) one may derive

$$v_0^p(\eta, \zeta) \leq \frac{M(p-1)}{\sqrt{\pi p}}e^{-\zeta^2/4} \quad \text{for all } (\eta, \zeta) \in S. \tag{7.3.110}$$

Using (7.3.103) in (7.3.110), one may estimate the magnitude of L as

$$L \leq \frac{2p}{3-p}\left(\frac{M(p-1)}{\sqrt{\pi p}}\right)^{(p-1)/p}. \tag{7.3.111}$$

The list of properties is concluded with two conjectures about the asymptotic properties of the solution near $\eta = 0$ and $\eta \uparrow L$; these conjectures are subsequently supported with the help of numerical solution. We note that (7.3.93) has a separable solution which vanishes at $\eta = 0$, namely,

$$v(\eta, \zeta) = \left(\frac{p-1}{p} \eta \right)^{1/(p-1)} f_0(\zeta) \quad \text{for } \eta > 0 \quad \text{and } -\infty < \zeta < \infty \quad (7.3.112)$$

where $f_0(\zeta)$ satisfies the ODE

$$f_0'' + \frac{\zeta}{2} f_0' + \frac{f_0}{p-1} - f_0^p = 0 \quad -\infty < \zeta < \infty \quad (7.3.113)$$

and the boundary conditions

$$f_0(\pm \infty) = 0. \quad (7.3.114)$$

The BVP (7.3.113)-(7.3.114) was studied by Brezis, Peletier, and Terman (1986). They proved the existence of a solution f_0 satisfying $\max_{\zeta \in \Re} f_0(\zeta) = f_0(0) < (1/(p-1))^{1/(p-1)}$, $\zeta f'(\zeta) < 0$ for $\zeta \neq 0$; $f_0(\zeta)$ decays to zero exponentially as $|\zeta| \to \infty$. Dawson et al. (1996) prove the following behaviour of v_0 near $\eta = 0$ in the appendix of their paper:

$$\frac{v_0(\eta, \zeta)}{\left(\frac{p-1}{p} \eta \right)^{1/(p-1)}} - f_0(\zeta) = O(\eta^\lambda) \text{ as } \eta \downarrow 0 \quad (7.3.115)$$

where λ is a positive constant.

The conjecture about the behaviour of v_0 as $\eta \uparrow L$ is that for any $\zeta \neq 0$,

$$\lim_{\eta \uparrow L} v_0(\eta, \zeta) = 0 \quad (7.3.116)$$

while at $\zeta = 0$

$$\lim_{\eta \uparrow L} v_0(\eta, 0) = v_0(L^-, 0) = \left(\frac{3-p}{2p} L \right)^{1/(p-1)}. \quad (7.3.117)$$

The behaviour of singularities at $\eta = L$ and $\eta = 0$ is discussed in detail. The qualitative features derived or conjectured above are validated by the numerical solution of the problem (7.3.93)-(7.3.94).

Dawson et al. (1996) compare their asymptotic results via balancing arguments with the more rigorous results of Escobedo, Vazquez, and Zuazua (1993) and others for $p > 1$ and find them in good agreement. The dominant balance approach is simpler and, when combined with numerical solution, can provide a good picture of the asymptotic structure of the solution.

Chapter 8

Series Solutions of Nonlinear PDEs

8.1 Introduction

From early times, exact similarity solutions were intuitively sought in the form $\vec{u} = t^m \vec{f}(xt^n)$ when the PDEs involved two independent variables; m and n were found either by dimensional argument or direct substitution so that PDEs reduced to ODEs. This class was more fully identified by the use of Lie group methods as described in Section 3.1, or intuitively by the direct similarity approach detailed in Section 3.4. In the latter, the solution was written in the form $u = \alpha(x,t) + \beta(x,t)U(\eta(x,t))$ where the functions α, β, and η were subsequently found by substitution into a PDE so that the resulting equation became an ODE in U with η as the independent variable. However, even this substitution (which again requires invariance properties of the original PDE) may not yield physically meaningful solutions. In such a circumstance, one attempts to write an infinite series for each of the dependent variables in powers of one of the independent variables, time, say, with coefficient functions depending on the similarity variable. The substitution of the infinite series in the PDEs and BCs results in an infinite system of ODEs with appropriate boundary conditions which must be solved analytically or otherwise. The series solution must then be summed up. The convergence of the series may be proved numerically or analytically. In the following sections we take up several problems for which closed form exact solutions are known for some special cases, say for plane geometry. These special exact solutions motivate infinite series form of the solution for the more general problems. The approach is best understood by discussing the solution of some specific problems taken mostly from gasdynamics.

8.2 Analysis of Expansion of a Gas Sphere (Cylinder) into Vacuum

Imagine a gas sphere or cylinder initially at rest in a state of uniform pressure p_0 and density ρ_0, surrounded by vacuum. At some initial time the gas is suddenly allowed to expand into the surrounding vacuum. We shall discuss the mathematical solution of this problem which, as we shall show, has to go beyond similarity solution and be expressed in an infinite series in time with coefficient functions depending on a similarity variable.

It may be noted that if there is no spherical or cylindrical geometry constraining the flow, and the problem is strictly one-dimensional (in this case the gas initially occupies an entire half-space), the solution of the problem can be found explicitly. This solution also suggests the form of the solution for other geometries. We therefore discuss this simpler case first (Stanyukovich (1960)).

If we let $l, c_0 = (\gamma p_0/\rho_0)^{1/2}$, $l/c_0, \rho_0$ and $\rho_0 c_0^2$ represent, respectively, the characteristic length, velocity, time, density, and pressure, then the equations of plane isentropic fluid motion in dimensionless form are

$$u_t + u u_x + \frac{2}{\gamma - 1} c c_x = 0 \tag{8.2.1}$$

$$\frac{2}{\gamma - 1} (c_t + u c_x) + c u_x = 0 \tag{8.2.2}$$

where u is the particle velocity, c is the local sound speed, and γ the ratio of specific heats; x and t are the space and time coordinates, respectively. Since the flow is isentropic, we have $c^2 = \rho^{\gamma-1}$.

Immediately after release, the only particles in motion lie between the gas vacuum interface and the rarefaction front which propagates into the stationary gas. The condition at the interface is $c = 0$ (pressure and density are also zero there). This interface moves with a velocity which must be found as part of the solution. The sound (characteristic) front moves into the undisturbed gas with unit dimensionless velocity. If the gas initially occupies the half space $x < 0$, the sound front has the locus $x = -t$ (obtained by solving $dx/dt = -1 + 0$) where $u = 0$. We have thus prescribed boundary conditions both at the vacuum front and the sound front. The system (8.2.1)-(8.2.2) must be solved subject to these BCs. The solution of this problem may be found by similarity approach or otherwise (Stanyukovich (1960)):

$$u = \frac{2}{\gamma + 1} (1 + \eta) \tag{8.2.3}$$

$$c = \frac{\gamma - 1}{\gamma + 1} \left(\frac{2}{\gamma - 1} - \eta \right) \tag{8.2.4}$$

where $\eta = x/t$ and $-1 < \eta \le 2/(\gamma - 1)$. For $\eta \le -1$, the solution is the uniform state $u = 0$ and $c = 1$, and for $\eta \ge 2/\gamma - 1$ we have vacuum where $c = 0$. The C^+ characteristics are obtained by integrating

$$\frac{dx}{dt} = u + c \qquad (8.2.5)$$

with u and c from (8.2.3)-(8.2.4), and the IC $x = 0$, $t = 0$; we thus obtain

$$x = kt^{(3-\gamma)/(\gamma+1)} + \frac{2}{\gamma - 1}t \qquad (8.2.6)$$

where k is the parameter denominating individual characteristics. The negative characteristics $dx/dt = u - c$ are similarly found to be

$$x = \eta t, \qquad (8.2.7)$$

different values of η giving different characteristics issuing from the origin. The negative characteristics (8.2.7) are straight lines in the (x, t) plane.

We now turn to the case of spherical or cylindrical gas mass expanding freely into the complementary vacuous state. The fronts separating the moving gas from the stationary one and from the vacuum are the same, namely, a sound wave and vacuum front, respectively. The Euler equations of motion in the present case are

$$u_t + uu_r + \frac{2}{\gamma - 1}cc_r = 0 \qquad (8.2.8)$$

$$2(\gamma - 1)^{-1}(c_t + uc_r) + cu_r + \sigma cu/r = 0 \qquad (8.2.9)$$

where $\sigma = 0, 1, 2$ for plane, cylindrical, and spherical symmetry, respectively. Here, we follow the work of Greenspan and Butler (1962), considerably improvised by Nagesawara Yogi (1995). The boundary conditions, as before, are $u = 0$ on the curve $c = 1$, the leading characteristic propagating into the stationary gas. The gas-vacuum front must be located by the requirement that $c = 0$ there.

Greenspan and Butler (1962) showed, by using the Euler and Lagrangian forms of the equations applied at the vacuum front, that it does not decelerate, and that, indeed, it must move with a constant velocity $2/(\gamma - 1)$. We shall show by our construction of the solution that there is a unique solution of the problem when the front is assumed to move with constant speed. For this purpose, appropriate series solution would be shown to converge; this will be accomplished both analytically and numerically.

A similarity solution of the form (8.2.3)-(8.2.4) does not exist for the nonplanar geometries ($\sigma \ne 0$). For the latter, we first introduce the Riemann invariants

$$\phi = u + 2c/(\gamma - 1) \qquad (8.2.10)$$

$$\psi = u - 2c/(\gamma - 1) \tag{8.2.11}$$

as the new dependent variables, and

$$\eta = \frac{r - 1}{t} \tag{8.2.12}$$

and time as the new independent variables, into (8.2.8) and (8.2.9) (observe that the initial radius is unity, hence the definition (8.2.12) of η). Equations (8.2.8)-(8.2.9) now become

$$(1 + \eta t)[t\phi_t - \eta\phi_\eta + \frac{1}{4}\{(\gamma + 1)\phi + (3 - \gamma)\psi\}\phi_\eta] + \frac{1}{8}\sigma(\gamma - 1)t(\phi^2 - \psi^2) = 0 \tag{8.2.13}$$

$$(1 + \eta t)[t\psi_t - \eta\psi_\eta + \frac{1}{4}\{(3 - \gamma)\phi + (\gamma + 1)\psi\}\psi_\eta] - \frac{1}{8}\sigma(\gamma - 1)t(\phi^2 - \psi^2) = 0 \tag{8.2.14}$$

It may be noted that the system (8.2.13)-(8.2.14) has certain symmetries. If (8.2.13) is written as $L(\phi, \psi) = 0$, then (8.2.14) is simply $L(\psi, \phi) = 0$; besides, if $\phi(\eta, t)$ and $\psi(\eta, t)$ are solutions of (8.2.13)-(8.2.14), then so are

$$\phi^*(\eta, t) = -\psi(-\eta, -t), \quad \psi^*(\eta, t) = -\phi(-\eta, -t). \tag{8.2.15}$$

These symmetries have important consequences. If one finds the solution for the expanding gas-vacuum interface, the solution for the problem of cavity collapse can be easily derived therefore (see Section 8.3), and vice versa.

The boundary conditions $u = 0$, $c = 1$ on the leading front, in view of (8.2.10) and (8.2.11), become $\phi = -\psi = 2/(\gamma - 1)$ on $\eta = -1$, that is, on $r = 1 - t$. The vacuum front must be located by the condition $c = 0$.

Restricting our attention to $t \leq 1$, that is, before the sound wave reaches the center or axis of symmetry (see (8.2.12)), we seek solution of (8.2.13)-(8.2.14) in the form

$$\phi = \sum_{n=0}^{\infty} A_n(\eta)t^n \tag{8.2.16}$$

$$\psi = \sum_{n=0}^{\infty} B_n(\eta)t^n. \tag{8.2.17}$$

Putting (8.2.16)-(8.2.17) into (8.2.13) and (8.2.14) and equating coefficients of different powers of t to zero, we get the following infinite system of ODEs for the coefficient functions $A_n(\eta)$ and $B_n(\eta)$:

$n = 0$.

$$\left(-\eta + \frac{1}{4}(\gamma + 1)A_0 + \frac{1}{4}(3 - \gamma)B_0\right)\frac{dA_0}{d\eta} = 0 \tag{8.2.18}$$

$$\left(-\eta + \frac{1}{4}(3 - \gamma)A_0 + \frac{1}{4}(\gamma + 1)B_0\right)\frac{dB_0}{d\eta} = 0 \tag{8.2.19}$$

$n \geq 1$.

$$S_n(\{A_i\}, \{B_i\}) + \eta S_{n-1}(\{A_i\}, \{B_i\}) = -T_{n-1}(\{A_i\}, \{B_i\}) \qquad (8.2.20)$$

$$S_n(\{B_i\}, \{A_i\}) + \eta S_{n-1}(\{B_i\}, \{A_i\}) = T_{n-1}(\{B_i\}, \{A_i\}) \qquad (8.2.21)$$

where

$$S_n(\{A_i\}, \{B_i\}) = nA_n - \eta\frac{dA_n}{d\eta} + \frac{1}{4}\sum_{k=0}^{n}((\gamma+1)A_{n-k} + (3-\gamma)B_{n-k})\frac{dA_k}{d\eta}$$
$$(8.2.22)$$

and

$$T_n(\{A_i, B_i\}) = \frac{1}{8}\sigma(\gamma-1)\sum_{k=0}^{n}(A_{n-k}A_k - B_{n-k}B_k). \qquad (8.2.23)$$

The boundary conditions at the sound wave interface become

$$A_n(-1) = -B_n(-1) = \frac{2}{\gamma-1}\delta_{0n} \qquad (8.2.24)$$

where δ_{0n} is the Kronecker's delta. The system (8.2.18)-(8.2.19) with BC from (8.2.24) for $n = 0$ is easily found to be

$$A_0 = \frac{2}{\gamma-1} \qquad (8.2.25)$$

$$B_0 = \frac{4}{\gamma+1}\left(\eta - \frac{1}{2}\frac{3-\gamma}{\gamma-1}\right). \qquad (8.2.26)$$

This, with (8.2.10)-(8.2.11) and (8.2.16)-(8.2.17), is just the exact solution of the planar problem, $\sigma = 0$. The series solution (8.2.16)-(8.2.17) builds upon this planar solution to take into account the geometrical effects for $\sigma = 1, 2$ via the coefficients A_n and $B_n (n \geq 1)$. To further facilitate the imposition of BCs it is convenient to introduce

$$z = \frac{\gamma-1}{\gamma+1}\left(\frac{2}{\gamma-1} - \eta\right) \qquad (8.2.27)$$

and t as the new independent variables instead of η and t (see Equation (8.2.4)). The coefficients in the solution (8.2.16)-(8.2.17) would now become $A_n = A_n(z), B_n = B_n(z)$ and $\phi = \phi(z,t)$ and $\psi = \psi(z,t)$, etc. The zeroth order solution (8.2.25)-(8.2.26) now becomes

$$A_0 = \frac{2}{\gamma-1}, B_0 = \frac{2}{\gamma-1}(1 - 2z). \qquad (8.2.28)$$

Equations (8.2.20)-(8.2.21) can be manipulated to the form

$$z\frac{dA_n}{dz} - \frac{n}{2}\left(\frac{\gamma+1}{\gamma-1}\right)A_n = -\frac{1}{\gamma}\sum_{k=1}^{n-1}[(\gamma+1)A_{n-k} + (3-\gamma)B_{n-k}]\frac{dA_k}{dz}$$

$$+\frac{1}{2}\frac{\gamma+1}{\gamma-1}\sum_{k=1}^{n}\left(\frac{2}{\gamma-1}\right)^{k-1}\left(\frac{\gamma+1}{2}z-1\right)^{k-1}$$

$$\times T_{n-k}(\{A_i\},\{B_i\}) \qquad\qquad (8.2.29)$$

$$(n+1)B_n + \frac{3-\gamma}{\gamma+1}A_n = \frac{1}{4}\frac{\gamma-1}{\gamma+1}\sum_{k=1}^{n-1}[(\gamma+1)B_{n-k} + (3-\gamma)A_{n-k}]\frac{dB_k}{dz}$$

$$+\sum_{k=1}^{n}\left(\frac{2}{\gamma-1}\right)^{k-1}\left(\frac{\gamma+1}{2}z-1\right)^{k-1}$$

$$\times T_{n-k}(\{A_i\},\{B_i\}), \qquad\qquad (8.2.30)$$

T_n is again given by (8.2.23).

Greenspan and Butler (1962), using an induction argument, prove that all the coefficient functions $A_n(z)$ and $B_n(z)$ satisfying (8.2.29)-(8.2.30) are members of a certain class of functions R, which consists of a linear finite sum of terms of the form $Kz^\nu \ln^m z$, where $\nu \geq 1$, where m is a positive integer and K is a constant. Thus, if $p(z)$ is a typical element of this set, then it has the form

$$p(z) = z^{\nu_0}[k_0^{(0)} + k_1^{(0)}\ln z + \cdots K_{m_0}^{(0)}\ln^{m_0} z] + \cdots$$

$$+z^{\nu_j}[K_0^{(\gamma)} + K_1^{(\gamma)}\ln z + \cdots + K_{m_j}^{(\gamma)}\ln^{m_j} z]. \qquad (8.2.31)$$

It is clear from (8.2.31) that every function $p(z)$ belonging to the class R has the following properties: (i) $p(0) = 0$, (ii) $|dp/dz| < \infty$ for all finite values of z. Since $A_0(0) = -B_0(0) = 2/(\gamma-1)$, it follows that $c = 0$, $u = 2/(\gamma-1)$ on the curve $z = 0$. That is, the gas-vacuum interface is given by $z = 0$, corresponding to $\eta = 2/(\gamma-1)$. It moves with constant speed $2/(\gamma-1)$.

The first five coefficients for $A_n(z)$ and $B_n(z)$ for $\gamma = 5/3, 3$ are given explicitly in Nagesawara Yogi (1995).

The following statements about $A_n(z)$ and $B_n(z)$ for different γ may be easily verified.

(a) For $1 < \gamma < 3$, the leading power of z in $A_n(z)$ and $B(z)$ is max $[1+n, nN]$, where $N = (\gamma+1)/(2(\gamma-1))$. There is a countable set of values of γ for which either N or nN is an integer.

(b) Three distinct forms of $A_n(z)$ and $B_n(z)$ may be identified.

(i) If $N \neq (n+1)/n$ is fractional, A_n and B_n are multinomials in z and z^{N-1}.

(ii) If $N \neq (n+1)/n$ is integer, A_n and B_n are polynomials in z for $\gamma = 7/5, 9/7, 11/9, \cdots$.

(iii) $N = 1, 2$, corresponding to $\gamma = 3$ and $5/3$, respectively. In these cases, A_n and B_n contain powers of both z and $z \log z$. Thus, logarithmic terms appear only for $\gamma = 5/3$ in the range $1 < \gamma < 2$ and for $\gamma = 3$ in $\gamma > 2$.

To give an idea of the specific forms for A_n and B_n for case (i) above, we have for $n \geq 1$,

$$A_n(z) = \sum_{i=1}^{n+1} a_{n,i} z^i + \sum_{k=1}^{n} \sum_{j=1}^{n} a_{n,k,j} z^{kN+j-k} \qquad (8.2.32)$$

$$B_n(z) = \sum_{i=1}^{n+1} b_{n,i} z^i + \sum_{k=1}^{n} \sum_{j=1}^{n} b_{n,k,j} z^{kN+j-k}. \qquad (8.2.33)$$

For the specific value $\gamma = 7/5$ in case (ii), we have

$$A_n(z) = \sum_{k=1}^{nN} A_{k,n} z^k, \quad B_n(q) = \sum_{k=1}^{nN} B_{k,n} z^k. \qquad (8.2.34)$$

For case (iii) above we have the following:
$\gamma = 5/3, n \geq 1$.

$$A_n(z) = \left[\sum_{i=1}^{2n} A_{n,i} z^i \right] + \left[\sum_{k=1}^{n-1} (z \ln z)^k \left(\sum_{j=1}^{2(n-k)} A_{n,k,j} z^j \right) \right]$$
$$+ (z \ln z)^n (A_{n,n,1} z) \qquad (8.2.35)$$

$$B_n(z) = \left[\sum_{i=1}^{2n} B_{n,i} z^i \right] + \left[\sum_{k=1}^{n-1} (z \ln z)^k \left(\sum_{j=1}^{2(n-k)} B_{n,k,j} z^j \right) \right]$$
$$+ (z \ln z)^n (B_{n,n,1} z) \qquad (8.2.36)$$

$\gamma = 3, n \geq 1$.

$$A_n(z) = \sum_{i=0}^{n+1} A_{n,i} z^i + \sum_{i=1}^{n} (\ln^i z) \left(\sum_{j=1}^{n} A_{n,i,j} z^j \right) \qquad (8.2.37)$$

$$B_n(z) = \sum_{i=0}^{n+1} B_{n,i} z^i + \sum_{i=1}^{n} (\ln^i z) \left(\sum_{j=1}^{n} B_{n,i,j} z^j \right)$$
$$B_{n,n,n} = 0. \qquad (8.2.38)$$

The solution of the problem under consideration was obtained by a direct summation of the series (8.2.16)-(8.2.17), using Pade sums and by direct numerical solution of the system of ODEs (8.2.29)-(8.2.30). The solution of the expansion of gas sphere (cylinder) into vacuum, with the vacuum front moving with constant speed, is thus shown to exist by actual construction.

An attempt to prove analytically the convergence of the series (8.2.16) and (8.2.17) was made by Greenspan and Butler (1962). They argued that, if the series for $\phi_z(z,t)$ and $\psi_z(z,t)$, the most highly differentiated terms evaluated at the most crucial physical positions $z = 0, 1$ can be shown to converge, the convergence elsewhere would follow. Indeed, they showed that the series in t at these locations can be summed up for $\gamma = 5/3$ both for convergent and divergent flows (see Section 8.3). This, in fact, is also true for $z = 1$ for all values of γ. Thus, we have

$$\phi_z(0,t) = -\frac{12t}{1+3t}, \quad \psi_z(0,t) = -6\frac{(1+2t)}{1+3t} \quad \text{for } \gamma = 5/3 \qquad (8.2.39)$$

$$\phi_z(1,t) = 0, \quad \psi_z(1,t) = \frac{4}{\gamma-1}\frac{t}{1-t}\frac{1}{\ln(1-t)} \quad \text{for all } \gamma. \qquad (8.2.40)$$

The series converge at either end of the rarefaction wave and Greenspan and Butler (1962) surmised that it would therefore converge throughout the interval $0 \leq z \leq 1$.

Sachdev, Gupta, and Ahluwalia (1992a) proved the local convergence of the series in the neighbourhood of $z = 0$ for the cavity collapse problem (see Section 8.3); the same argument holds for the expansion front. Our computations clearly bear out the conjecture of Greenspan and Butler (1962) for the entire flow for all γ: $1 < \gamma < 5/3$ (see Nagesawara Yogi (1995)).

8.3   Collapse of a Spherical or Cylindrical Cavity

The present problem is entirely analogous to the one treated in the previous section, namely the expansion of gas sphere (cylinder) into surrounding vacuum; however, it has considerable interest — both physical and mathematical — and merits discussion. Imagine an ideal homogeneous polytropic gas at rest, surrounding a spherical or cylindrical vacuous space. At time $t = 0$, the surface containing the vacuous space is instantaneously removed and a one-dimensional flow of the medium ensues inward. The flow region is bounded by the free surface separating the gas from the vacuum (hereafter called gas-vacuum interface) and a sound front, a characteristic, adjoining the medium at rest. This characteristic is given by $dx/dt = c_0$, the speed of sound in the undisturbed gas. The gas-vacuum interface moves, starting initially with a constant speed $2/(\gamma - 1)$ corresponding to the planar case (see Section 8.2) and continues to do so until it collapses to the center or axis of asymmetry for $1 < \gamma < 1 + 2/(1 + \nu)$, as we shall demonstrate (Sachdev, Gupta, and Ahluwalia (1992a)). Here, $\nu = 0, 1, 2$ for plane, cylindrical, and spherical symmetry, respectively (see Equation (8.3.2) below). The Eulerian equations of motion for an isentropic flow in these geometries are

$$u_t + uu_r + \beta cc_r = 0 \tag{8.3.1}$$

$$c_t + uc_r + \beta^{-1}\left(cu_r + \frac{\nu uc}{r}\right) = 0 \tag{8.3.2}$$

where $\beta = 2/(\gamma - 1)$, $\gamma = c_p/c_v$, being the ratio of specific heats at constant pressure and volume, respectively. The spatial variable r and time t have been rendered nondimensional by reference to the initial cavity radius R_0 and R_0/c_0, respectively. The radial velocity u and the sound speed c have also been made nondimensional by the undisturbed sound speed c_0.

The boundary condition on the leading characteristic separating the undisturbed gas from the disturbed one is

$$u = 0 \quad \text{on the curve} \quad c = 1. \tag{8.3.3}$$

The movement of the gas vacuum interface is given by the condition

$$c = 0. \tag{8.3.4}$$

This is the second boundary condition and locates the interface.

The gas-vacuum interface, on initiation of the collapse, accelerates instantaneously to move subsequently with a constant speed β. It may appear that the spherical or cylindrical cavity would accelerate as it converges. This, however, is not the case. It was shown by Sachdev et al. (1992a) using a wave-fronts analysis on the back of the interface, that infinite gradients develop before the collapse only if $\gamma > \gamma_*$, where $\gamma_* = 1 + 2/(1 + \nu)$. This is also confirmed by the convergence argument given in Section 8.2. We therefore construct the global solution of the problem for $1 < \gamma < \gamma_*$, for which the flow behind the interface remains free from infinite gradients.

As in Section 8.2, we introduce in (8.3.1)-(8.3.2) the Riemann invariants

$$\phi = u + \beta c \tag{8.3.5}$$
$$\psi = u - \beta c \tag{8.3.6}$$

as the dependent variables, and

$$z = \frac{1}{1 + \beta}\{\beta + (x - 1)/t\} \tag{8.3.7}$$

and t as the independent variables. We obtain

$$\{1 + ((1 + \beta)z - \beta)t\} \left[t\frac{\partial \phi}{\partial t} + \frac{1}{1 + \beta} \right.$$

$$\times \left. \left\{ \frac{1}{2\beta}((\beta + 1)\phi + (\beta - 1)\psi) + \beta - (1 + \beta)z \right\} \frac{\partial \phi}{\partial z} \right]$$

$$+ \frac{\nu t}{4\beta}(\phi^2 - \psi^2) = 0 \tag{8.3.8}$$

$$\{1 + ((1 + \beta)z - \beta)t\} \left[t\frac{\partial \psi}{\partial t} + \frac{1}{1 + \beta} \right.$$

$$\times \left. \left\{ \frac{1}{2\beta}((\beta + 1)\psi + (\beta - 1)\phi) + \beta - (1 + \beta)z \right\} \frac{\partial \psi}{\partial z} \right]$$

$$- \frac{\nu t}{4\beta}(\phi^2 - \psi^2) = 0. \tag{8.3.9}$$

Here, $0 \le z \le 1$; $z = 0$ corresponds to gas-vacuum interface starting from $x = 1$ and moving with constant speed $-\beta < 0$, while $z = 1$ represents the sound front. On the latter, the condition (8.3.3), in terms of ϕ and ψ, becomes

$$\phi(1, t) = \beta, \quad \psi(1, t) = -\beta. \tag{8.3.10}$$

We seek solution of (8.3.8)-(8.3.9) subject to (8.3.10) in the form

$$\phi(z, t) = \sum_{i=1}^{\infty} f_i(z)t^i, \quad \psi(z, t) = \sum_{i=0}^{\infty} g_i(z)t^i. \tag{8.3.11}$$

We restrict ourselves to the interval $0 < t < 1$, which covers the time $t = (\gamma - 1)/2$ for a uniformly moving cavity surface to collapse to center (axis) of symmetry provided $1 < \gamma < 3$. Substituting (8.3.11) into (8.3.8) and (8.3.9) and equating coefficients of different powers of t to zero, we arrive at the following system of ODEs for f_i, g_i:

$$\left[\beta - (1 + \beta)z + \frac{1}{2\beta}\{(1 + \beta)f_0 + (\beta - 1)g_0\} \right] \frac{df_0}{dz} = 0 \tag{8.3.12}$$

$$\left[\beta - (1 + \beta)z + \frac{1}{2\beta}\{(1 + \beta)g_0 + (\beta - 1)f_0\} \right] \frac{dg_0}{dz} = 0 \tag{8.3.13}$$

and

$$z\frac{dg_i}{dz} - \frac{i(1 + \beta)}{2}g_i - \frac{1}{4\beta}G_i = -\frac{(1 - \beta)}{2}\sum_{k=1}^{i}\{\beta - (1 + \beta)z\}^{k-1}T_{i-k} \tag{8.3.14}$$

$$(1 + i)f_i + \frac{(\beta - 1)}{(\beta + 1)}g_i + \frac{1}{2\beta(1 + \beta)}F_i = -\sum_{k=1}^{i}\{\beta - (1 + \beta)z\}^{k-1}T_{i-k} \tag{8.3.15}$$

for $i = 0$ and $i \geq 1$, respectively. Here

$$T_i = \frac{\nu}{4\beta} \sum_{p=0}^{i} (f_{i-p}f_p - g_{i-p}g_p)$$

$$G_i = \begin{cases} 0 & i = 1 \\ \sum_{k=1}^{i-1} \{(1+\beta)g_{i-k} + (\beta - 1)f_{i-k}\} \dfrac{dg_k}{dz}, & i \geq 2 \end{cases} \qquad (8.3.16)$$

and

$$F_i = \begin{cases} 0 & i = 1 \\ \sum_{k=1}^{i-1} \{(1+\beta)f_{i-k} + (\beta - 1)g_{i-k}\} \dfrac{df_k}{dz}, & i \geq 2. \end{cases} \qquad (8.3.17)$$

The boundary conditions on f_i and g_i become

$$f_0(1) = \beta, \ g_0(1) = -\beta \qquad (8.3.18)$$

and

$$f_i(1) = 0, \ g_i(1) = 0, i \geq 1. \qquad (8.3.19)$$

The solution of zeroth order (nonlinear) system (8.3.12)-(8.3.13) subject to (8.3.18), namely

$$f_0(z) = \beta(2z - 1), \ g_0(z) = -\beta, \qquad (8.3.20)$$

corresponds just to the planar solution of the problem for $\nu = 0$, namely the escape of a slab of gas into vacuum. The functions $g_i(z), i \geq 1$ are governed by a system of inhomogeneous linear, first-order, ordinary differential equations (8.3.14) and (8.3.15), subject to the homogeneous boundary conditions (8.3.19). If $g_i, i \geq 1$ is found by solving (8.3.14), $f_i, i \geq 1$ is found from the algebraic relation (8.3.15). $f_i, g_i (i \geq 1)$ embody the effect of spherical or cylindrical contraction.

The coefficient functions $f_i(z)$ and $g_i(z)$ enjoy the same properties as the functions named $A_n(z)$ and $B_n(z)$ in the solution of the expansion problem detailed in Section 8.2. The series (8.3.11) was shown to be locally convergent near $z = 0$ by Sachdev et al. (1992a).

The series solution for c was computed for spherical geometry ($\nu = 2$) for $\gamma = 1.4, 5/3$, and for cylindrical geometry ($\nu = 1$) for $\gamma = 1.999$. These values represent distinct structures of the series solution (see Section 8.2). Table 8.1 shows the rate of convergence for the series solution for c for spherical symmetry for the case $\gamma = 1.4$ at different times before the point of focusing as the number of terms increases. It is clear that the rate of convergence decreases as t tends to $(\gamma - 1)/2$, the time of collapse of the cavity. It may also be observed that the convergence is slower near the gas-vacuum interface ($z = 0$). The computations for $\gamma = 5/3$ show that the series for c converges in the closed domain between the vacuum front and the characteristic front for the entire time to the point of collapse, that is, $t \leq 0.3332$. For the cylindrical case, the series solution converges for $\gamma \lesssim 2$ all the way to the time of collapse. For $\gamma > 2$, the solution series diverges at times earlier than the collapse time,

Table 8.1. Partial sums c_i for the series solutions for c as given by (8.3.5), (8.3.6), and (8.3.11) for $\gamma = 1.4$ and $\nu = 2$ at $t = 0.149, 0.199$.

			$t = 0.149$			
z	0.0	0.01	0.021	0.61	0.81	1.0
c_0	0.00000	0.01000	0.21000	0.61000	0.81000	1.00000
c_2	0.00000	0.01366	0.26142	0.65537	0.83018	1.00000
c_2	0.00000	0.01557	0.28105	0.66144	0.83014	1.00000
c_3	0.00000	0.01670	0.28968	0.66243	0.83016	1.00000
c_4	0.00000	0.01740	0.29373	0.66261	0.83016	1.00000
c_5	0.00000	0.01785	0.29570	0.66264	0.83016	1.00000
c_6	0.00000	0.01815	0.29669	0.66264	0.83016	1.00000
c_7	0.00000	0.01835	0.29719	0.66265	0.83016	1.00000
c_8	0.00000	0.01848	0.29745	0.66265	0.83016	1.00000
c_9	0.00000	0.01857	0.29758	0.66265	0.83016	1.00000
$c_1 0$	0.00000	0.01864	0.29765	0.66265	0.83016	1.00000
			$t = 0.199$			
z	0.0	0.01	0.21	0.61	0.81	1.0
c_0	0.00000	0.01000	0.21000	0.61000	0.81000	1.00000
c_2	0.00000	0.01489	0.27867	0.67060	0.83695	1.00000
c_2	0.00000	0.01830	0.31370	0.68143	0.83688	1.00000
c_3	0.00000	0.02098	0.33425	0.68379	0.83692	1.00000
c_4	0.00000	0.02321	0.34713	0.68434	0.83693	1.00000
c_5	0.00000	0.02512	0.35552	0.68447	0.83693	1.00000
c_6	0.00000	0.02681	0.36112	0.68450	0.83693	1.00000
c_7	0.00000	0.02832	0.36492	0.68451	0.83693	1.00000
c_8	0.00000	0.02969	0.36753	0.68451	0.83693	1.00000
c_9	0.00000	0.03094	0.36934	0.68452	0.83693	1.00000
$c_1 0$	0.00000	0.03210	0.37061	0.68452	0.83693	1.00000

the greater the value of γ, the earlier the divergence. The series structure for the solution of the system (8.3.14)-(8.3.15) for $\gamma = 2$ is extremely complicated, involving as it does powers of z, z^α, and $z^\alpha \ln z$, $\alpha > 1$. This case was not computed. Instead, the solution was found for $\gamma = 1.999$, for which we have a double series in z and $z^{\alpha-1}$.

Now we compare the results of the global solution thus found with the numerical solution of the problem by Thomas, Pais, Gratton, and Diez (1986), and the asymptotic self-similar solution of Sachdev et al. (1992a), which holds close to the time of cavity collapse. The series solution for a uniformly moving spherical or cylindrical cavity exists for

$0 < \gamma < 1 + 2/(1 + \nu)$ up to the time of collapse. This is in agreement
with the numerical results of Thomas et al. (1986) for a spherical cavity
for $\gamma < \gamma_p$ where $3/2 < \gamma_p \lesssim 5/3$. For $\gamma > \gamma_p$, the numerical results show
that the spherical cavity moves initially with a constant velocity equal to
$-2/(\gamma-1)$, but then it begins to accelerate as the center is approached. The
solution in the limiting situation assumes a self-similar form, as discussed
by Lazarus (1981). Lazarus (1982) also carried out the stability analysis
of the self-similar solutions for uniformly moving cavity surfaces. He concluded that these solutions are unstable for $\gamma < 1 + 2/(1 + \nu)$, confirming
that the self-similar solution for the uniformly moving cavity does not form
intermediate asymptotics.

8.4   Converging Shock Wave from a Spherical or Cylindrical Piston

In Section 8.3 we considered the global solution of the problem describing
spherical or cylindrical cavity collapse; we also summarised the results with
respect to the self-similar solutions of this problem, which were shown to
be unstable for most ranges of γ. A related problem is that of converging
shock waves, which also possesses a similarity solution. Indeed, self-similar
solutions of both these problems belong to the class called the "second
kind" (see Zel'dovich and Raizer (1967)) for which dimensional analysis or
group properties of the PDEs do not fully determine the self-similar form
of the problem; they require a global solution of an eigenvalue problem
for the reduced system of ODEs. Typically, for this class of problems the
exponent in the definition of the similarity variable turns out, in general,
to be an irrational number. For the converging shock problem, which was
first studied by Guderley (1942), this exponent α in the similarity variable
$\xi = rt^{-\alpha}$ was found to be 0.717 for the spherical converging shock for $\gamma(= c_p/c_v) = 1.4$. This value was later refined by several other investigators.
    Indeed, the cavity collapse and the converging shock phenomena have
some interesting physical closeness (Greenspan and Butler (1962)). The
gas-vacuum interface is related, in a sense, to an infinitely strong shock.
Imagine one gas escaping into another gas at rest; the resultant flow may
be divided into three regions, including a shock proceeding into the rest
gas and a stationary rarefaction wave into the escaping gas. If, now, the
density of the rest gas tends to zero, the entire shock regime, including
the stationary wave, becomes meaningless inasmuch as the density also
vanishes in this domain. The shock, however, becomes infinitely strong,
i.e., the density ratio approaches $(\gamma + 1)/(\gamma - 1)$, even though the density
itself is zero. The gas-vacuum front, in short, behaves in certain respects
like an infinitely strong shock. These ideas are illustrated by Greenspan
and Butler (1962) by considering the reflection of a vacuum front off a wall;

the front does indeed reflect as an infinitely strong shock.

As in Sections 8.2 and 8.3, we are not concerned with local behaviour of the solution in the close vicinity of the point (axis) of convergence of the shock. Instead, we describe a global problem arising out of the motion of a spherical or cylindrical piston that collapses with uniform inward speed (Van Dyke and Guttmann (1982)). As for the motions of expanding and contracting vacuum fronts, the base flow here is also assumed to be produced by impulsive motion of a plane piston. This piston motion is assumed to be so strong that the shock generated is of infinite strength. Even though this motion seems rather contrived, it can be shown that several other similar piston motions asymptotically lead to Guderley's self-similar solution near the center (axis).

Consider a spherical or cylindrical container of initial radius, R_0, containing a perfect gas at rest with uniform density ρ_0 and adiabatic constant γ. At time $t = 0$, the container suddenly contracts with a very large velocity V, emitting ahead of it a shock wave of radius $R(t)$, whose trajectory must be found as part of the solution.

The equation governing this flow in different geometries are

$$\rho_t + (\rho v)_r + j\frac{\rho v}{r} = 0 \tag{8.4.1}$$

$$v_t + vv_r + \frac{1}{\rho}p_r = 0 \tag{8.4.2}$$

$$(p\rho^{-\gamma})_t + v(p\rho^{-\gamma})_r = 0 \tag{8.4.3}$$

where ρ, p, and v are density, pressure, and particle velocity at any point r and time t; $j = 0, 1, 2$ for planar, cylindrical, and spherical symmetry, respectively. The Rankine-Hugoniot relations connecting the states ahead of and behind an infinitely strong shock are

$$v = \frac{2}{\gamma + 1}\dot{R} \tag{8.4.4}$$

$$\rho = \frac{\gamma + 1}{\gamma - 1}\rho_0 \tag{8.4.5}$$

$$p = \frac{2}{\gamma + 1}\rho_0\dot{R}^2 \tag{8.4.6}$$

where $\dot{R}$ is the velocity of the shock.

The boundary condition at the piston generating the shock is

$$v = -V \quad \text{at} \quad r = R_0 - Vt \tag{8.4.7}$$

where R_0 is the initial position of the piston. It is convenient to introduce the coordinate $x = R_0 - r$, measuring the distance (inward) from the initial position of the piston. The particle velocity v then becomes $-u$, say.

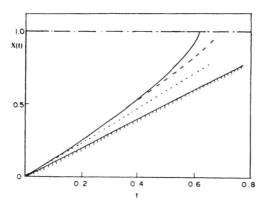

Figure 8.1. History of converging shock wave in (x,t) plane for spherical piston with $\gamma = 7/5$. ⊤⊤⊤⊤, path of piston; ....., 1-term planar approximation to shock wave; - - - - -, 3-term approxima- tion (8.4.23); ———, full solution.

The flow in the (x,t) plane is shown in Figure 8.1 for a short initial time; when the effect of geometry is insignificant, the flow between the uniformly moving strong shock with velocity $(\gamma + 1)V/2$ and the (uniformly) moving piston is simply $u = V$, the density $\rho = [(\gamma+1)/(\gamma-1)]\rho_0$, and the pressure $p = \frac{1}{2}(\gamma + 1)\rho_0 V^2$, in accordance with (8.4.4)-(8.4.6).

With this short-time plane flow in view, it is convenient to introduce the "similarity" variable

$$\xi = \frac{2}{\gamma - 1}\left(\frac{x}{Vt} - 1\right) \tag{8.4.8}$$

and time as new independent variables so that, for this basic flow, the position of the piston is given simply by $\xi = 0$, and that of the shock by $\xi = 1$. (cf. the transformation (8.3.7)). The initial radius of the position R_0, its velocity V, time R_0/V, initial density ρ_0, and the pressure $\rho_0 V^2$ may be used to render all the variables dimensionless. The governing system of PDEs (8.4.1)-(8.4.3) and BC (8.4.4)-(8.4.7) may be rewritten in terms of the variables ξ and t as follows:

$$\left[1 - \left(1 + \frac{1}{2}(\gamma - 1)\xi\right)t\right]\left\{\rho u_\xi + \left(u - 1 - \frac{1}{2}(\gamma - 1)\xi\right)\rho_\xi + \frac{1}{2}(\gamma - 1)t\rho_t\right\}$$

$$= \frac{1}{2}(\gamma - 1)jt\rho u \tag{8.4.9}$$

$$\rho\left(u - 1 - \frac{1}{2}(\gamma - 1)\xi\right)u_\xi + \frac{1}{2}(\gamma - 1)t\rho u_t + p_\xi = 0 \tag{8.4.10}$$

$$\left(u - 1 - \frac{1}{2}(\gamma - 1)\xi\right)(\rho p_\xi - \gamma p\rho_\xi) + \frac{1}{2}(\gamma - 1)t(\rho p_t - \gamma p\rho_t) = 0 \tag{8.4.11}$$

$$u = \frac{2}{\gamma+1}\dot{X}, \rho = \frac{\gamma+1}{\gamma-1}, p = \frac{2}{\gamma+1}\dot{X}^2 \text{ at } \xi = \frac{2}{\gamma-1}\left[\frac{X(t)}{t} - 1\right] \quad (8.4.12)$$

and

$$u = 1 \quad \text{at} \quad \xi = 0 \quad (8.4.13)$$

where $x = X(t)$ is the locus of the shock for the nonplanar flow. We may quite reasonably assume that the shock trajectory $X(t)$ is analytic in time so that

$$X(t) = \sum_{n=1}^{\infty} X_n t^n. \quad (8.4.14)$$

The other flow variables are expanded as

$$u = \sum_{n=1}^{\infty} U_n(\xi)t^{n-1}, \quad \rho = \sum_{n=1}^{\infty} R_n(\xi)t^{n-1}, \quad p = \sum_{n=1}^{\infty} P_n(\xi)t^{n-1}. \quad (8.4.15)$$

Here, we take the basic flow U_1, R_1, and P_1 to be that given by the planar solution mentioned earlier, which in nondimensional form is

$$U_1 = 1, R_1 = \frac{\gamma+1}{\gamma-1}, P_1 = \frac{1}{2}(\gamma+1), X_1 = \frac{1}{2}(\gamma+1). \quad (8.4.16)$$

Substituting (8.4.15) into (8.4.9)-(8.4.11) and equating coefficients of different powers of t to zero, we get a sequence of triads of first-order, linear ordinary differential equations for U_n, R_n, and P_n. The equations for U_2, R_2, and P_2, for example, are

$$\frac{\gamma+1}{\gamma-1}U_2' - \frac{1}{2}(\gamma-1)\xi R_2' + \frac{1}{2}(\gamma-1)R_2 = \frac{1}{2}(\gamma+1)j \quad (8.4.17)$$

$$-\xi U_2' + U_2 + \frac{2}{\gamma+1}P_2' = 0 \quad (8.4.18)$$

$$\xi\left(P_2' - \frac{1}{2}\gamma(\gamma-1)R_2'\right) - \left(P_2 - \frac{1}{2}\gamma(\gamma-1)R_2\right) = 0. \quad (8.4.19)$$

The boundary condition (8.4.13) on the piston gives $U_n(0) = 0$ for all $n > 1$. The BCs at the shock must be found by substituting (8.4.14) and (8.4.15) in (8.4.12) and then putting $\xi = 1$. For the second-order terms we get

$$U_2(1) = \frac{4}{\gamma+1}X_2, R_2(1) = 0, P_2(1) = 4X_2. \quad (8.4.20)$$

The form of this problem suggests that U_2, R_2, and P_2 are linear in ξ. On using the BCs (8.4.20), the solution of (8.4.17)-(8.4.19) comes out to be

$$U_2 = \frac{\gamma(\gamma-1)}{2(2\gamma-1)}j\xi, \quad R_2 = \frac{\gamma+1}{2\gamma-1}j(1-\xi)$$

$$(8.4.21)$$

$$P_2 = \frac{\gamma(\gamma+1)(\gamma-1)}{2(2\gamma-1)}j, \quad X_2 = \frac{\gamma(\gamma+1)(\gamma-1)}{8(2\gamma-1)}j.$$

Indeed, for this problem the higher order coefficients U_n, R_n, and P_n are simply polynomials in ξ of degree $n - 1$:

$$U_n(\xi) = \sum_{k=2}^{n} U_{nk}\xi^{k-1}, R_n(\xi) = \sum_{k=1}^{n} R_{nk}\xi^{k-1},$$

(8.4.22)

$$P_n(\xi) = \sum_{k=1}^{n} P_{nk}\xi^{k-1}.$$

We substitute (8.4.22) into the corresponding ODE, for $U_n(\xi), R_n(\xi)$, and $P_n(\xi)$, and expand the shock conditions (8.4.12) about $\xi = 1$, using (8.4.14). We then equate like powers of ξ as well as of t, and obtain for each approximation a system of $3n$ inhomogeneous algebraic equations for U_{nk}, R_{nk}, P_{nk}, and X_n, with right-hand sides depending on the previous approximations.

Solving these equations in the third order, we get the position of the shock as

$$X(t) = \frac{1}{2}(\gamma + 1)t + \frac{\gamma(\gamma + 1)(\gamma - 1)}{8(2\gamma - 1)}jt^2 + \frac{(\gamma + 1)(\gamma - 1)}{48(7\gamma - 5)}$$
$$\times \left[(\gamma + 1)(3\gamma + 1)j + \frac{\gamma(13\gamma^3 - 21\gamma^2 + 13\gamma - 1)}{(2\gamma - 1)^2}j^2\right]t^3 + \cdots$$

(8.4.23)

This result for $\gamma = 7/5$ for a spherical shock is shown in Figure 8.1 The results obtained from (8.4.23) were also compared with the numerical evaluation of Lee (1968) to this order. The second and third coefficients in (8.4.23) agree with his to three significant places.

Van Dyke and Guttmann (1982) wrote a computer programme to generate the general term in (8.4.23). Table 8.2 gives 40 coefficients in the series for shock trajectory, obtained in the manner described earlier for three terms, for spherical symmetry for $\gamma = 7/5, 5/3, 3$ and for cylindrical symmetry for $\gamma = 7/5$. Since all the coefficients in the series are positive, the singularity (if one arises) on the shock trajectory must lie on the positive t-axis. It is also observed that the coefficients in this series increase steadily in magnitude, implying that the radius of convergence must be less than unity. This is sensible since the piston itself would reach the axis with unit velocity at $t = 1$. It is also found that the coefficients grow faster for the spherical case than for the cylindrical case, indicating that the focusing is more intense for the former.

For estimating the radius of convergence of the series (8.4.14), Van Dyke and Guttmann (1982) used the Domb and Sykes (1957) approach. If the series in the neighbourhood of the nearest singularity (assuming there is a finite one) has the form

Table 8.2. Coefficients X_n in series (8.4.14) for shock waves.

n	Spherical, $\gamma = 7/5$	Spherical, $\gamma = 5/2$	Spherical, $\gamma = 3$	Cylindrical, $\gamma = 7/5$
1	1.200000000000	1.333333333333	2.00000000000	1.2000000000000
2	0.186666666667	0.317460317460	1.20000000000	0.0933333333333
3	0.188345679012	0.330964978584	1.83333333333	0.0730864197531
4	0.172851981806	0.351087328915	3.40035087719	0.0577257959714
5	0.172147226896	0.428702976041	7.24262900585	0.0497185254748
6	0.195748089820	0.581262688522	16.7325356185	0.0473867537972
7	0.239592510180	0.833416073327	40.8212062145	0.0487020337051
8	0.303219524757	1.24182040572	103.538798073	0.0525457596193
9	0.394337922617	1.90667020627	270.351164204	0.0586078973893
10	'0.525663995528	2.99573095341	721.973134446	0.0670385267585
11	0.714271423746	4.79335492559	1962.93555769	0.0782473038694
12	0.985060389731	7.78505460535	5415.71134591	0.0928536362648
13	1.37561449412	12.8028036868	15125.3041521	0.111712634840
14	1.94193338406	21.2785506061	42681.0787588	0.135973603154
15	2.76700088699	35.6880234991	121509.247882	0.167161791445
16	3.97437632751	60.3290517468	348589.799633	0.207289223128
17	5.74887230925	102.690500277	1006783.95686	0.259004706586
18	8.36757126135	175.866803349	2925043.77126	0.325795437783
19	12.2467590407	302.827404305	8543150.61409	0.412256433109
20	18.0133655273	523.983733067	25069946.9513	0.524449933038
21	26.6137638895	910.630204719	73881275.4824	0.670384889712
22	39.4795522908	1588.86850668	218567708.399	0.860657162621
23	58.7805913213	2782.27435391	648869068.945	1.10930506416
24	87.8118838110	4888.12883923	1932484742.18	1.43495376450
25	131.585889835	8613.85327622	5772286224.84	1.86234753012
26	197.740487538	15221.5900368	17288250591.8	2.42440315613
27	297.032522044	26967.3254176	51908194965.1	3.16496440419
28	449.979223858	47890.4406560	156214990411	4.14250004943
29	681.152558004	85235.2928220	471129305758	5.43507307638
30	1033.25274612	152014.101220	1.42371888519 12	7.14702352524
31	1570.42985951	271633.152889	4.31039960943 12	9.41796318002
32	2391.25395142	486253.291668	1.30727810033 13	12.4348912563
33	3647.35908070	871915.946437	3.97126124722 13	16.4485262673
34	5572.26775610	1565938.77695	1.20824798683 14	21.7953372274
35	8525.99348990	2816584.06448	3.68138936143 14	28.9272839088
36	13064.1157676	5073195.16260	1.12320678078 15	38.4519908447
37	20044.8405790	9149924.83352	3.43136709434 15	51.1870510605
38	30795.0631275	16523403.6091	1.04955212172 16	68.2334756229
39	47368.1399675	29874379.4238	3.21397456855 16	91.0751000913
40	72944.3025390	54074091.6579	9.85273540521 16	121.713200768

$$X(t) = \sum_{n=1}^{\infty} X_n t^n \sim A_1 \left(1 - \frac{t}{t_c}\right)^{\alpha_1} \quad \text{as} \quad t \to t_c, \qquad (8.4.24)$$

then

$$\frac{X_n}{X_{n-1}} \sim \frac{1}{t_c}\left(1 - \frac{1+\alpha_1}{n}\right) \quad \text{as} \quad n \to \infty. \qquad (8.4.25)$$

Figure 8.2 shows $1/n$ versus X_n/X_{n-1} for the spherical converging shock for $\gamma = 7/5$. A linear fit with the exponent $\alpha_1 = 0.717$, as given by Guderley, yields $1/t_c = 1.61$ or $t_c = 0.62$ to graphical accuracy. A more accurate fit by a polynomial in $1/n$ gave a value of $1/t_c$ as 1.609021, which agrees with Guderley's result to three significant figures.

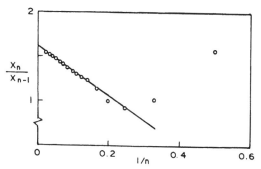

Figure 8.2. Graphical ratio test of Domb and Sykes for the series (8.4.14) for position of shock wave. ———, $1.61\ (1-1.717/n)$.

To verify that the nearest singularity for the above case does not occur before collapse, the series (8.4.14) was solved for t_0 such that $X(t_0) = 1$. For $\gamma = 1.4$, this value was found to be 0.62149604. Similar figures were obtained for spherical symmetry for $\gamma = 5/3$ and 3, and for cylindrical symmetry for $\gamma = 1.4$.

A series equivalent of (14) in the form

$$R(\tau) = \sum_{i=1}^{\infty} \frac{A_i}{1+\alpha_i \tau}, \tau = \ln\left(1 - \frac{t}{t_c}\right)^{-1} \qquad (8.4.26)$$

was also constructed using a Padé approximation. For $\gamma = 7/5$, the values of A_1 and α_1 for the spherical converging shock were found to be 0.71717450 and 0.981706, respectively. The value of α_1 thus calculated is in excellent agreement with that obtained from the precise numerical solution of the governing PDEs and boundary conditions by Lazarus and Richtmyer (1977). Indeed, it was found that the three-term series (8.4.23) for the shock trajectory gives an excellent description of the trajectory of the converging shock in its entire course, the error never exceeding 0.5%.

Regarding the intermediate asymptotic character of the Guderley's similarity solution, there are conflicting views in Russian and Western literature (see Van Dyke and Guttmann (1982) for a discussion). To the author's knowledge, this matter has not yet been fully resolved.

For other investigations related to this problem, which are partly analytic, reference may be made to Lee (1968) and Nakamura (1983). The latter work is close to that of Van Dyke and Guttmann (1982); however, here the piston velocity was assumed to be quadratic in time. The first three-term solution (similar to that of Van Dyke and Guttmann (1982)) was used to determine the starting conditions for the numerical solution. The numerical method was based on characteristics and the transition of the nonself-similar motion of the shock to its self-similar asymptotic regime was analysed.

We conclude this section with a summary of the work of Kozmanov (1977), which assumes a general piston motion

$$x(t) = \xi_1 t + \xi_2 t^2 + \cdots \quad + \xi_n t^n, \quad \xi_1 > 1, \qquad (8.4.27)$$

but does not quite carry the work to its completion. The shock trajectory was written out as

$$x = c_1 t + c_2 t^2 + \cdots \quad + c_n t^n \qquad (8.4.28)$$

where c_i are constants. The flow between the piston and the shock was to be found, leading in the process to the determination of the unknown constants c_i in (8.4.28). The series form of the solution was assumed as

$$u = \sum_{k=0}^{\infty} u_k(t)\phi^k(x,t), \rho = \sum_{k=0}^{\infty} \rho_k(t)\phi^k(x,t)$$

$$S = \sum_{k=0}^{\infty} S_k(t)\phi^k(x,t) \qquad (8.4.29)$$

where

$$\phi(x,t) = x - c_1 t - c_2 t^2 - \ldots - c_n t^n. \qquad (8.4.30)$$

$\phi(x,t)$ is similar to the variable ξ in the analysis of Van Dyke and Guttmann (1982): $\phi(x,t) = 0$ is the shock trajectory. Kozmanov (1977) considered plane, cylindrical, and spherical geometries. Explicit results were found for the case for which $x(t)$ in (8.4.27) is a quadratic and the geometry is planar. For $\xi_1 = 10$ and $\xi_2 = 5$, the shock trajectory was found to be $X = 15.132t + 4.241t^2$. A comparison with the numerical solution of the problem showed a discrepancy in the shock trajectory to be less than 0.1% for $t < 0.3$.

References

Acheson, D. J. (1972) The critical level for hydromagnetic waves in a rotating fluid, *J. Fluid Mech.*, 53, 401-415.

Ardavan -Rhad, H. (1970) The decay of a plane shock wave, *J. Fluid Mech.*, 43, 737-751.

Barenblatt, G. I. (1979) *Similarity, Self-similarity, and Intermediate Asymptotics*, Consultants Bureau, New York.

Barenblatt, G. I. (1996) *Scaling, Self-similarity, and Intermediate Asymptotics*, Cambridge University Press, New York.

Bender, C. M. and Orszag, S. A. (1978) *Advanced Mathematical Methods for Scientists and Engineers*, McGraw-Hill, New-York.

Benton, E. R. and Platzman, G. W. (1972) A table of solutions of one dimensional Burgers equation, *Quart. Appl. Math.*, 30, 195-212.

Bluman, G. W. and Cole, J. D. (1974) *Similarity Methods for Differential Equations*, Springer-Verlag, New York.

Bluman, G. W. and Kumei, S. (1987) On invariance properties of the wave equation, *J. Math. Phys.*, 28, 307-318.

Bluman, G. W. and Kumei, S. (1989) *Symmetries and Differential Equations*, Springer-Verlag, New York.

Brezis, H., Peletier, L. A., and Terman, D. (1986) A very singular solution of the heat equation with absorption, *Arch. Rat. Mech. Anal.*, 95, 185-209.

Bukiet, B., Pelesko, J., Li, X. L., and Sachdev, P. L. (1996) A characteristic based numerical method with tracking for nonlinear wave equations, *Comput. Math. Applic.*, 31, 75-99.

Calegaro, F. (1991) Why are certain nonlinear PDEs both widely applicable and integrable? in *What is Integrability*, 1-62. V.E. Zakharov and F. Calegaro (Eds.), Springer Verlag, New York.

Canosa, J. (1973) On a nonlinear diffusion equation describing population growth, *IBM J. Res. & Dev.*, 17, 307-313.

Carbonaro, P. (1997) Group analysis for the equations describing the one-dimensional motion of an ideal gas in the hodograph plane, *Int. J. Non-Linear Mechanics*, 32, 455-464.

Carrier, G. F. and Greenspan, H. P. (1958) Water waves of finite amplitude on a sloping beach, *J. Fluid Mech.*, 4, 97-109.

Case, K. M. and Chiu, S. C. (1969) Burgers' turbulence models, *Phys. Fluids*, 12, 1799-1808.

Clarkson, P. A. and Kruskal, M. D. (1989) New similarity reductions of the Boussinesq equation, *J. Math. Phys.*, 30, 2201-2213.

Cole, J.D. (1951) On a quasi-linear parabolic equation occurring in aerodynamics, *Quart. Appl. Math.*, 9, 225-236.

Colton, D. (1976) The approximation of solutions to initial boundary value problems for parabolic equations in one space variable, *Quart. Appl. Math.*, 34, 377-386.

Courant, R. and Friedrichs, K. O. (1948) *Supersonic Flow and Shock Waves*, Interscience, New York.

Crighton, D. G. (1979) Model equations of nonlinear acoustics, *Ann. Rev. Fluid. Mech.*, 11, 11-33.

Crighton, D. G. and Scott, J. F. (1979) Asymptotic solution of model equations in nonlinear acoustics, *Phil. Trans. Roy. Soc. Lond.*, A 292, 101-134.

Dawson, C. N. (1991) Godunov - mixed methods for advective flow problems in one space dimension, *SIAM J. Numer. Anal.*, 28, 1282-1309.

Dawson, C. N., Van Duijn, C. J., and Grundy, R. E. (1996) Large time asymptotics in contaminant transport in porous media, *SIAM J. Appl. Math.*, 56, 965-993.

Domb, C. and Sykes, M. F. (1957) On the susceptibility of a ferromagnetic above the Curie point, *Proc. Roy. Soc. Lond.*, A 240, 214-228.

Earnshaw, S. (1858) On the mathematical theory of sound, *Phil. Trans. Roy. Soc. Lond.*, A 150, 133-148.

Escobedo, M. and Grundy, R. E. (1996) Asymptotic profiles of a decaying contaminant in transport through a porous medium, *Euro. J. Appl. Math.*, 7, 395-416.

Escobedo, M., Vazquez, J. L., and Zuazua, E. (1993) A diffusion - convection equation in several space dimensions, *Indiana Univ. Math. J.*, 42, 1413-1440.

Fisher, R. A. (1937) The wave of advance of advantageous genes, *Ann. Eugenics*, 7, 355-369.

Freeman, N. C. (1972) Simple waves on shear flows: similarity solutions, *J. Fluid Mech.*, 56, 257-263.

Friedman, A. (1964) Partial Differential Equations of Parabolic Type, Prentice-Hall, Englewood Cliffs, NJ.

Goard, J. M., Broadbridge, P., and Arrigo, D. J. (1996) The integrable nonlinear degenerate diffusion equation $u_t = [f(u)u_x^{-1}]_x$ and its relatives, *Z. Angew. Math. Phys.*, 47, 926-942.

Goldstein, S. and Murray, J. D. (1959) On the mathematics of exchange processes in fixed columns. III The solution for general entry conditions, and a method of obtaining asymptotic expressions. IV Limiting values, and correction terms, for the kinetic -theory solution with general entry conditions. V The equilibrium -theory and perturbation solutions, and their connexion with kinetic -theory solutions, for general entry conditions, *Proc. Roy. Soc. Lond.*, A 252, 334-375.

Gratton, J. and Minotti, F. (1990) Self-similar viscous gravity currents: phase plane formalism, *J. Fluid Mech.*, 210, 155-182.

Greenspan, H. P. and Butler, D. S. (1962) On the expansion of a gas into vacuum, *J. Fluid Mech.*, 13, 101-119.

Grundy, R. E., Sachdev, P. L., and Dawson, C. N. (1994) Large time solution of an initial value problem for a generalised Burgers equation, in *Nonlinear Diffusion Phenomenon*, P. L. Sachdev and R. E. Grundy (Eds.) pp.68 -83. Narosa Publishing House, New Delhi.

Grundy, R. E., Van Duijn, C. J., and Dawson, C. N. (1994) Asymptotic profiles with finite mass in one dimensional contaminant transport through porous media: The fast reaction case, *Q. Jl. Mech. Appl. Math.*, 47, 69 -106.

Guderley, G. (1942) Starke kugelige und zylindrische Verdichtungsstösse in der Nähe des Kugelmittelpunktes bzw.der Zylinderachse, Luftfahrtforschung 19, 302-312.

Haberman, R. (1987) *Elementary Applied Partial Differential Equations*, Prentice-Hall, Englewood Cliffs, NJ.

Harris, S. E. (1996) Sonic shocks governed by the modified Burgers' equation, *Euro. J. Appl. Math.*, 7, 201 - 222.

Hille, E. (1970) Some aspects of Thomas-Fermi equation, *J. Analyse Math.*, 23, 147-170.

Hille, E. (1970a) Aspects of Emden's equation, *J. Fac. Sci. Tokyo. Sect.*, 1, 17, 11-30.

Hood, S. (1995) New exact solutions of Burgers equation - an extension to the direct method of Clarkson and Kruskal, *J. Math. Phys.*, 36, 1971-1990.

Hopf, E. (1950) The partial differential equation $u_t + uu_x = \mu u_{xx}$, *Comm. Pure Appl. Math.*, 3, 201-230.

Huppert, H. E. (1982) The propagation of two-dimensional viscous gravity currents over a rigid horizontal surface, *J. Fluid Mech.*, 121, 43-58.

Joseph, K. T. and Sachdev, P. L. (1994) On the solution of the equation $u_t + u^n u_x + H(x, t, u) = 0$, *Quart. Appl. Math.*, 52, 519-527.

Keener, J. P. (1988) *Principles of Applied Mathematics: Approximation and Transformation*, Addison-Wesley, New York.

Kevorkian, J. and Cole, J. D. (1996) *Multiple Scale and Singular Perturbation Methods*, Springer-Verlag, New York.

Kolmogoroff, A. N., Petrovsky, I. G., and Piskunov, N. S. (1937) Investigation of the diffusion equation connected with an increasing amount of matter and its application to a biological problem, *Bull. MGU*, A1 (6), 1-26.

Kozmanov, M. Iu. (1977) On the motion of piston in a polytropic gas, *J. Appl. Math. Mech.*, 41, 1152-1156.

Landau, L. D. (1945) On shock waves at large distances from the place of their origin, *Soviet J. Phys.*, 9, 496-500.

Larsen, D. A. (1978) Transient bounds and time asymptotic behaviour of solutions of nonlinear equations of Fisher type, *SIAM J. Appl. Math.*, 34, 93-103.

Lazarus, R. B. and Richtmyer, R. D. (1977) Similarity solutions for converging shocks, *Los Alamos Scientific Lab., Rep.*, LA-6823-MS.

Lazarus, R. B. (1981) Self-similar solutions for converging shocks and collapsing cavities, *SIAM J. Num. Anal.*, 18, 316-371.

Lazarus, R. B. (1982) One -dimensional stability of self-similar converging flows, *Phys. Fluids*, 25, 1146-1155.

Lee, B.H.K. (1968) The initial phases of a collapse of an imploding shock wave and the application to hypersonic internal flow. *C.A.S.I. Trans.*, 1, 57-67.

Lee-Bapty, I. P. and Crighton, D. G. (1987) Nonlinear wave motion governed by the modified Burgers equation, *Phil. Trans. R. Soc. Lond.*, A 323, 173-209.

Lefloch, P. (1988) Explicit formula for scalar nonlinear conservation laws with boundary condition, *Math. Methods Appl. Sci.*, 10, 265-287.

Lehnigk, S. H. (1976) Conservative similarity solutions of the one-dimensional autonomous parabolic equation, *J. Appl. Math. Phys. (ZAMP)*, 27, 385-391.

Lehnigk, S. H. (1976a) A class of conservative diffusion processes with delta function initial conditions, *J. Math. Phys.*, 17, 973-976.

Leibovich, S. and Seebass, A. R. (Eds.) (1974) in *Nonlinear Waves*, pp.103-138, Cornell University Press, Ithaca, New York.

Levi, D. and Winternitz, P. (1989) Nonclassical symmetry reductions: example of the Boussinesq equation, *J. Phys. A: Math. Gen.*, 22, 2915-2924.

Levinson, N. (1969) Asymptotic behaviour of solutions of nonlinear differential equations, *Stud. Appl. Math.*, 49, 285-297.

Lighthill, M. J. (1956) Viscosity effects in sound waves of finite amplitude, in *Surveys in Mechanics*, G. K. Batchelor and R. M. Davies, (Eds.), pp.250-351, Cambridge University Press, New York.

Logan, J. D. and Pérez, J. D. J. (1980) Similarity solutions for reactive shock hydrodynamics, *SIAM J. Appl. Math.*, 39, 512-527.

Logan, J. D. (1987) *Applied Mathematics: A Contemporary Approach*, John Wiley & Sons, New York.

Logan, J. D. (1994) *An Introduction to Nonlinear Partial Differential Equations*, John Wiley & Sons, New York.

Ludlow, D.K., Clarkson, P.A., and Bassom, A.P. (1999) Similarity reduction and exact solutions for the two-dimensional incompressible Navier-Stokes equations, *Stud. Appl. Math.*, 103, 183-240.

Manoranjan, V. S. and Mitchell, A. R. (1983) A numerical study of the Belousov-Zhabotinskii reaction using Galerkin finite element methods, *J. Math. Biol.*, 16, 251-260.

Mayil Vaganan, B. (1994) Exact Analytic Solutions for Some Classes of Partial Differential Equations, Ph. D. thesis, Indian Institute of Science, Bangalore, India.

McKean, H. P. (1975) Application of Brownian motion to the equation of Kolmogorov-Petrovskii-Piskunov, *Comm. Pure. Appl. Math.*, 28, 323-331.

Merkin, J. H. and Sadiq, M. A. (1996) The propagation of travelling waves in an open cubic autocatalytic chemical system, *IMA J. Appl. Math.*, 57, 273-309.

Mollison, D. (1977) Spatial contact models for ecological and epidemic spread, *J. Roy. Stat. Soc.*, B39, 283-326.

Murray, J. D. (1968) Singular perturbations of a class of nonlinear hyperbolic and parabolic equations, *J. Math. Phys.*, 47,111-133.

Murray, J. D. (1970) Perturbation effects on the decay of discontinuous solutions of nonlinear first order wave equations, *SIAM J. Appl. Math.*, 19, 273-298.

Murray, J. D. (1973) On Burgers' model equation for turbulence, *J. Fluid Mech.*, 59, 263-279.

Murray, J. D. (1977) *Nonlinear Differential Equation Models in Biology*, Clarendon Press, Oxford.

Murray, J. D. (1989) *Mathematical Biology*, Springer-Verlag, New York.

Nagesawara Yogi, A.M. (1995) On the analytic theory of explosions, Ph. D. thesis, Indian Institute of Science, Bangalore, India.

Nakamura, Y. (1983) Analysis of self-similar problems of imploding shock waves by the method of characteristics, *Phys. Fluids.*, 26, 1234-1239.

Nerney, S., Schmahl, E. J., and Musielak, Z. E. (1996) Analytic solutions of the vector Burgers' equation, *Quart. Appl. Math.*, 54, 63-71.

Nimmo, J. J. C. and Crighton, D. G. (1982) Bäcklund transformations for nonlinear parabolic equations: the general results. *Proc. R. Soc. Lond.*, A 384, 381-401.

Nimmo, J. J. C. and Crighton, D. G. (1986) Geometrical and diffusive effects in nonlinear acoustic propagation over long ranges, *Phil. Trans. Roy. Soc. Lond.*, A 320, 1-35.

Nye, A. H. and Thomas, J. H. (1976a) Solar magneto-atmospheric waves. I. An exact solution for a horizontal magnetic field, *Astrophys. J.*, 204, 573-581.

Nye, A. H. and Thomas, J. H. (1976b) Solar magneto-atmospheric waves. II. A model for running Penumbral waves, *Astrophys. J.*, 204, 582-588.

Odulo, Al. B., Odulo, An. B., and Chusov, M. A. (1977) On the class of nonlinear stationary waves in the ocean, *Izv. Acad. Sci. USSR Atmospher. Ocean. Phys.*, 13(8), 584-587.

Oleinik, O. A. (1957) Discontinuous solutions of nonlinear differential equations, *Uspekhi Mat. Nauk.*, 12, 3(75), 3-73.

Ovsiannikov, L. V. (1982) *Group Analysis of Differential Equations*, Academic Press, New York.

Peletier, L. A. (1998) Self-similar solutions of the second kind, in *Nonlinear Analysis and Continuum Mechanics*, G. Buttazzo, G. P. Galdi, E. Lanconelli., and P. Pucci (Eds.), Springer, New York.

Rothe, F. (1978) Convergence to travelling fronts in semilinear parabolic equations, *Proc. R. Soc. Edin.*, A 80, 213-234.

Sachdev, P. L. (1978) A generalised Cole - Hopf transformation for nonlinear parabolic and hyperbolic equations. *J. Appl. Math. Phys. (ZAMP)*, 29, 963 -970.

Sachdev, P. L. (1980) Exact, self-similar, time-dependent, free surface flows under gravity, *J. Fluid Mech.*, 96, 797-802.

Sachdev, P. L. (1987) *Nonlinear Diffusive Waves*, Cambridge University Press, Cambridge.

Sachdev, P. L. (1991) *Nonlinear Ordinary Differential Equations and Their Applications*, Marcel Dekker, Inc., New York.

Sachdev, P. L., Dowerah, S., Mayil Vaganan, B., and Philip, V. (1997) Exact analysis of a nonlinear partial differential equation of gasdynamics, *Quart. Appl. Math.*, 55, 201-229.

Sachdev, P. L. and Gupta, N. (1990) Exact traveling-wave solutions for model geophysical systems, *Stud. Appl. Math.*, 82, 267-289.

Sachdev, P. L., Gupta, N., and Ahluwalia, D. S. (1992) Exact analytic solutions describing unsteady plane gas flows with shocks of arbitrary strength, *Quart. Appl. Math.*, 50, 677-726.

Sachdev, P. L., Gupta, N., and Ahluwalia, D. S. (1992a) Global solutions describing the collapse of a spherical or cylindrical cavity, *J. Appl. Math Phys. (ZAMP)*, 43, 856-874.

Sachdev, P. L. and Joseph, K. T. (1994) Exact representations of N-wave solutions of generalised Burgers equations, in *Nonlinear Diffusion Phenomenon*, P. L. Sachdev and R. E. Grundy (Eds.), Narosa Publishing House, New Delhi.

Sachdev, P. L., Joseph, K. T., and Mayil Vaganan, B. (1996) Exact N-wave solutions of generalized Burgers equations, *Stud. Appl. Math.*, 97, 349 -367.

Sachdev, P. L., Joseph, K. T., and Nair, K. R. C. (1994) Exact N-wave solutions for the non-planar Burgers equation, *Proc. Roy. Soc. Lond.*, A 445, 501-517.

Sachdev, P. L. and Mayil Vaganan, B. (1992) Exact solutions of linear partial differential equations with variable coefficients, *Stud. Appl. Math.*, 87, 213-237.

Sachdev, P. L. and Mayil Vaganan, B. (1994) Exact free surface flows for shallow water equations I: The incompressible case. *Stud. Appl. Math.*, 93, 251-274.

Sachdev, P. L. and Mayil Vaganan, B. (1995) Exact free surface flows for shallow water equations II: The compressible case. *Stud. Appl. Math.*, 94, 57-76.

Sachdev, P. L. and Nair, K. R. C. (1987) Generalised Burgers equations and Euler - Painlevé transcendents II, *J. Math. Phys.*, 28, 997 -1004.

Sachdev, P. L. and Narasimha Chari, M V. (1982) Multinomial solutions of the beach equation, *J. Appl. Math. Phys. (ZAMP)*, 33, 534-539.

Sachdev, P. L. and Philip, V. (1986) Invariance group properties and exact solutions of equations describing time-dependent free surface flows under gravity, *Quart. Appl. Math.*, 43, 463-480.

Sachdev, P. L. and Philip, V. (1988) Exact simple waves on shear flows in a compressible barotropic medium, *Stud. Appl. Math.*, 79, 193-203.

Sachdev, P. L. and Reddy, A. V. (1982) Some exact solutions describing unsteady plane gas flows with shocks, *Quart. Appl. Math.*, 40, 249-272.

Sachdev, P. L. and Seebass, A. R. (1973) Propagation of spherical and cylindrical N- waves, *J. Fluid Mech.*, 58, 197-205.

Sachdev, P. L. and Srinivasa Rao, Ch. (1999) N-wave solution of modified Burgers equation, To appear in *Appl. Math. Letters.*

Sachdev, P. L., Srinivasa Rao, Ch., and Joseph, K. T. (1999) Analytic and numerical study of N-waves governed by nonplanar Burgers equation, *Stud. Appl. Math.*, 103, 89-120.

Schindler, G. M. (1970) Simple waves in multidimensional gas flow, *SIAM J. Appl. Math.*, 19, 390-407.

Sedov, L. I. (1959) *Similarity and Dimensional Methods in Mechanics*, Academic Press, New York.

Seshadri, V. S. and Sachdev, P. L. (1977) Quasi-simple wave solutions for acoustic gravity waves, *Phys. Fluids*, 20, 888-894.

Seymour, B. and Varley, E. (1984) A Bäcklund transformation for a nonlinear telegraph equation, in *Wave Phenomena: Modern Theory and Applications*, C. Rogers and T.B. Moodie (Eds.), Elsevier Science Publishers B. V., Amsterdam.

Sharma, V. D., Ram, R., and Sachdev, P. L. (1987) Uniformly valid analytical solution to the problem of a decaying shock wave, *J. Fluid Mech.*, 185, 153-170.

Sherratt, J. A. and Marchant, B. P. (1996) Algebraic decay and variable speeds in wave front solutions of a scalar reaction-diffusion equation, *IMA Jl. Appl. Math.*, 56, 289-302.

Shih, C. C. (1974) Attenuation characteristics of nonlinear pressure waves propagating in pipes, Finite Amplitude Wave Effects in Fluids, L. Bjorno (Ed.). IPC Science and Technology Press, Guildford.

Smith, S. H. (1969) On initial value problems for the flow in a thin sheet of viscous liquid, *J. Appl. Math. Phys. (ZAMP)*, 20, 556-560.

Spielvogel, L. Q. (1975) Single-wave run-up on sloping beaches, *J. Fluid Mech.*, 74, 685-694.

Stanyukovich, K. P. (1960) *Unsteady Motion of Continuous Media*, Pergamon Press, New York.

Steketee, J. A. (1976) Transformations of the equations of motion for the unsteady rectilinear flow of a perfect gas, *J. Engg. Math.*, 10, 69-94.

Stikker, U. O. (1970) Numerical simulation of the coil annealing process, in Mathematical Models in Metallurgical Process Development, *Iron and Steel Institute, Special Report*, 123, pp. 104-113.

Stoker, J. J. (1948) The formation of breakers and bores, *Comm. Pure. Appl. Math.*, 1, 1-87.

Swan, G. W. (1977) Exact fundamental solutions of linear parabolic equations with spatially varying coefficients, *Bull. Math. Biol.*, 39, 435-451.

Taylor, G. I. (1910) The conditions necessary for discontinuous motion in gases, *Proc. Roy. Soc. Lond.*, A 84, 371-377.

Taylor, G. I. (1950) The formation of a blast wave by a very intense explosion. I, Theoretical discussion. *Proc. Roy. Soc. Lond.*, A 201, 159-174.

Thomas, L. P., Pais, V., Gratton, R., and Diez, J. (1986) A numerical study on the transition to self-similar flow in collapsing cavities, *Phys. Fluids*, 29, 676-679.

Tuck, E. O. and Hwang, L. (1972) Long wave generation on a sloping beach, *J. Fluid Mech.*, 51, 449-461.

Ustinov, M. D. (1982) Approximate solution to nonself-similar problem of motion of a piston after an impact, *Izv. Akad. Nauk. SSSR, Mekh. Zhid. Gaza.*, 6, 167-171.

Ustinov, M. D. (1984) Motion of a piston under the influence of gas pressure in the presence of an initial temperature gradient, *Izv. Akad. Nauk. SSSR, Mekh. Zhid. Gaza.*, 2, 177-180.

Ustinov, M. D. (1986) Some one-dimensional unsteady adiabatic gas flows with plane symmetry, *Izv. Akad. Nauk. SSSR, Mekh. Zhid. Gaza.*, 5, 96-104.

Van Duijn, C. J., Grundy, R. E., and Dawson, C. N. (1997) Large time profiles in reactive solute transport, *Transport in Porous Media*, 27, 57-84.

Van Dyke, M. and Guttmann, A. J. (1982) The converging shock wave from a spherical or cylindrical piston, *J. Fluid Mech.*, 120, 451-462.

Varley, E. and Seymour, B. R. (1988) A method for obtaining exact solutions to partial differential equations with variable coefficients, *Stud. Appl. Math.*, 78, 183-225.

Varley, E., Kazakia, J. Y. and Blythe, P. A. (1977) The interaction of large amplitude barotropic waves with an ambient shear flow: Critical flows, *Phil. Trans. Roy. Soc. Lond.*, A 287, 189-236.

Venkatachalappa, M., Rudraiah, N., and Sachdev, P. L. (1992) Exact nonlinear travelling hydromagnetic wave solutions, *Acta Mech.*, 93, 1-11.

Watson, G. N. (1962) *A Treatise on the Theory of Bessel Functions*, Cambridge University Press, Cambridge.

Whitham, G. B. (1950) The propagation of spherical blast, *Proc. Roy. Soc. Lond.*, A 203, 571-581.

Whitham, G. B. (1952) The flow pattern of a supersonic projectile, *Comm. Pure. Appl. Math.*, 5, 301-348.

Whitham, G. B. (1974) *Linear and Nonlinear Waves*, John Wiley & Sons, New York.

Williams, W. E. (1980) *Partial Differential Equations*, Clarendon Press, Oxford.

Willms, A. R. (1995) An exact solution of Stikker's nonlinear heat equation, *SIAM J. Appl. Math.*, 55, 1059-1073.

Zel'dovich, Ya. B. and Raizer, Yu. P. (1967) *Physics of Shock Waves and High-Temperature Hydrodynamic Phenomena*, Vol. 2, Academic Press, New York.

INDEX

Printed and bound by CPI Group (UK) Ltd, Croydon, CR0 4YY

23/10/2024

01778238-0008

Printed and bound by CPI Group (UK) Ltd, Croydon, CR0 4YY